Viruses

Viruses

Molecular Biology, Host Interactions, and Applications to Biotechnology

Paula Tennant
The University of the West Indies, Mona, Jamaica

Gustavo Fermin
Universidad de Los Andes, Mérida, Venezuela

Jerome E. Foster
The University of the West Indies, St. Augustine, Trinidad

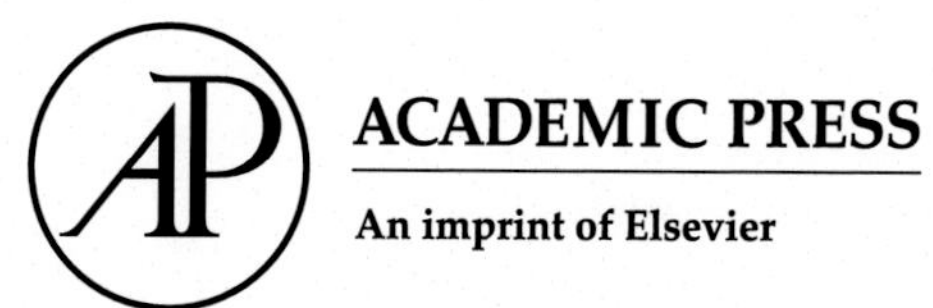

Academic Press is an imprint of Elsevier
125 London Wall, London EC2Y 5AS, United Kingdom
525 B Street, Suite 1800, San Diego, CA 92101-4495, United States
50 Hampshire Street, 5th Floor, Cambridge, MA 02139, United States
The Boulevard, Langford Lane, Kidlington, Oxford OX5 1GB, United Kingdom

Notices

Knowledge and best practice in this field are constantly changing. As new research and experience broaden our understanding, changes in research methods, professional practices, or medical treatment may become necessary.

Practitioners and researchers must always rely on their own experience and knowledge in evaluating and using any information, methods, compounds, or experiments described herein. In using such information or methods they should be mindful of their own safety and the safety of others, including parties for whom they have a professional responsibility.

To the fullest extent of the law, neither the Publisher nor the authors, contributors, or editors, assume any liability for any injury and/or damage to persons or property as a matter of products liability, negligence or otherwise, or from any use or operation of any methods, products, instructions, or ideas contained in the material herein.

British Library Cataloguing-in-Publication Data
A catalogue record for this book is available from the British Library

Library of Congress Cataloging-in-Publication Data
A catalog record for this book is available from the Library of Congress

ISBN: 978-0-12-811257-1

Publisher: John Fedor
Acquisition Editor: Linda Versteeg-Buschman
Editorial Project Manager: Pat Gonzalez
Production Project Manager: Stalin Viswanathan
Cover Designer: Mark Rogers

Typeset by MPS Limited, Chennai, India

Contents

List of Contributors

Gustavo Fermin Universidad de Los Andes, Mérida, Venezuela

Jerome E. Foster The University of the West Indies, St. Augustine, Trinidad

Augustine Gubba University of KwaZulu-Natal, Pietermaritzburg, South Africa

Sudeshna Mazumdar-Leighton University of Delhi, Delhi, India

José A. Mendoza Universidad de Los Andes, Mérida, Venezuela

Sephra Rampersad The University of the West Indies, St. Augustine, Trinidad and Tobago

Marcia Roye The University of the West Indies, Mona, Jamaica

Janine Seetahal Eric Williams Medical Sciences Complex, Champs Fleurs, Trinidad

Paula Tennant The University of the West Indies, Mona, Jamaica

Preface

It is an exciting time to study virology! The world of viruses is much greater and more diverse than previously recognized. Discoveries such as the Mimivirus and its virophage, have stimulated new discussions on the definition of viruses, their place in the current view of the tree of life, and their inherent and derived "interactomics" defined as the molecules and the processes by which virus gene products interact with themselves and with the host's cellular gene products to bring about changes in phenotypes, and of their own evolution. Viruses are also facilitating the development and creation of new materials and powerful tools for gene and genome engineering. However, viral diseases continue to plague us, our animals and crop plants; sadly, in the last decade, diseases such as SARS, H1N1 influenza, Nipah virus disease have wreaked havoc around the world as have Cassava Mosaic and Brown Streak virus diseases in Eastern and Southern Africa. Ebola, Chikungunya, and West Nile viruses have reemerged and the AIDS epidemic continues to sweep across sub-Saharan Africa and parts of Asia.

This book explores several concepts fundamental to Virology. While most texts focus on the pathogenesis and clinical aspect of viral diseases or the molecular biology of viral replication, we present viruses not only as formidable foes, but also as entities that can be beneficial to their hosts and humankind, and as entities that are helping to shape the Tree of Life. A complete overview of modern virology is intended, not a comprehensive or encyclopedia treatment of these topics. It is hoped that this overview with sufficient documentation for more indepth study will be of value to anyone learning Virology at any stage; a novice trying to understand the basic principles for the first time, an intermediate student of Virology preparing to study more advanced areas in the discipline, or anyone who has ended their formal education, but has maintained an interest in the discipline. An elementary knowledge of molecular biology is assumed, especially of the basic structures of nucleic acids and proteins, of the genetic code, and the processes involved in transcription and protein synthesis.

The volume comprises 13 chapters. The first chapter provides a general introduction on the history of the discipline. Subsequent four chapters cover the basic aspects of virology including the structure of viruses, the organization of their genomes, and basic strategies in replication and expression.

Principles are emphasized in these chapters, along with the diversity and versatility of viruses. The five chapters that follow examine virus examples, from a cross section of hosts, to illustrate the principles as well as the diversity of viruses, how they cause disease and how their hosts react to such disease, exploring developments in the field of host–microbe interactions in recent years. We then explore the medical and agricultural importance of viruses and how viruses can be surprisingly beneficial to their hosts.

The book is the product of a corporative effort. We wish to express our appreciation to all the contributing authors. We are especially grateful to members of the editorial team, Jill Leonard, Linda Versteeg-Buschman, Halima Williams, Joslyn Paguio, Pat Gonzalez, and Stalin Viswanathan, for providing support, feedback, and guidance during the process. Gratitude is extended to many of our colleagues for their advice and helpful discussions. Finally, we are indebted to all who have helped shape our careers over the years, especially our students.

Paula Tennant, Gustavo Fermin, and Jerome E. Foster

Chapter 1

Introduction: A Short History of Virology

Gustavo Fermin[1] and Paula Tennant[2]
[1]*Universidad de Los Andes, Mérida, Venezuela,* [2]*The University of the West Indies, Mona, Jamaica*

[Physicists] feel that the field of bacterial viruses is a fine playground for serious children who ask ambitious questions.

Max Delbrück

As masterly expressed by the French microbiologist André Lwoff in his Nobel Lecture of 1965, "*For the philosopher, order is the entirety of repetitions manifested, in the form of types or of laws, by perceived objects. Order is an intelligible relation. For the biologist, order is a sequence in space and time. However, according to Plato, all things arise out of their opposites. Order was born of the original disorder, and the long evolution responsible for the present biological order necessarily had to engender disorder. An organism is a molecular society, and biological order is a kind of social order. Social order is opposed to revolution, which is an abrupt change of order, and to anarchy, which is the absence of order. I am presenting here today both revolution and anarchy, for which I am fortunately not the only one responsible. However, anarchy cannot survive and prosper except in an ordered society, and revolution becomes sooner or later the new order. Viruses have not failed to follow the general law. They are strict parasites which, born of disorder, have created a very remarkable new order to ensure their own perpetuation.*" In this introductory chapter, we present a short history of the main breakthroughs in the history of virology along with some of the reasons why viruses are considered among the most intriguing and fascinating creations of nature.

Viruses. DOI: https://doi.org/10.1016/B978-0-12-811257-1.00001-2

$C_{332,652}H_{492,388}N_{98,245}O_{131,196}P_{7,501}S_{2,340}$

The empirical formula, $C_{332,652}H_{492,388}N_{98,245}O_{131,196}P_{7,501}S_{2,340}$, represents the chemical composition of the poliovirus, a virus that has earned the reputation of being one of the world's most feared pathogens. In most infections this virus is limited to the alimentary tract but paralytic poliomyelitis occurs in less than 1% of cases when the virus enters the central nervous system and replicates in the motor neurons of the spinal cord. Describing the poliovirus as a formula portrays the virus as a chemical, a particle of high symmetry containing all the properties required for survival. Even so, one cannot recreate the virus by mixing all its components—even in exact amounts. At most, one could synthesize a template free, biologically meaningful array of logically ordered nucleotides based on prior *knowledge*—since a virus is more than a chemical entity, and an evolved product of biological information. Indeed viruses were first defined as simple entities, lacking the mechanisms necessary for metabolic function, consisting of a single type of nucleic acid encased in a protein coat. Since some virus particles can form aggregates and bond to each other to form a crystal, viruses were considered as molecular, and not cellular, entities. Many definitions of life, being based on the cellular theory, excluded viruses as living organisms.

The recent discovery of a new group of virus species is, however, challenging the classification of viruses as nonliving entities and has reignited debates on the definition of life. These viruses, designated as mimiviruses, were isolated in 1992 from a water sample collected in an air conditioning system during investigations of a pneumonia outbreak in England. A bacterial etiology was first suspected because of the resemblance to Gram-positive cocci. Electron microscopy in 2003 revealed icosahedral virus particles; the dimensions of which rival those of many microbes. These viruses are three times larger than any virus known at that time and carry DNA genomes consisting of 1.2 million base pairs encoding 1260 genes, 7 of which are common to all cellular life: eukaryotes, bacteria, and archaea. The genome of a relative, a megavirus, was later isolated from a marine sample and shown to be 6.5% larger than that of reported mimiviruses, and unusually packed with DNA repair enzymes. Further analysis of this relative suggested that the complexity of these giant DNA viruses has not yet been uncovered, and perhaps these virus lineages that are as old as those of other microbes on Earth could contribute to the reconstruction of the evolutionary history of viruses. Some researchers are of the view that the giant viruses are the origin of the eukaryotic nucleus; others speculate that they assisted in the emergence of DNA from RNA precursors. These are all hypotheses and much work is needed to understand the origin and evolution of life on Earth.

FROM FILTERABLE AGENT TO GENETIC PARASITE

Much of the initial attention of virologists was focused on viruses as disease causing agents. Certainly, attempts of treating virus infections were recorded long before there was an understanding of the concept of a virus as a distinct entity. Case in point, the development of vaccines and vaccination. Vaccine development began with attempts to prevent infections of the dreaded smallpox disease from as early as the 15th century. In 1796 the Physician Edward Jenner tested his hypothesis that secretions from the wounds that occurred on the udder of milking cows contained material that could protect against smallpox, an acute contagious disease caused by the *Variola virus*. His hypothesis was based on the observation that milkmaids very rarely contracted the rash that appears on the face and body of an infected person. Instead they often developed pustules on their hands, which were later shown to be caused by the closely related *Cowpox virus*. A method of intentional infection or variolation was used earlier in China and the Middle East sometime in the 15th and 16th centuries as it was quickly realized that individuals who survived the disease were immune to subsequent infections. The practice involved the application of pustular secretions onto superficial scratches on the arm or leg of an uninfected person, sometimes with grievous consequences and not the mild infection hoped for.

Almost a century later, the French chemist and microbiologist Louis Pasteur, in honoring Jenner's discovery, coined the term vaccination to refer to the use of a weakened pathogen or "*vaccine*" to defend against infectious diseases. With the young physician, Emil Roux, Pasteur developed a vaccine against rabies and techniques for attenuating materials he used for his live vaccines. The rabies virus (*Rabies lyssavirus*) causes acute infection of the central nervous system. In humans the infection is characterized by a neurologic period, coma, and death. By the time the signs of rabies become apparent the disease is nearly always fatal. Pasteur's vaccine was developed from dried spinal cord tissues collected from rabbits that had died from the infection. The material, initially tested in dogs, was administered to a young boy who had been bitten by a rabid dog and was certain to die. He received multiple shots of the vaccine and survived not only the bite wounds of the attack, but also the experimental vaccine. Pasteur, already a respected scientist, was elevated to the level of an idol—but he never attempted to identify the parasitic agent responsible for the disease. Others after Pasteur, such as Mayer, Ivanovsky, Beijerinck, Loeffler, Frosch, and Reed, showed that infectious agents smaller than bacteria were associated with many of the prevalent diseases at the time. This opened the door to a separate discipline, that of Virology, which came to later contribute to a greater understanding of the most intricate life processes.

Pasteur is also credited for bringing acceptance to the germ theory of disease which states that some diseases are caused by parasitic agents. Prior to the 19th century, disease was thought of as either a divine intervention and punishment or the result of noxious odors. Through Pasteur's work and the development of microscopic techniques by the German physician and microbiologist Robert Koch, microorganisms became visible and identifiable; and gave credence to the germ theory of disease. In 1840 it was Jacob Henle, Koch's mentor, who proposed that infectious diseases were caused by living organisms capable of reproducing outside the infected individual. Koch went on to isolate the bacterium responsible for tuberculosis, and from this developed four criteria for the identification of the causative agent of a disease; the pathogen is always associated with a given disease, the pathogen can be isolated from the diseased host and grown in pure culture, the cultured pathogen causes disease when transferred to a healthy susceptible host and the same pathogen is recoverable from the experimentally infected host.

The first evidence that showed the existence of viruses came from experiments set up to fulfill Koch's postulates and determine the cause of a disease that was plaguing tobacco. In 1879 Adolf Mayer, the director of the Agricultural Experimental Station in Wageningen, Holland, initiated work on tobacco mosaic disease. He reported "*the harm done by this disease is often very great and it has caused the cultivation of tobacco to be given up entirely*" in the Netherlands and certain parts of Germany. In attempting to follow Koch's postulates, Mayer used sap extracted from diseased tobacco plants as inoculum to infect healthy tobacco plants. He was successful in reproducing the original disease and subsequently launched a microbiological study to identify the causative agent. Samples were examined microscopically. Samples were passed through filter paper. Samples were cultured on medium devised for growing bacteria. None of these tests were successful in identifying or isolating the etiological agent. Nonetheless, Mayer concluded that the agent was a bacterium that had probably lost activity upon filtration. Dmitri Ivanovsky, a Russian botanist, and the soil microbiologist Martinus Beijerinck, conducted further investigations into the relationship between this etiological agent and the disease of tobacco.

In 1890 Dmitri Ivanovsky was commissioned to study the mosaic disease that was destroying tobacco plants in Crimea. As did Mayer, Ivanovsky showed that sap from diseased tobacco remained infectious to healthy tobacco plants after filtration. Similar observations were obtained when filtrate derived from porcelain filters was used to inoculate healthy tobacco plants. These filters were invented in the 19th and early 20th centuries to retain bacteria and purify water and other liquids. Ivanovsky also reported that it was impossible to grow an organism in pure culture. He came to the conclusion that the pathogenic agent was a minuscule, unculturable bacterium. Beijerinck, on the other hand, proposed a revolutionary idea based on similar observations and three others; the agent could be precipitated by alcohol, it could diffuse through a solid agar medium and it was not able to reproduce outside of a host. Beijerinck then posited that the filterable agent was a unique type of pathogen and coined the term *contagium vivum fluidum* to convey his concept of a new type of infectious agent that exists in a fluid, noncellular state.

Similar filterable agents too small to be observed by visible light microscopy, but capable of causing disease in animals, were subsequently recognized. Koch's disciples, Friedrich Loeffler and Paul Frosch (1898), isolated the first agent from animals associated with an extremely contagious disease of cloven footed animals. Later designated the *Foot-and-mouth disease virus*, the agent that was found filterable, could not be grown on medium used for the cultivation of bacterial pathogens, and it was shown, by dilution, to be infectious. In 1901 Walter Reed and his team isolated the first filterable agent from humans, the *Yellow fever virus*. Yellow fever is a mosquito-borne viral-spread hemorrhagic fever with a high case–fatality rate. It is endemic to tropical regions of South America and sub-Saharan Africa.

Also around this time a link between filterable or cell-free agents and cancer was proposed. Two Danish scientists in 1908, Vilhelm Ellerman and Oluf Bang, successfully used a cell-free filtrate from chickens with avian erythroblastosis to transmit the disease to healthy chickens. Similarly, 3 years later Peyton Rous in the United States showed that a filterable agent extracted from a sarcoma in chickens was infectious. These findings, however, went unrecognized until interest in the involvement of filterable agents in tumor pathogenesis revived some 20 years later with the discovery of agents responsible for murine tumors. Finally, in 1964, the first filterable agent linked to human cancer was discovered by the United Kingdom scientists Anthony Epstein and Yvonne Barr. The herpesvirus-like Epstein-Barr virus, derived from African Burkitt's lymphoma tissue. It was years later before it was appreciated that infection is generally not sufficient for cancer, and additional events and host factors, including immunosuppression, somatic mutations, genetic predisposition, and exposure to carcinogens also play a role in the development of cancers.

Concurrently, two independent investigations led to the discovery of filterable agents that infect bacteria, confirming that all organisms can harbor these agents. While trying to grow the bacterium *Staphylococcus aureus*, Frederick Twort, an English medical bacteriologist, noted the development of small, clear areas, which on further investigation were found to represent zones of lysed bacteria. Twort did not, however, pursue this finding and it was another 2 years before an explanation was presented by Felix d'Herelle, a French-Canadian microbiologist at the Institut Pasteur in Paris. d'Herelle's 1911 studies on the cause of dysentery in locust populations in Mexico led to the detection of a bacterium as the causal agent of the epizootic. Like Twort, d'Herelle observed small, clear regions of lysed bacteria during his attempts at culturing the bacterium. During subsequent investigations with human dysentery in 1917, d'Herelle noticed his bacterial culture was completely destroyed when he happened to mix a filtrate of the clear areas with a culture of dysentery bacteria and attributed the phenomenon to an unknown agent. And he aptly named the unknown agent in the filtrate a bacteriophage or "*bacteria eater*." By the start of the 20th century the concept of viruses was firmly established, though entirely in negative terms; they were associated with many diseases, they could not be seen, could not be cultivated in the absence of cells and they were not retained by bacteria-proof filters.

Beijerinck's concept of the *contagium vivum fluidum* was virtually forgotten until 1935, when Wendell Stanley reported on the crystallization of *Tobacco mosaic virus* (TMV). The crystallization of TMV suggested to Stanley that TMV was a pure protein with a regular structure. Stanley later demonstrated that the virus, apparently lifeless in a crystallized form, multiplied when dissolved and reintroduced into tobacco plants. This was a shocking finding at the time that blurred the lines between the living organisms studied by biologists and the nonliving molecules studied by chemists. Even more surprising was the discovery by two British scientists, the plant pathologist Frederick Bawden and the biochemist and virologist, Norman Pirie. Bawden and Pirie in 1936 demonstrated that TMV was not a pure protein, but contained about 6% **r**ibo**n**ucleic **a**cid (RNA) indicating that a virus is more complex than a mere chemical, even if not quite an organism. John Bernal and Isidor Fankuchen "*visualized*" the virus for the first time by X-ray crystallography in 1941. And the first virus particle (virion) ever visualized (and photographed) using an electron microscopy was accomplished later by Helmuth Ruska in 1940.

Stanley's crystallization of TMV marked the beginning of an understanding of the molecular organization of viruses. Viruses by their nature are biochemical entities consisting of two major elements and a common "*body plan*" composed of large number of protein subunits and other components assembled to form symmetrical, reproducible structures that encase the genome (DNA or RNA). Viruses can be classified as rod-like or spherical, with the capsid proteins of rod-like viruses arranged with helical symmetry around the nucleic acid, and the capsids of most spherical viruses arranged with icosahedral symmetry. The principles of the natural engineering utilized during the formation of a virus are ubiquitous in biology. Together the early studies with bacteriophages, cell culture methods, technological breakthrough in DNA sequencing and the development of polymerase chain reaction, contributed to the realization of a range of viral genome sizes that spans three orders of magnitude. Later in the early 1970s the Baltimore classification recognized seven types of nucleic acid genomes and seven replication strategies within infected cells. This followed on from an appreciation that viruses are totally dependent on cells for multiplication and exploit all possible DNA and RNA interconversions. Viruses are genetic parasites.

In this book, viruses are regarded as macromolecular complexes that, through biological evolution, came into existence and acquired the capacity to replicate themselves by using the genetic instructions they encode and the hosts' cell machinery. We continue our short account, unraveling the mystery of the virus, beginning in the 1930s with the convergence of previously distinct biological and physical disciplines: Biochemistry, Genetics, Microbiology, and Physics.

VIRUSES HAVE CONTRIBUTED TO OUR UNDERSTANDING OF A GENE AND HOW IT WORKS

Quantum mechanics, the branch of physics that describes the behavior of particles in the subatomic realm, made its debut in the late 1930s. The revolutionary science permeated to and has influenced all fields of knowledge, including biology and the worrisome issue of genes and heredity. Of course, the work initiated by Mendel with peas continued by Morgan and colleagues with *Drosophila* and many others with different organisms, formulated the basis of heredity and a foundation for efforts in the coming century to understand the underlying mechanisms. But the search for the gene transcended the boundaries of a single science subject and a single method of research. The German-American physicist, Max Delbrück, played a crucial role in explaining the nature of genes. Initially a student of Astronomy in the 1920s, Delbrück turned to quantum mechanics and went on to use the advances in quantum physics to determine how genes operated and how genetic information was replicated over generations.

At the start, Delbrück had plans of developing his theories of the gene at the California Institute of Technology in 1938 using fruit flies. Instead he joined in the study investigating the biology of bacteriophages (i.e., viruses that infect bacteria) as a model for oncogenic virus research. His focus was on the multiplication of bacteriophages. This was because bacteriophages provided a more tractable experimental system, and heredity could be studied in terms of the multiplication of virus particles. Working with Emory Ellis, it was determined that bacteriophages reproduce in one step and not exponentially as do cells. Later collaborations with Salvador Luria (1943) provided the first real evidence that bacterial inheritance, like that of the cells of higher organisms, is mediated by genes and not by mechanisms of adaptation as was widely held at the time. Their classical experiment is commonly referred to as the fluctuation test. Essentially, Luria and Delbrück attempted to distinguish between the two prevailing hypotheses; whether bacterial mutants arise before any selection or only after bacteria are subjected to selection. They measured the number of mutants resistant to the bacteriophage T1 in a large number of replicate cultures of *Escherichia coli*. Delbrück then worked out the expected statistical distribution of the number of mutant cells per culture. Together, their data supported the hypothesis that phage resistant mutations had a constant probability of occurring in each cell division, and that bacteria did not become resistant after exposure to the virus. In subsequent studies with strains of the bacteriophage T2 in 1946, Alfred Hershey showed that different strains of the bacteriophage could exchange genetic material when both have infected the same bacterial cell, resulting in the generation of a bacteriophage that is a hybrid of the two. Hershey referred to the process as genetic recombination. His successive discovery that genetic damage caused by radiation in bacteriophages could be repaired by gene exchange following infection of the same host bacterium with several damaged phages gave further credence that recombination is not limited to life forms that reproduce sexually. Later on, Delbrück and other scientists were able to dissect all stages of the replication cycles of other bacteriophages. And by 1945 the phage group consisting of Delbrück, Luria and Hershey, and others, set up courses on bacteriophage genetics at Cold Spring Harbor (Long Island, New York). A number of geneticists consequently turned to studying bacteria and viruses rather than the higher order organisms for a better understanding of genetic behavior. The field of Molecular Biology arose from the convergence of work by geneticists and physicists on a common problem: the nature of inheritance.

During this time the British bacteriologist Frederick Griffith (1928) demonstrated "*genetic transformation*." Griffith found that strains of *Streptococcus pneumoniae* took up genetic material released by other cells and became phenotypically "*transformed*" by the new genetic information. The chemical nature of this material was DNA, as later demonstrated by Avery, McLeod, and McCarthy in 1944. But there was still some hesitation within the scientific community in accepting this finding, as there was the general assumption that the genetic material must be a protein. This set the stage for the Hershey-Chase experiment. In their elegant experiment performed in 1952, Alfred Hershey and Martha Chase demonstrated that during the infection of bacterial host cells, the DNA component of the viruses was transferred to bacteria to form bacteriophage progeny, and not the protein component, thus confirming that DNA contained the genetic information, and is the hereditary material. In a way, all these discoveries paved the way for the later work on the structure of DNA by Francis Crick and James Watson—who, by the way, was the first PhD student of Luria and a member of the Phage Group.

In their landmark 1961 *Nature* paper entitled "*General nature of the genetic code for proteins*," Francis Crick, Sydney Brenner, and collaborators, described a series of genetic experiments in which they showed that the genetic code for protein is a triplet code, the code is degenerate, triplets are not overlapping, and each nucleotide sequence is read from a specific starting point. Their conclusions derived from the previously determined fine structure genetic mapping using mutants with alterations in a specific gene, namely the rII region of bacteriophage T4. In the early 1950s Seymour Benzer provided the first detailed fine structure map of a genetic region through extensive genetic experiments with the rII region of bacteriophage T4. Brenner and coworkers used mutagenesis and genetic recombination with the T4 rII system to map altered sites in this genetic region and to establish the general nature of the genetic code. Later evidence for the existence of stop codons, which terminate protein synthesis, was reported in 1965 by Brenner from his work with amber mutants of the T4 bacteriophage. Taken together, viruses have helped to elucidate the nature and structure of genes that later was demonstrated to be a universal characteristic of all organisms.

VIRUSES HAVE ALLOWED US TO UNDERSTAND HOW GENES ARE REGULATED

The bacteriophage lambda was one of the first biological entities whose transcriptional regulation was studied and understood in detail. It was found early on that lambda, and other phages, possess a temporally controlled pattern of transcription and gene expression, commonly referred to as immediate early, early, and late transcription. Gene expression in phage lambda has contributed, almost as equally as bacterial gene regulation, to an understanding of the many facets of gene expression.

Besides the discovery of lytic and lysogenic cycles in the 1960s, that in itself is the outcome of specific gene regulatory circuits in action, the study of repressors (namely, cI and Cro) allowed for the analysis of not only the factors involved in the expression (activation or repression) of genes but also their kinetics. Beyond the roles of promoters and operators (which act in *cis*) as well as the repressors (which act in *trans*), phage lambda provided insight into antitermination regulation (proteins N and Q), the action of genetic switches—as defined by Mark Ptashne—the intimate relationship between a prophage and its host, the SOS response, and much more. Since the 1980s Ptashne has focused on applying insights gained from the study of the lambda bacteriophage to eukaryotic cells, in particular yeast. He wrote that they "*had no way of knowing, at the start, that studying the lambda repressor and its action would yield a coherent picture of a regulatory switch and even less an indication that the principles of protein–DNA interaction and gene regulation, gleaned from the lambda studies, would apply even in eukaryotes.*"

VIRUSES HAVE CONTRIBUTED TO OUR UNDERSTANDING OF HOW GENOMES ARE ORGANIZED

The bacteriophage lambda was also one of the earliest model systems for studying the physical nature of DNA and organization of genes. A substantial amount of early research examined the hydrodynamic properties of the lambda DNA with the goal of determining its absolute molecular weight. Methods were developed for separating the two halves of lambda DNA and for mapping genes identified genetically to these halves and to smaller physical intervals. In 1971 electron microscopic analysis led to the construction of a detailed physical map of lambda DNA—a gene map in base pairs (bps) rather than recombination frequency units. Ultimately, lambda contributed greatly to the mapping of DNA. The 12 bp lambda cohesive ends were the subject of the first direct nucleotide sequencing of DNA. Fourteen years following the determination of the exact sequences making up 12 bp nucleotide sequence (1968), the complete genome of lambda was determined by Sanger (1982). It was then realized that sequence data could give profound insights into genetic organization. Analysis of the lambda replication cycle also contributed to the dissection of the mechanics by which DNA molecules recombine (by site-specific recombination between lambda and the host chromosome, and by the generalized recombination pathway mediated by the host RecBCD complex between lambda DNA molecules) and replicate (in this case, by the rolling circle model). Altogether lambda phage experiments laid much of the groundwork for the science of Molecular Biology. Genome circularization, recombination, gene expression, and genome replication, combined, illustrated that the genes in lambda's genome are *organized* in terms of function, and hence, timing of their use and expression. We can deem genome organization as a different way of regulating gene expression: in eukaryotes, e.g., gene and genome organization, along with nucleus architecture, determine when and how genetic information is expressed in order to comply with the commands of the organism's developmental program. Nonetheless, recent research indicates that the historical contributions of bacteriophage studies might have overshadowed their key roles in global ecology and bacterial pathogenicity.

VIRUSES HAVE CONTRIBUTED TO OUR UNDERSTANDING OF THE MECHANISMS UNDERLYING RNAi

RNA interference (RNAi) has over the years been known under a number of different names; cosuppression, **p**ost**t**ranscriptional **g**ene **s**ilencing (PTGS), and quelling. Only after these seemingly unrelated processes were fully understood did it become clear that they all described a similar phenomenon. RNAi is a nucleotide sequence-specific process that induces mRNA degradation or translation inhibition at the posttranscriptional level. The two-step mechanism uses short RNA species generated (by the RNase III-like nuclease Dicer) from dsRNA precursors to target corresponding mRNAs for cleavage. Pioneering observations on RNAi were reported in plants and plant viruses, and later on described in almost all eukaryotic organisms, including mammals. It is surprising that the molecular basis of RNAi, one of the oldest and most ubiquitous systems, was first reported little more than 20 years ago.

The events leading to the elucidation of the biochemical mechanisms underlying RNAi began with the 1980s investigations on pathogen derived resistance. Roger Beachy and coworkers transformed tobacco with the coat protein gene of a plant virus, *Tobacco mosaic virus*, and showed that transgenic plants expressing the gene were resistant to infection by the homologous virus. Multiple strategies were rapidly developed to engineer virus resistant plants. Over the 1990s, David Baulcombe and colleagues, working with transgenic tobacco plants carrying *Potato virus X* derived sequences or nonvirus derived sequences (β glucuronidase or green fluorescence protein), discovered that small RNA species were associated with the resistant phenotype exhibited by transgenic plants and also illustrated the sequence specificity, RNA degradation, and the posttranscriptional nature of the resistance mechanism. By 1998 Andrew Fire and Craig Mello investigated the nature of the short RNA species in the nematode worm, *Caenorhabditis elegans*. They confirmed the involvement of double-stranded RNA in silencing genes in the animal, the specificity of the mechanism, and that RNA interference can spread between cells and is inherited. Waterhouse and colleagues in 1999 provided additional support of the association of RNA duplexes and gene silencing in tobacco plants transformed with protease gene constructs derived from *Potato virus Y*.

Although initially recognized as a handy tool to regulate gene expression and develop plant varieties resistant to viruses, RNAi is now recognized as a mechanism for cellular protection. The mechanism defends the genome against viruses and transposons, while removing abundant but aberrant nonfunctional mRNAs. Similar genes and RNA intermediates are required in RNAi pathways in protozoa, plants, fungi, and animals, thus indicating they are ancient strategies of genome defense. RNAi is being considered as an important tool for functional genomics and for gene-specific therapeutic activities that target the mRNAs of disease-related genes. Increasing knowledge of the interaction between virus and host RNAi machinery should not only lead to the development of effective and durable RNAi-based antiviral strategies but also insights into the escape strategies exploited by viruses.

VIRUSES THAT CAUSE DISEASES REPRESENT A SMALL FRACTION OF THE VIRAL COMMUNITY

Diseases such as smallpox, rabies, AIDS, bird flu, swine flu, herpes, hepatitis, Japanese encephalitis, cassava mosaic disease as well as some of the more common, chickenpox, and the ever prolific common cold have helped perpetuate the bias toward viewing viruses as major challenges to humans, either directly through affecting our health or through affecting the health of livestock and crops. Indeed, much of the historical and current research resolved around determining the causative agent of virus diseases that often negatively affect their hosts, whether by causing disease in humans, plants, or animals or by killing their microbial hosts. Not only can virus diseases take a large toll on human life, some infections are seen as development concerns affecting education, income, productivity, and economic growth. Economic losses due to viruses derive from the treatment of the diseases they cause, their prevention, control of their vectors, as well as a myriad of varying consequences of social, economic and political impact, and the research aimed at developing control strategies—including vaccines and their administration. In agriculture viruses can lead to a complete crop loss due to reductions in plant growth and vigor, decreases in product quality (and hence, market value), investments in the development of prevention and control strategies, the implementation of quarantine programs, development of detection protocols, and more.

Although some virus diseases, such as smallpox and poliomyelitis, have been eradicated or almost wiped out, many diseases persist with limited little success at management and containment. In addition, new infectious diseases are emerging and old ones are reappearing after a significant decline in incidence. The term emerging disease refers to the appearance of an as yet unrecognized infection, or a previously recognized infection that has expanded into a new ecological niche or geographical zone. Emerging or re-emerging diseases are typically zoonotic, i.e., the infection can be spread between animals and humans and is often accompanied by some change in pathogenicity. HIV/AIDS is an example. HIV crossed into humans from chimpanzees in the 1920s presumably because of the bush meat trade—the hunting and killing of chimpanzees and other animals for human consumption. Severe acute respiratory syndrome (SARS), avian influenza, and Zika are more recent emerging zoonotic diseases. Zoonoses have been known since early historical times, but their incidence has quadrupled in the last half-century, mainly because of increasing human encroachment into wildlife habitats, air travel, and wildlife trafficking.

As this book will present diseases of viral etiology no further details will be provided in this section. Suffice to say that viruses can inflict damage in ways not caused by other pathogens mainly because virus infections are not always easy to control or prevent. However, it now appears that viruses with bad intent represent only a small fraction of a massive viral community and that a large number of viruses are unknown to science. Technologies of DNA and RNA deep sequencing, as well as genomics and metagenomics, are rapidly uncovering new species of viruses from seemingly healthy hosts. A diverse, abundant, and underappreciated viral community exists on and within us, even within our own genomes. Unlike the influenza- and Ebola-like viruses, these viruses establish a balanced coexistence by regulating their gene expression (possibly involving the use of noncoding RNAs such as microRNAs) thus allowing them to exist for the host's entire lifetime under the radar of the host's defense systems.

VIRUSES CAN BE BENEFICIAL TO THEIR HOSTS

Amazingly, viruses can also protect us from diseases. Bacteriophages, e.g., defend mucosal surfaces against bacterial infections. These surfaces form vast interfaces of the host organism with the environment and thus are the main ports of entry for pathogenic microorganisms. Also present in the mucus environment are bacteriophages that prey on specific bacterial hosts. Using a novel hunting strategy known as "*subdiffusive motion*" to increase their chances of finding bacteria within a mucosal surface, the phages adhere to mucus, create a layer of viruses that can then infect and lyse invading, pathogenic bacteria. This surveillance mechanism not only helps in the fight against pathogenic bacteria, but also to modulate and regulate the bacteriome associated with the animal.

Further studies of symbiotic organisms and their hosts in complex tripartite interactions reveal that viruses are key players in these associations. A well-studied example is the colony collapse disorder involving the parasitic mite (*Varroa destructor*) and the *Deformed wing virus*. On its own, the *Deformed wing virus* (*Iflaviridae*) infects healthy bee hives at low levels, but when the mites are present, the entire hive can be infected. The bees not only suffer direct damage from the mites, but they are also efficiently inoculated with viruses. Recent research suggests another layer in the association: that the virus weakens the bees' immune system by affecting the expression of proteins of the NF-κB family of immune response genes, presumably making the bees more tolerant of the invading mites. It was already known that the virus benefitted from its association with the mite; now it appears that the mite gains reproductive success from serving as a viral vector. Mutualistic interactions between viruses and plant hosts have also been documented. For example, infection of beets with *Cucumber mosaic virus* (*Bromoviridae*) enhanced the plants' thermal tolerance as well as water stress tolerance. The underlying mechanisms driving enhanced tolerance to water stress in wheat plants following infection with Barley yellow dwarf virus, and its many variants (*Luteoviridae*), appears to involve the induction of phytohormones in a time-dependent manner.

Contributions of viruses to aquatic ecosystems cannot be overlooked. In these environments bacteriophages mediate a myriad of mechanisms that influence gene flow, niche expansion, adaptation, speciation, population control, and others, among its members. In oceans, viruses play a key role in the dynamics and structure of the ocean's host populations and communities. They also impact geochemical changes in the oceans; e.g., viruses can alter the pathways of carbon cycling as a result of the lysis of the cells they infect. On the other hand, viruses help to free essential nonorganic elements derived from cell lysis (e.g., iron), or from the viruses themselves (mainly phosphorus and nitrogen).

VIRUSES CAN FUNCTION AS TOOLS FOR BIOTECHNOLOGY

Since the discovery of bacteriophages, scientists have rapidly harnessed the biology of these viruses for the development of new tools and applications in Molecular Biology. Transduction, e.g., has been invaluable in two areas of bacterial Molecular Biology: genetic mapping and strain construction. The process is defined as the bacteriophage-mediated transfer of genetic material between bacterial cells. That is, in transduction, DNA of a donor bacterium packaged in a bacteriophage virion is introduced into a recipient bacterium upon infection. Transduction occurs in nature, where bacteriophages are numerous, and is possibly a strong driving force shaping bacterial evolution. There are two types, generalized and specialized. In the first case, transducing viruses harbor any portion of the donor bacteria DNA, and thus, are able to transfer any bacterial genes in fragments limited by the physical capacity of the virion. These are lytic phages that carry segments of the donor bacterium DNA that is accidentally packaged. Recombination occurs upon release of the donor DNA and infection of the recipient bacterium. The best example of a generalized transducing phage is the *Escherichia virus P1* (*Caudovirales*: *Myoviridae*), which has also been used for the development of the artificial genetic vectors called pacmids and the development of the Cre-*Lox* site-specific recombination strategy of artificial genetic engineering. Generalized transduction by P1 was first described by Edward Lennox in 1955. Specialized transducing phages, on the other hand, are lysogenic phages that integrate into only one or very few sites in the donor host genome. Hence they can carry only a few specific bacterial genes limited to the markers flanking the inserted virus genome. The mechanism was discovered in *Escherichia virus Lambda* by Esther Lederberg in 1950.

Other bacteriophages, notably Lambda, have contributed to the development of genetic manipulation technology. Derivatives of the genome of bacteriophage Lambda have been constructed to serve as cloning vectors, including cosmids. More recently, many other viruses of plant and animal origin are being used as vectors for genetic engineering and/or heterologous expression of different genes in diverse organisms, as well as the silencing of specific genes. Apart from these studies with bacteriophages, researchers have found several other uses for viruses; not only in genetic engineering but also in medicine and agriculture. The most obvious of these is vaccines and vaccine vectors. Other ways involve the use of viruses as vectors or carriers that take material required for the treatment of a disease to various target cells, viruses in nanotechnology as carriers for genetically modified sequences of genomes to the host cells, and viruses used to target cancer cells. Viruses are elegant macromolecular assemblies that display geometrically sophisticated architecture not seen in other biological assemblies. They use simple general principles, a very limited number of viral components, extensive symmetry and built in conformational flexibility to complete their replication cycles in their host cells. Although they come in various sizes and shapes, there are only two major types of symmetric assemblies, icosahedral and helical. Viruses are considered attractive for materials science and nanotechnology because of their regular structures and homogeneity of particle size, potential for mutagenesis to manipulate virus surface proteins, stability of the particles, accessibility of the particle interior, ease of production, and dynamic structural properties. As one scientist puts it, these protein structures are evolution tested, multifaceted systems with highly ordered spatial arrangements and natural cell targeting and genetic material storing capabilities. And finally, other uses for viruses surround the control of damaging agricultural pests including viruses themselves.

Taken together viruses long have been the subjects of scientific study. Much of the early studies were focused on viruses as disease causing agents and involved basic studies of viral replication, transmission, and pathogenesis. The 1970s to the present signify an era of major discoveries and technological breakthroughs in virology, mostly driven by advances in Molecular Biology. The advent of next generation sequencing technologies has expedited the process of novel virus discovery, identification, and viral genome sequencing. It is now clear that viruses are the most diverse and uncharacterized components of the major ecosystems on Earth. As briefly outlined in this chapter, viruses have played a major role in 20th-century biology and still continue to serve as ideal tools to dissect the most intricate life processes; the numerous Nobel prizes awarded to virologists is one measure of their impact (Table 1.1).

TABLE 1.1 Discoveries in Virology Recognized by a Nobel Prize

Year	Recipient(s)	Achievement
2008	H. zur Hausen (Physiology or Medicine)	Discovery of human papilloma viruses
2008	F. Barré-Sinoussi and L. Montagnier (Physiology or Medicine)	Discovery of *Human immunodeficiency virus*
1996	P. C. Doherty and R. M. Zinkernagel (Physiology or Medicine)	Discoveries concerning the specificity of the cellular immune system against virus antigens
1993	R. J. Roberts and P. A. Sharp (Physiology or Medicine)	Discovery of splicing, firstly demonstrated in adenoviruses introns
1989	J. M. Bishop and H. E. Varmus (Physiology or Medicine)	Discovery of the cellular origin of retroviral oncogenes
1980	F. Sanger (and W. Gilbert) (Chemistry)	First genome sequence ever determined (bacteriophage φX174)
1980	P. Berg (Chemistry)	For creating the first recombinant DNA molecule including the use of SV40, among others

(Continued)

TABLE 1.1 (Continued)

Year	Recipient(s)	Achievement
1976	B. S. Blumberg (and D. C. Gadjuseck) (Physiology or Medicine)	Discoveries concerning *Hepatitis B virus*
1975	D. Baltimore, R. Dulbecco and H. M. Temin (Physiology or Medicine)	Discoveries concerning the interaction between tumor viruses and the cell genetic material
1969	M. Delbrück, A. Hershey, and S. Luria (Physiology or Medicine)	Discoveries concerning the mechanism of replication and the genetic structure of viruses
1966	P. Rous (Physiology or Medicine)	Discovery of tumor-inducing viruses
1965	F. Jacob, A. Lwoff, and J. Monod (Physiology or Medicine)	Discoveries concerning genetic control of enzymes and virus synthesis
1954	J. F. Enders, T. H. Weller, and F. Chapman Robbins (Physiology or Medicine)	Discovery of the ability of poliomyelitis viruses to grow in cultures of various types of tissues
1951	M. Theiler (Physiology or Medicine)	Discoveries concerning yellow fever and how to combat it
1946	J. H. Northrop and W. M. Stanley (Chemistry)	Crystallization of *Tobacco mosaic virus*

FURTHER READING

Brock, T.D., 1990. The Emergence of Bacterial Genetics. Cold Spring Harbor Laboratory Press, Cold Spring Harbor, NY, p. 346.

Casjens, S.R., Hendrix, R.W., 2015. Bacteriophage lambda: early pioneer and still relevant. Virology 479–480, 310–330. https://doi.org/10.1016/j.virol.2015.02.010.

Lustig, A., Levine, A.J., 1992. One hundred years of Virology. J. Virol. 66, 4629–4631.

Murphy, F.A., 2016. The Foundations of Virology. Discoverers and Discoveries, Inventors and Inventions, Developers and Technologies. INFINITY Publishing, West Conshohocken, PA, p. 570, Extended Online Edition 2014–2016.

Taylor, M.W., 2014. Viruses and Man: A History of Interactions. Springer International Publishing, Switzerland, p. 430.

Chapter 2

Virion Structure, Genome Organization, and Taxonomy of Viruses

Gustavo Fermin
Universidad de Los Andes, Mérida, Venezuela

All living habitats . . . have and must operate in a virosphere.

Luis Villareal

WHAT IS A VIRUS?—PERHAPS THERE IS A SIMPLE ANSWER, *NOT AN EASY ANSWER*

The notion of what a virus is, and the nature of its relationship with the cellular world, has changed considerably over the past few decades. Viruses were initially distinguished from other organisms by their small size and their ability to pass through filters. They were described as "*filterable disease agents*": in other words, infectious agents that are small enough to pass through bacterial filters. Since viruses are quintessential parasites that depend on a host for most of their sustaining functions, logic dictated that viruses also be defined and classified on the basis of the host they infect. However, the host range has proven to be a very complex biological phenomenon with intricacies that are yet to be fully understood. Additionally, not all viruses can be defined as strict parasites given the benefits derived from their association with certain hosts.

The recent discovery of giant viruses that infect protists and bacteria has initiated a major shift in the thinking on how to define and classify viruses. Indeed both the particle and genome sizes of these viruses overlap significantly with those of bacteria, some eukaryotes, and archaea. The discovery called for a definition of viruses based on their essential nature rather than on their size. To this end, one proposal called for the revision of the living world into two major groups of organisms, ribosome-encoding organisms

Viruses. DOI: https://doi.org/10.1016/B978-0-12-811257-1.00002-4

that include all archaea, bacteria and eukarya, and capsid-encoding organisms, the viruses. Clearly, this division distinguishes viruses from cellular organisms based on genome content; in particular, the genes that encode ribosomal proteins and ribosomal RNAs. These genes are among the few genes whose sequences are conserved in all cellular organisms. The last universal common ancestor probably possessed a sophisticated ribosome that contained at least 34 ribosomal proteins that are shared by all archaeal, bacterial, and eukaryotic organisms. Unlike cellular organisms, viruses lack ribosomes and must use the ribosomes of their host cells for the translation of their mRNA into proteins.

While there is no single protein that is common to all viruses the expression of a capsid is a necessary structure that is used to disseminate viruses between ribosome-encoding hosts. One striking feature of viral capsids lies in the folded topology of the protein monomers that make up the capsid. The jellyroll β barrel appears to be the most prevalent structural core motif among capsid proteins, but it has not been found in any cellular protein. However, numerous groups of viruses share a common evolutionary history with genetic elements that lack a capsid (and its coding gene) and are never encapsidated or, in some cases, encapsidated in the virions of "*host*" viruses. Hence another proposal that has been put forward is a scheme that does not focus on the presence of a capsid or any particular gene, but rather is rooted in the concept of genetic, informational parasitism. That is, viruses are parasitic genetic information; they possess a varied repertoire of replication strategies and establish a range of relationships with cellular hosts that exhibit various degrees of reliance on the information processing systems of the host. In this chapter we first examine the range of biological features of viruses. We later look at the taxonomic classification schemes for viruses, along with specific pitfalls. For a list of all recognized virus families and unassigned genera, the reader is referred to Table 2.1.

VIRION MORPHOLOGIES

A virion is the physical entity that encompasses all that a virus represents in terms of its genome and the encapsidating protein shell, which maintains structural integrity and contribute to functional properties such as transmission. A virion is only one of the physical manifestations of a virus during a defined stage of its replication cycle; indeed, the most complex and complete of its physical manifestations. The most abundant forms of virions are based on spheres and rods with icosahedral and helical symmetries, respectively, and all their elegant variations. Some bacteriophages possess a unique morphology and are tadpole shaped. They possess a head-tail morphology consisting of a combination of icosahedral and helical symmetries. The recent discovery of archaeal viruses has widened our knowledge of viral morphologies. Although most resemble bacteriophages, some virions of

TABLE 2.1 List of Virus Families and Unassigned Genera as Recognized by the ICTV (2017), and Some Selected Features

Order	Viral Family	Genome Type	Genomic Molecule(s)	Host	Species	Env.	Virion Morphology
Unassigned	*Adenoviridae*	dsDNA	1 L	A	62	N	Icosahedral
Herpesvirales	*Alloherpesviridae*	dsDNA	1 L	A	12	Y	Spherical to pleomorphic
Tymovirales	*Alphaflexiviridae*	ssRNA(+)	1 L	F-P	50	N	Filamentous, flexible
Unassigned	*Alphatetraviridae*	ssRNA(+)	1 or 2 L	A	10	N	Icosahedral
Unassigned	*Alvernaviridae*	ssRNA(+)	1 L	Pr	1	N	Icosahedral
Unassigned	*Amalgaviridae*	dsRNA	1 L	P	4	NA	Capsid-less
Unassigned	*Ampullaviridae*	dsDNA	1 L	Ar	1	Y	Bottle-shaped, helical nucleoprotein
Unassigned	*Anelloviridae*	ssDNA(−)	1 C	A	68	N	Icosahedral
Unassigned	*Arenaviridae*	ssRNA(+/−)	2 L	A	36	Y	Spherical to pleomorphic
Nidovirales	*Arteriviridae*	ssRNA(+)	1 L	A	17	Y	Spherical to pleomorphic
Unassigned	*Ascoviridae*	dsDNA	1 C	A	4	Y	Bacilliform
Unassigned	*Asfarviridae*	dsDNA	1 L	A	1	Y	Spherical to pleomorphic
Unassigned	*Astroviridae*	ssRNA(+)	1 L	A	22	N	Icosahedral
Unassigned	*Avsunviroidae*	ssRNA	1 C	P	4	Vrd	Capsid-less
Unassigned	*Baculoviridae*	dsDNA	1 C	A	66	Y	Rod-shaped, helical nucleocapsid
Unassigned	*Barnaviridae*	ssRNA(+)	1 L	F	1	N	Bacilliform
Unassigned	*Benyviridae*	ssRNA(+)	1 L	P	4	N	Rod shaped

(*Continued*)

TABLE 2.1 (Continued)

Order	Viral Family	Genome Type	Genomic Molecule(s)	Host	Species	Env.	Virion Morphology
Tymovirales	*Betaflexiviridae*	ssRNA(+)	1 L	F-P	89	N	Filamentous, flexible
Unassigned	*Bicaudaviridae*	dsDNA	1 C	Ar	1	N	Lemon shaped
Unassigned	*Bidnaviridae*	ssDNA	2 L	A	1	N	Icosahedral
Unassigned	*Birnaviridae*	dsRNA	2 L	A	6	N	Icosahedral
Mononegavirales	*Bornaviridae*	ssRNA(−)	1 L	A	8	Y	Spherical
Unassigned	*Bromoviridae*	ssRNA(+)	3 L	P	33	N	Icosahedral or Bacilliform
Unassigned	*Caliciviridae*	ssRNA(+)	1 L	A	7	N	Icosahedral
Unassigned	*Carmotetraviridae*	ssRNA(+)	1 L	A	1	N	Icosahedral
Unassigned	*Caulimoviridae*	dsDNA-RT	1 C	A-P	62	N	Bacilliform or Icosahedral
Unassigned	*Chrysoviridae*	dsRNA	4 L	F	9	N	Icosahedral
Unassigned	*Circoviridae*	ssDNA(+/−)	1 C	A	70	N	Icosahedral
Unassigned	*Clavaviridae*	dsDNA	1 C	Ar	1	N	Bacilliform
Unassigned	*Closteroviridae*	ssRNA(+)	1-3 L	P	49	N	Filamentous, flexible
Nidovirales	*Coronaviridae*	ssRNA(+)	1 L	A	39	Y	Bacilliform or Spherical
Unassigned	*Corticoviridae*	dsDNA	1 C	B	1	N	Icosahedral
Unassigned	*Cystoviridae*	dsRNA	3 L	B	1	Y	Spherical
Picornavirales	*Dicistroviridae*	ssRNA(+)	1 L	A	15	N	Icosahedral
Unassigned	*Endornaviridae*	dsRNA	1 L	C-F-P	22	NA	Capsid-less

Bunyavirales	*Feraviridae*	ssRNA(−)	3 L	A	1	Y	Spherical, helical nucleocapsid
Mononegavirales	*Filoviridae*	ssRNA(−)	1 L	A	7	Y	Filamentous, flexible, helical nucleocapsid
Bunyavirales	*Fimoviridae*	ssRNA(−)	4-8 L	P	9	Y	Spherical, nucleocapsid
Unassigned	*Flaviviridae*	ssRNA(+)	1 L	A	82	Y	Spherical
Unassigned	*Fuselloviridae*	dsDNA	1 C	Ar	9	Y	Spindle-shaped
Tymovirales	*Gammaflexiviridae*	ssRNA(+)	1 L	F	1	N	Filamentous, flexible
Unassigned	*Geminiviridae*	ssDNA(+ / −)	1-2 C	P	369	N	Icosahedral
Unassigned	*Genomoviridae*	ssDNA(+ / −)	1 C	A-F-P	73	N	Icosahedral
Unassigned	*Globuloviridae*	dsDNA	1 L	Ar	2	Y	Spherical, pleomorphic
Unassigned	*Guttaviridae*	dsDNA	1 C	Ar	2	Y	Droplet shaped
Bunyavirales	*Hantaviridae*	ssRNA(−)	3 L	A	41	Y	Spherical, nucleocapsid
Unassigned	*Hepadnaviridae*	dsDNA-RT	1 C	A	12	Y	Spherical
Unassigned	*Hepeviridae*	ssRNA(+)	1 L	A	5	N	Icosahedral
Herpesvirales	*Herpesviridae*	dsDNA	1 L	A	89	Y	Spherical
Unassigned	*Hypoviridae*	ssRNA(+)	1 L	F	4	NA	Capsid-less
Unassigned	*Hytrosaviridae*	dsDNA	1 C	A	2	Y	Helical nucleocapsid
Picornavirales	*Iflaviridae*	ssRNA(+)	1 L	A	15	N	Icosahedral
Unassigned	*Inoviridae*	ssDNA(+)	1 C	B	31	N	Filamentous to rod shaped
Unassigned	*Iridoviridae*	dsDNA	1 L	A	12	Y/N	Icosahedral or Spherical
Bunyavirales	*Jonviridae*	ssRNA(−)	3 L	A	1	Y	Tubular or Spherical

(*Continued*)

TABLE 2.1 (Continued)

Order	Viral Family	Genome Type	Genomic Molecule(s)	Host	Species	Env.	Virion Morphology
Unassigned	*Lavidaviridae*	dsDNA	1 C	Pr	3	N	Icosahedral
Unassigned	*Leviviridae*	ssRNA(+)	1 L	B	4	N	Icosahedral
Ligamenvirales	*Lipothrixviridae*	dsDNA	1 L	Ar	8	Y	Filamentous
Unassigned	*Luteoviridae*	ssRNA(+)	1 L	P	34	N	Icosahedral
Herpesvirales	*Malacoherpesviridae*	dsDNA	1 L	A	2	Y	Icosahedral to pleomorphic
Picornavirales	*Marnaviridae*	ssRNA(+)	1 L	C	1	N	Icosahedral
Unassigned	*Marseilleviridae*	dsDNA	1 C	Pr	4	N	Icosahedral
Unassigned	*Megabirnaviridae*	dsRNA	2 L	F	1	N	Icosahedral
Nidovirales	*Mesoniviridae*	ssRNA(+)	1 L	A	7	Y	Spherical
Unassigned	*Metaviridae*	ssRNA-RT	1 L	A-F-P	39	Y	Ovoidal
Unassigned	*Microviridae*	ssDNA(+)	1 C	B	21	N	Icosahedral
Unassigned	*Mimiviridae*	dsDNA	1 L	Pr	2	N	Icosahedral (Pseudo)
Mononegavirales	*Mymonaviridae*	ssRNA(−)	1 L	F	1	Y	Filamentous, flexible, helical nucleocapsid
Caudovirales	*Myoviridae*	dsDNA	1 L	Ar-B	271	N	Icosahedral head with tail
Bunyavirales	*Nairoviridae*	ssRNA(−)	3 L	A	12	Y	Spherical
Unassigned	*Nanoviridae*	ssDNA(+)	8 or 6 C	P	12	N	Icosahedral
Unassigned	*Narnaviridae*	ssRNA(+)	1 L	C-F-Pr	7	NA	Capsid-less

Unassigned	*Nimaviridae*	dsDNA	1 C	A	1	Y	Helical nucleocapsid
Unassigned	*Nodaviridae*	ssRNA(+)	2 L	A	9	N	Icosahedral
Unassigned	*Nudiviridae*	dsDNA	1 C	A	3	Y	Helical nucleocapsid
Mononegavirales	*Nyamiviridae*	ssRNA(−)	1 L	A	5	Y	Spherical, helical nucleocapsid
Unassigned	*Ophioviridae*	ssRNA(−)	3 or 4 L	P	7	N	Filamentous, flexible nucleocapsids
Unassigned	*Orthomyxoviridae*	ssRNA(−)	6-8 L	A	9	Y	Spherical, Filamentous, Pleomorphic
Unassigned	*Papillomaviridae*	dsDNA	1 C	A	116	N	Icosahedral
Mononegavirales	*Paramyxoviridae*	ssRNA(−)	1 L	A:	49	Y	Spherical, Filamentous, Pleomorphic
Unassigned	*Partitiviridae*	dsRNA	2 L	F-P	60	N	Icosahedral
Unassigned	*Parvoviridae*	ssDNA(+/−)	1 L	A	62	N	Icosahedral
Bunyavirales	*Peribunyaviridae*	ssRNA(−)	3 L	A	52	Y	Spherical, nucleocapsid
Unassigned	*Permutotetraviridae*	ssRNA(+)	1 L	A	2	N	Icosahedral
Bunyavirales	*Phasmaviridae*	ssRNA(−)	3 L	A	6	Y	Helical nucleocapsid (?)
Bunyavirales	*Phenuiviridae: Goukovirus*	ssRNA(−)	3 L	A	4	Y	Spherical, icosahedral arrangement of glycoproteins
Bunyavirales	*Phenuiviridae: Phasivirus*	ssRNA(−)	3 L	A	3	Y	Spherical, icosahedral arrangement of glycoproteins
Bunyavirales	*Phenuiviridae: Phlebovirus*	ssRNA(+/−)	3 L	A	10	Y	Spherical, icosahedral arrangement of glycoproteins

(*Continued*)

TABLE 2.1 (Continued)

Order	Viral Family	Genome Type	Genomic Molecule(s)	Host	Species	Env.	Virion Morphology
Bunyavirales	*Phenuiviridae: Tenuivirus*	ssRNA(−)	4–5 C	P	7	N	Helical nucleocapsid (also they are circular)
Unassigned	*Phycodnaviridae*	dsDNA	1 L	A-C-P	33	Y	Icosahedral
Unassigned	*Picobirnaviridae*	dsRNA	2 L	A	2	N	Icosahedral
Picornavirales	*Picornaviridae*	ssRNA(+)	1 L	A	80	N	Icosahedral
Unassigned	*Plasmaviridae*	dsDNA	1 C	B	1	Y	Quasipherical to pleomorphic
Unassigned	*Pleolipoviridae*	dsDNA; ssDNA	1 C	Ar	5	Y	Pleomorphic
Unassigned	*Pleolipoviridae*	dsDNA	1 C	Ar	3	Y	Pleomorphic
Mononegavirales	*Pneumoviridae*	ssRNA(-)	1 L	A	5	Y	Spherical, helical nucleocapsid
Caudovirales	*Podoviridae*	dsDNA	1 L	B	132	N	Icosahedral head with tail
Unassigned	*Polydnaviridae*	dsDNA	1 C	A	53	Y	Prolate ellipsoid; helical nucleocapsid
Unassigned	*Polyomaviridae*	dsDNA	1 C	A	80	N	Icosahedral
Unassigned	*Pospiviroidae*	ssRNA	1 C	P	28	Vrd	Capsid-less
Unassigned	*Potyviridae*	ssRNA(+)	1 or 2 L	P	195	N	Filamentous, flexible
Unassigned	*Poxviridae*	dsDNA	1 L	A	71	Y	Pleomorphic
Unassigned	*Pseudoviridae*	ssRNA-RT	1 L	A-F-P-Pr	34	N	Spherical (Isometric to quasiisometric)
Unassigned	*Quadriviridae*	dsRNA	4 L	F	1	N	Icosahedral
Unassigned	*Reoviridae*	dsRNA	9–12 L	A-F-P	90	N	Icosahedral, double layered

Unassigned	*Retroviridae*	ssRNA-RT	1 L	A	55	Y	Spherical to pleomorphic
Mononegavirales	*Rhabdoviridae*	ssRNA(−)	1 L	A-P	131	Y	Bullet-shaped or Bacilliform; helical nucleocapsid
Nidovirales	*Roniviridae*	ssRNA(+)	1 L	A	1	Y	Bacilliform, helical nucleocapsid
Ligamenvirales	*Rudiviridae*	dsDNA	1 L	Ar	3	N	Rod-shaped
Unassigned	*Sarthroviridae*	ssRNA(+)	1 L	A	1	Sat	Icosahedral
Picornavirales	*Secoviridae*	ssRNA(+)	1 or 2 L	P	81	N	Icosahedral
Caudovirales	*Siphoviridae*	dsDNA	1 L	Ar-B	553	N	Icosahedral head with tail
Unassigned	*Solinviviridae*	ssRNA(+)	1 L	A	2	N	Spherical
Unassigned	*Sphaerolipoviridae*	dsDNA	1 L or 1 C	Ar-B	7	N	Icosahedral
Unassigned	*Spiraviridae*	ssDNA(+)	1 C	Ar	1	N	Coil-shaped, helical nucleoprotein
Mononegavirales	*Sunviridae*	ssRNA(−)	1 L	A	1	Y	Helical nucleocapsid (?)
Unassigned	*Tectiviridae*	dsDNA	1 L	B	5	N	Icosahedral
Unassigned	*Togaviridae*	ssRNA(+)	1 L	A	32	Y	Spherical, icosahedral
Unassigned	*Tolecusatellitidae*	ssDNA	1 C	P	72	Sat	Icosahedral
Unassigned	*Tombusviridae*	ssRNA(+)	1 or 2 L	P	74	N	Icosahedral
Bunyavirales	*Tospoviridae*	ssRNA(+ / −)	3 L	P	11	Y	Spherical, nucleocapsid
Unassigned	*Totiviridae*	dsRNA	1 L	F-Pr	28	N	Icosahedral
Unassigned	*Tristromaviridae*	dsDNA	1 L	Ar	2	Y	Rod-shaped helical nucleocapsid
Unassigned	*Turriviridae*	dsDNA	1 L	Ar	2	N	Icosahedral
Tymovirales	*Tymoviridae*	ssRNA(+)	1 L	A-P	39	N	Icosahedral

(*Continued*)

TABLE 2.1 (Continued)

Order	Viral Family	Genome Type	Genomic Molecule(s)	Host	Species	Env.	Virion Morphology
Unassigned	*Virgaviridae*	ssRNA(+)	1-3 L	P	59	N	Rod-shaped
Unassigned	Unassigned: *Albetovirus*	ssRNA(+)	1 L	P	3	Sat	Icosahedral
Mononegavirales	Unassigned: *Anphevirus*	ssRNA(−)	1 L	A	1	Y	Helical nucleocapsid (?)
Mononegavirales	Unassigned: *Arlivirus*	ssRNA(−)	1 L	A	1	Y	Helical nucleocapsid (?)
Unassigned	Unassigned: *Aumaivirus*	ssRNA(+)	1 L	P	1	Sat	Icosahedral
Unassigned	Unassigned: *Bacilladnavirus*	ssDNA(+/−)	1 C	C	1	N	Icosahedral
Picornavirales	Unassigned: *Bacillarnavirus*	ssRNA(+)	1 L	C	3	N	Icosahedral
Unassigned	Unassigned: *Blunervirus*	ssRNA(+)	4 L	P	1	N	Baciliform or filamentous?
Unassigned	Unassigned: *Botybirnavirus*	dsRNA	2 L	F	1	?	Icosahedral
Mononegavirales	Unassigned: *Chengtivirus*	ssRNA(−)	1 L	A	1	Y	Helical nucleocapsid (?)
Unassigned	Unassigned: *Cilevirus*	ssRNA(+)	2 L	P	2	N	Bacilliform
Mononegavirales	Unassigned: *Crustavirus*	ssRNA(−)	1 L	A	1	Y	Helical nucleocapsid (?)
Unassigned	Unassigned: *Deltavirus*	ssRNA(−)	1 C	A	1	Y	Spherical, helical nucleocapsid
Unassigned	Unassigned: *Dinodnavirus*	dsDNA	1 C	C	1	?	Icosahedral
Unassigned	Unassigned: *Higrevirus*	ssRNA(+)	3 L	P	1	N	Bacilliform

Unassigned	Unassigned: *Idaeovirus*	ssRNA(+)	2 L	P	1	N	Icosahedral
Picornavirales	Unassigned: *Labyrnavirus*	ssRNA(+)	1 L	C	1	N	Icosahedral
Unassigned	Unassigned: *Ourmiavirus*	ssRNA(+)	3 L	P	3	N	Bacilliform or Icosahedral
Unassigned	Unassigned: *Papanivirus*	ssRNA(+)	1 L	P	1	Sat	Icosahedral
Unassigned	Unassigned: *Polemovirus*	ssRNA(+)	1 L	P	1	N	Icosahedral
Unassigned	Unassigned: *Rhizidiovirus*	dsDNA	1 L	F-C	1	N	Icosahedral
Unassigned	Unassigned: *Salterprovirus*	dsDNA	1 L	Ar	1	Y	Spindle shaped
Unassigned	Unassigned: *Sinaivirus*	ssRNA(+)	1 L	A	2	?	Icosahedral
Unassigned	Unassigned: *Sobemovirus*	ssRNA(+)	1 L	P	19	N	Icosahedral
Unassigned	Unassigned: *Tilapinevirus*	ssRNA(+)	10 L	A	1	Y	Spherical, Filamentous, Pleomorphic (?)
Unassigned	Unassigned: *Virtovirus*	ssRNA(+)	1 L	P	1	Sat	Icosahedral
Mononegavirales	Unassigned: *Wastrivirus*	ssRNA(−)	1 L	A	1	Y	Helical nucleocapsid (?)

Notes: (1) This table was organized in terms of virus family names (second column) ordered alphabetically. Since there are many genera not assigned to a particular family, they appear under the designation "*Unassigned*" in Column 2 after the family *Virgaviridae*. (2) Although the main criterion for listing virus families here was type of genome molecule, the families *Phenuiviridae* and *Pleolipoviridae* appear more than once since not all members of these families possess the same kind of genome molecule. (3) In Column 4 the number of genome molecules are listed followed by the kind of molecule, be linear (L) or circular (C). (4) In Column 5, we followed the proposal that all living beings can be classified into seven kingdoms: Animalia (A), Archea (Ar), Bacteria (B), Chromista (C), Fungi (F), Plantae (P) or Protozoa (Pr). We are aware that members in some virus families use as hosts a much defined group of species from a given kingdom, but for the purposes of simplicity, we decided not to be more specific than we inform here. (5) Finally we tried to be consistent with the designation of virion morphology, but we also understand that this is not an easy task given the extreme variability of virion forms. *Env.*, Envelope; *Sat*, satellite; *Vrd*, viroid.

archaeal viruses are bottle-, spindle-, or droplet-shaped. It is argued that there are at least 20 unrelated types of capsid proteins in all viruses (and hence, it might be assumed that they evolved independently on different occasions), and that most, if not all, originated from ancestral proteins of cellular organisms. A survey of the protein folds of capsid and nucleocapsid (protein associated with nucleic acids) structures reveals strong similarities in viruses infecting species of different domains of life, which is an indication of their antiquity and of the evolutionary connections among the viruses that encode them. For instance, seven major structural classes encompass more than 60% of the known virosphere. The most prevalent folds in viral capsid proteins are the single jellyroll and the double jellyroll. They can be found in more than a third of the capsid protein folds studied so far.

The unique folds of viral capsid proteins present a distinctive geometry and allow effective packaging and virion assembly. In general self-assembly guides the generation of new virions consisting of proteins coded in the virus genome that are synthesized *de novo* in and by the host. The virion can be formed by a defined number of the same protein (capsomers or nucleocapsid proteins), or by a few different proteins held together by hydrogen bonding and hydrophobic interactions. The virion can also be covered by a lipid envelope, derived from the very host. In some other cases the virion carries an internal membrane situated between the genome and the viral coat. Virions exhibit an amazing diversity in size, and range from 18 to 19 nm (*Nanoviridae*) to c. 900 nm (*Mimiviridae*), or even more (i.e., the unrecognized genera Pandoravirus and Pithovirus; the latter are more than 1000 nm in diameter). Fig. 2.1 provides a general, yet incomplete catalog of virion forms.

Icosahedral/Spherical Virions

The majority of viruses adopt the form of a sphere. In this form the capsid is made up of 60 replicas of the same protein(s) that interact spontaneously through template domains at their edges, resulting in an icosahedral structure. This is the simplest icosahedral capsid with a diameter of 20–25 nm. The capsid is composed of 12 pentameric capsomeres, i.e., a total of 60 capsomeres are required to complete the capsid. In larger viruses, more than 60 capsomeres are required to completely encapsidate the genome. The icosahedral capsid of the *Rubella virus*, e.g., is composed of 12 pentameric and 30 hexameric capsomeres for a total of 240 capsomeres. The diameter of these virions is in the range of 65–70 nm. Some viruses are easily recognizable by the unique arrangements of their capsomeres at the vertices of the icosahedron. For example, virions of members in the family *Hepeviridae* show a slightly lumpy daisy shape, those of *Caliciviridae* cup-like indentations, while those of *Astroviridae* display a remarkable star shape. Other viruses exhibit a multilayered protein capsid morphology, among these

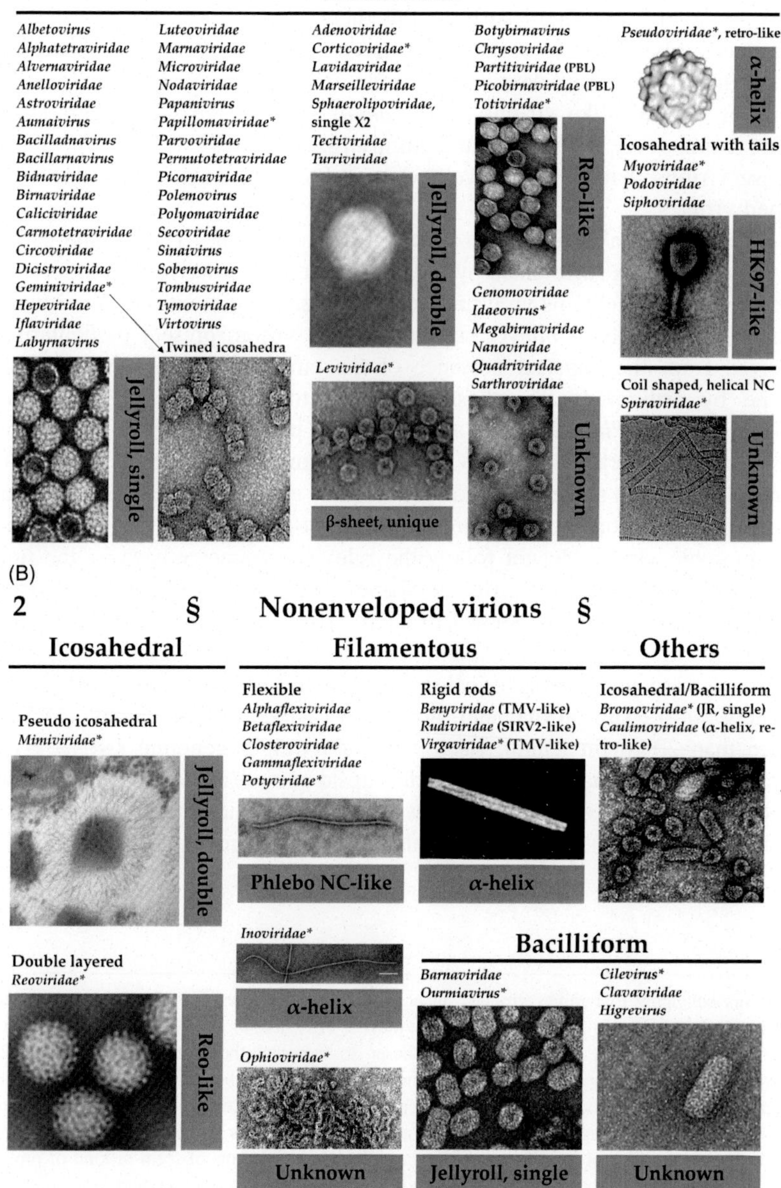

FIGURE 2.1 Morphological variability among viruses. In representing the diversity in virion forms, viruses were first grouped according to the possession, or not, of an envelope. Then, hierarchical groups were formed following the most accepted criterion of form (*spherical/icosahedral, helical/bacilliform/filamentous, etc.*) and where known, the prevalent structural core motif among the capsid proteins (*jellyroll, α helix, β sheet, etc.*). The latter is indicated below or at the right side of the photographs; when a particular fold belongs to a class not yet defined, the word "*unknown*" is used. In instances where another protein fold is also found within the group, this information is added beside the family or genus name. For example, filamentous rigid rods of

(*Continued*)

are species belonging to the family *Reoviridae*. Here, the capsids consist of 1–3 layers of protein surrounding the genome core of 9–12 dsRNA segments. Quite a few viruses also have fiber-like structures extending from their surfaces. Adenoviruses have long fibers emanating from its vertices, Sputnik-like virions (e.g., *Adenoviridae*) have short fibers stemming from every capsomer, and mimiviruses possess a forest of fibers. These virions that lack an envelope are referred to as naked or nonenveloped viruses.

Many spherical viruses also have an outer envelope. That is, the capsid of these viruses is surrounded by a lipid bilayer that is derived from host cell membranes and contains viral proteins. These enveloped virions, by virtue of the host-derived lipid bilayer surrounding the virion, present particularities derived from the physical presence of the envelope. Arboviruses (viruses borne by arthropods like those belonging to the families *Flaviviridae*, *Togaviridae*, or *Phenuiviridae*, among others) present as fuzzy spheres possessing a peripheral fringe. HIV-1 carries prominent projections because of viral glycoprotein spikes that protrude from its envelope. The latter virus and other members of the *Retroviridae* are also described as pleomorphic. Pleomorphic viruses do not follow the rules of symmetry because the lipid envelope readily adopts different shapes and sizes, making each virion unique. Their virions are known to adopt spherical or polygonal structures. The influenza virion is an example. The virions generally take on an irregular shape or may appear spherical or elliptical in shape, ranging from approximately 80–120 nm in diameter, and are occasionally filamentous, reaching more than 20 μm in length. Each copy of its eight genomic ssRNA (-) is folded into a rod-shaped, helical nucleocapsid complex referred to as **ribonu**cleoprotein **c**omplex (RNP). Along with multiple copies of the viral-encoded **n**ucleo**p**rotein (NP), each RNP contains a heterotrimeric viral polymerase

◀ members belonging to families grouped under nonenveloped viruses (Part 2 of the figure) may present coat proteins with α-helical motifs, that may be TMV-like or SIRV2-like. An asterisk indicates the specific taxon (family or genus) chosen to represent the morphology of the group. Unknown (*not in a gray box*) is also used when the presence or absence of an envelope, virion shape or kind of fold has not been reported for a particular family or genus of viruses. Part 6 of the figure presents viruses that do not possess a capsid or viruses for which limited information is available; i.e., families or genera with no further information in terms of form or kind of protein fold, the recently described unique group of viruses (e.g., *Tristromaviridae*), and a genus that is not fully accepted yet by ICTV (Faustovirus). *JR*, jellyroll; *NC*, nucleocapsid; *PBL*, picobirna-like. Source: Andrew King, Elliot Lefkowitz, Michael Adams and Eric Carstens. Ninth Report of the International Committee on Taxonomy of Viruses. Elsevier 2012. *Apart from the pending genus* Faustovirus, *taxa in the groups represented in Part 6 are not described in Krupovic and Koonin (2017) https://doi.org/10.1073/pnas.1621061114.*

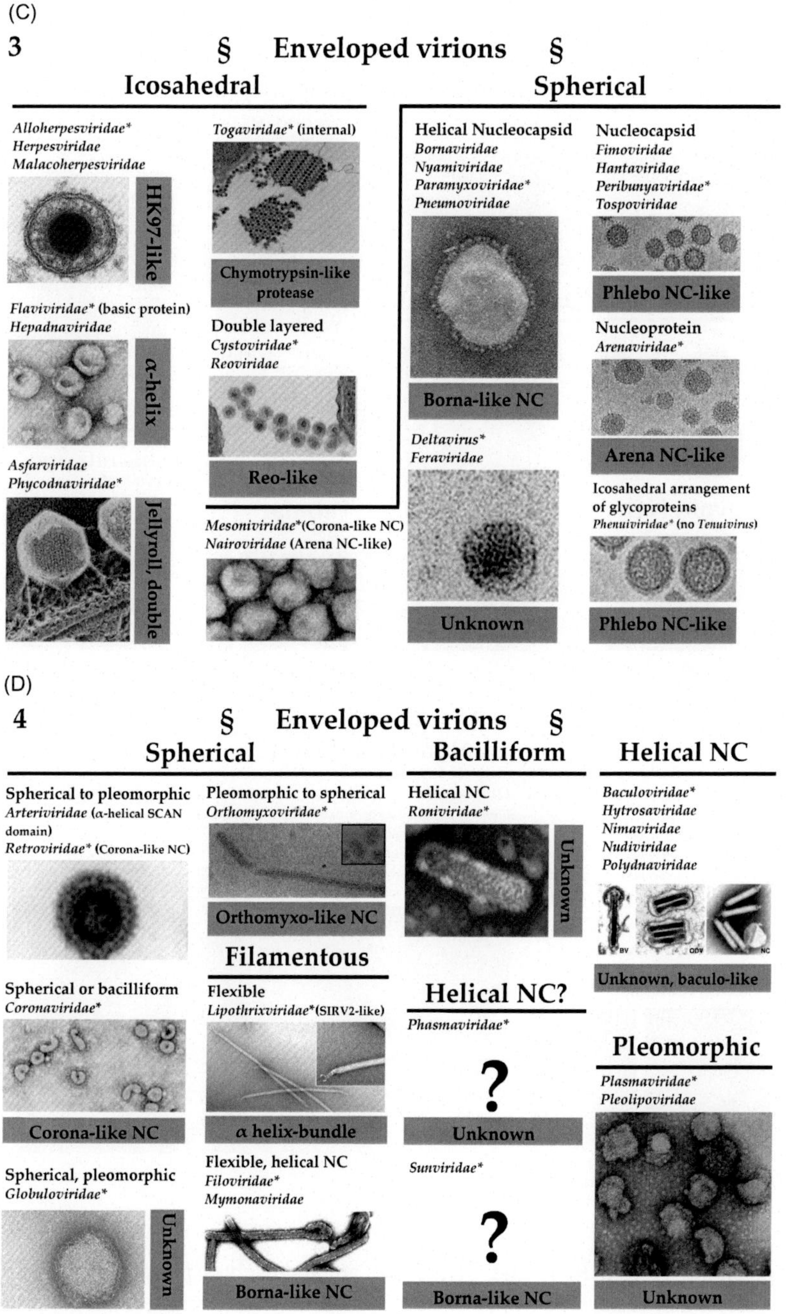

(C)

3 § Enveloped virions §

Icosahedral

*Alloherpesviridae**
Herpesviridae
Malacoherpesviridae
HK97-like

*Flaviviridae** (basic protein)
Hepadnaviridae
α-helix

Asfarviridae
*Phycodnaviridae**
Jellyroll, double

*Togaviridae** (internal)
Chymotrypsin-like protease

Double layered
*Cystoviridae**
Reoviridae
Reo-like

*Mesoniviridae**(Corona-like NC)
Nairoviridae (Arena NC-like)

Spherical

Helical Nucleocapsid
Bornaviridae
Nyamiviridae
*Paramyxoviridae**
Pneumoviridae
Borna-like NC

*Deltavirus**
Feraviridae
Unknown

Nucleocapsid
Fimoviridae
Hantaviridae
*Peribunyaviridae**
Tospoviridae
Phlebo NC-like

Nucleoprotein
*Arenaviridae**
Arena NC-like

Icosahedral arrangement of glycoproteins
*Phenuiviridae** (no *Tenuivirus*)
Phlebo NC-like

(D)

4 § Enveloped virions §

Spherical

Spherical to pleomorphic
Arteriviridae (α-helical SCAN domain)
*Retroviridae** (Corona-like NC)

Spherical or bacilliform
*Coronaviridae**
Corona-like NC

Spherical, pleomorphic
*Globuloviridae**
Unknown

Pleomorphic to spherical
*Orthomyxoviridae**
Orthomyxo-like NC

Filamentous

Flexible
*Lipothrixviridae**(SIRV2-like)
α helix-bundle

Flexible, helical NC
*Filoviridae**
Mymonaviridae
Borna-like NC

Bacilliform

Helical NC
*Roniviridae**
Unknown

Helical NC?

*Phasmaviridae**
?
Unknown

*Sunviridae**
?
Borna-like NC

Helical NC

*Baculoviridae**
Hytrosaviridae
Nimaviridae
Nudiviridae
Polydnaviridae
BV ODV NC
Unknown, baculo-like

Pleomorphic

*Plasmaviridae**
Pleolipoviridae
Unknown

FIGURE 2.1 Continued

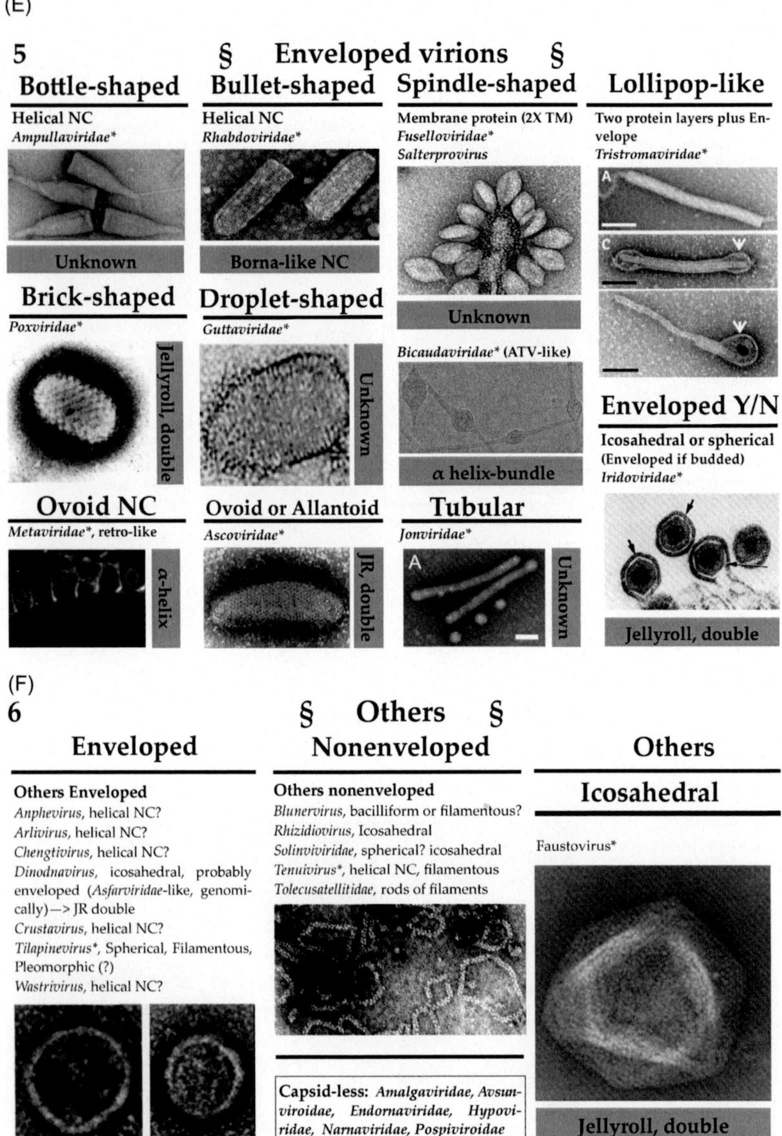

FIGURE 2.1 Continued

(consisting of PA, PB1, and PB2), and are themselves packaged in a lipid envelope that is derived from the host cell surface membrane. Two glycoproteins encoded by the virus are found in the envelope: hemagglutinin (HA) and neuraminidase (NA).

The most complex architecture of all spherical virions is shown by viruses of the families *Myoviridae*, *Podoviridae* and *Siphoviridae*, of the order *Caudovirales*, which infect bacteria and archaea. Some species of these three families show a head-tail structure where the icosahedral "*head*" is connected to a specialized host cell attachment structure, the tail, that may be contractile (*Myoviridae*) or not (the other two families). Three or four subterminal fibers, but more frequently six, may be present on these virions. Members belonging to the *Siphoviridae*, however, possess a noncontractile tail and never tail fibers. Virions with this type of structure are never enveloped.

Filamentous/Bacilliform Virions

The other simple way of constructing a virion is by making use of helical symmetry, in which the protein subunits and the nucleic acid are arranged in a helix. This gives a filament with flexibility that depends on the strength of the protein–protein interactions. That is, some filaments are rigid because of strong bonds between the protein monomers in successive turns of the helix (for instance *Tobacco mosaic virus*, *Virgaviridae*), while others are flexuous (*Potyviridae*) due to weaker bonds. Some filaments are so flexible that they can assume very different shapes. In *Filoviridae*, e.g., enveloped virions may be simple cylinders (the less frequent form) or form branches or loops. The loops can also vary in form and include the thread-like form (the most general morphotype) as well as U-, 9-, eye-bolt- or Shepperd's crook-shapes, among others. In the case of the family *Ophioviridae*, the highly flexuous, nonenveloped virions can form open circles.

Other virions with helical symmetry have a wider diameter so they look a little thicker than the typical filamentous virions. These are bacilliform (hemispherical at both ends) or bullet-shaped (with one rounded and one flattened end) filaments. They are more frequently found in plant and animal viruses (*Caulimoviridae*, e.g., with hosts in both kingdoms), but some archaeal virions (*Clavaviridae*) are also bacilliform.

Amazing Forms Among Old and New Comers

Viruses that exist in extreme environments display a wide range in morphotypical variations that are not observed with those entities found in less harsh habitats. A filamentous dsDNA virus that infects members of the archaeal genus *Pyrobaculum*, e.g., has its linear genome enclosed in a tripartite shell consisting of two protein layers and an envelope, which is unusual for dsDNA viruses. Other structures found with archaeal viruses include, as mentioned earlier, bottle-, lemon-, droplet-, and spindle-shapes. Virions of unusual forms are not exclusive to archaea, though, since bullet-shaped virions can be found in rhabdoviruses. Ovoid forms, on the other hand, are common among viruses belonging to the families *Ascoviridae* (which can also have a bacilliform or allantoid forms), *Metaviridae*, *Nimaviridae* (which also show a tail-like appendage at one end), and *Poxviridae* (which can also be brick shaped). Additionally, other virions (Pandoraviruses) resemble the shape of an amphora (Greek vase), have a prolate ellipsoid form (*Polydnaviridae*) or are twined icosahedra (*Geminiviridae*).

DO VIRUSES REALLY NEED A COAT?

The vast majority of viruses possess a capsid and form virions. The capsid plays multiple roles in the infection cycle of viruses:

1. A coat provides the virus genome protection against environmentally damaging agents. The virion serves as a physical barrier for the genome and proteins it may harbor. Indeed the structural integrity of the capsid is essential for viral replication and the expression of viral genes.
2. The coat facilitates interaction with host cells in terms of recognition, processing (mostly uncoating) and tropism—after host recognition, a virus displays its full biological potential in specific tissues of the host.
3. Coat proteins assist in virus movement between cells in multicellular organisms, thus allowing for systemic infection of the host.
4. Coat proteins, in addition, facilitate vector-mediated transmission of viruses between organisms. In the vector, coat proteins along with other viral proteins, allow for retention of virions to vector mouth parts, movement to other parts of the vector body, replication, and sometimes the production of more virions.

Nonetheless, viruses appear to have evolved from capsid-less selfish elements. We know of viruses that never have (nor code for) a capsid. For instance the families *Endornaviridae*, *Hypoviridae*, and *Narnaviridae* are comprised of members where a capsid is absent. Interestingly, members of the family *Amalgaviridae* encode for but do not form virions. The likely existence of capsid-less ancestors, along with the existence of capsid-less viruses today, suggests that the viral capsid was a late acquisition during their evolutionary history. Further, it appears that coat proteins not only accommodated the formation of capsids and their associated functions, but their acquisition seemingly widened opportunities for viruses to become the most numerous, conspicuous, and highly evolving biological entities known.

TYPES AND FUNCTION OF THE VIRUS ENVELOPE

As alluded to earlier, slightly less than a quarter of all viruses possess an envelope surrounding their protein capsid. The envelope typically consists of a host-derived lipid bilayer membrane plus glycoproteins of viral origin. Its structure varies in terms of size, composition, morphology, and complexity. Viral-derived glycoproteins are embedded in the lipid bilayer and in some cases, other nonglycosylated proteins of viral origin form part of the envelope. The number of virus glycoproteins varies among viral groups; in members of the family *Herpesviridae*, e.g., more than 10 glycoproteins have been identified. In simplier cases (e.g., members of *Togaviridae* and *Orthomyxoviridae*), there are one or two multimeric proteins. Other viral-encoded glycoproteins are the ion channel proteins, viroporins. Viroporins have at least one transmembrane domain and sometimes an extracellular membrane region that interacts with viral or host proteins. Viroporins have been discovered in *Influenza A virus*, *Hepacivirus C* (previously known as Hepatitis C virus), *Human immunodeficiency virus 1*, and Coronaviruses. Enveloped viruses may also contain a layer of protein between the envelope and the capsid, known as the matrix (*Orthomyxoviridae, Retroviridae)*, or there is no such layer and the capsid interacts directly with the internal tails of the membrane proteins (*Togaviridae, Bunyavirales*).

Recognition, attachment, and entry into cells through fusion with host cell membranes are probably the main roles played by the envelope and associated proteins. The envelope helps viruses to evade the mammalian host's immune system. Additionally, from an evolutionary point of view, virus envelope proteins, with variations, are well represented in vertebrate genomes and apparently were a potent force in placentation in different species. The syncytin genes represent a dramatic example of convergent evolution via the cooption of a retroviral envelope gene for a key biological function in placental morphogenesis.

GENOMES: THE NUCLEIC ACIDS SPACE EXPLOITED BY VIRUSES

All cellular organisms possess only one type of genomic molecule: double stranded DNA, along with a myriad of informational molecules represented by different types of RNA, and instances of ssDNAs. Viruses, on the other hand, have exploited the genomic space to its limits. Either DNA or RNA is used by viruses to encode their genomes, with many variations. In the case of RNA viruses, the genome can be single stranded in minus (−) or plus (+) orientations, or a combination of both (+/ −), or double stranded. The vast majority of viruses have RNA genomes. When it comes to DNA genomes, more than a third of all recognized virus species possess a dsDNA genome. DNA viruses can also have ssDNA(−) or ssDNA(+) molecules; ssDNA (+/ −) viruses also exist. In some cases the virus genome is only partially double stranded and partially single stranded (*Pleolipoviridae*), while other dsDNA genomes are described as open circular dsDNA due to nicks at specific sites along the DNA (*Caulimoviridae* and others). In general, viruses' genomes are either in linear or circular conformations.

The genome of a virus exists in the form of one or more molecules of nucleic acid. The most common number of viral genomic molecule is one (monopartite viruses). Other viruses are bipartite or tripartite, and yet others are so complex that their complete genome is represented by up to 12 different molecules. Although one genome molecule per virion is the "*rule,*" some viruses can carry all the genomic molecules in one particle while other viruses have their segmented genomes portioned in different particles (*Nanoviridae*); yet others encapsidate replicas of the same genome in a single particle (e.g., virus species of the family *Polydnaviridae*). A very important family of viruses, *Geminiviridae*, consists of two incomplete icosahedra joined together to form twinned particles; they are monopartite or bipartite. Monopartite geminiviruses encode all their genes on a single circular ssDNA. Bipartite geminiviruses (genus *Begomovirus*), on the other hand, consist of two virions containing different genomic components that are required for productive infection (i.e., genes are distributed on two separate ssDNA molecules that are packaged separately). Additional details on the categorization of virus' nucleic acids are given in the following section. A summary of the virosphere, in terms of nucleic acids, is provided in Table 2.2, and in Fig. 2.2.

dsDNA viruses

Viruses with a dsDNA genome amount to more than a third (38.6%) of all recognized viruses. These genomes can be found as circular molecules (25.3%) or as single linear molecules (74.3%). Typically, viral genomic DNA of members of a dsDNA viral family all have the same topological form. However, species of the family *Sphaerolipoviridae* can have either a circular or a linear genome. Some members of the family *Pleolipoviridae* have a monopartite circular genome, while others have a bipartite genome composed of a circular dsDNA and another which is circular, but single stranded. In both cases the dsDNA genomic molecules is interrupted by short runs of ss linear DNA. Members of the family *Caulimoviridae* possess a monopartite, circular dsDNA genome with gaps (discontinuities) in each strand of the genomic molecule. Specific nicks can be found in the transcribed strand, along with 1–3 in the nontranscribed strand.

Roughly 77% of viruses with a dsDNA genome are nonenveloped. Of note, members of the family *Iridoviridae* can be enveloped or not, depending on their route of exit from the host cell (budding or lysis). Other dsDNA viruses may possess a lipid membrane between the coat and the genome (*Turriviridae*). Although dsDNA viruses can infect members of all kingdoms

TABLE 2.2 Summary of Virus Families (and Unassigned Genera) Based on Genomic Molecule (Regardless Strategies of Expression), 2017

Genome	Type: (#)	Host(s)	Env: (#)	Species
dsDNA, including dsDNA-RT viruses (74 species)	1 L: 1265	A, Ar, B, C, F, P, Pr	Y: 376	38 families (2 RT)
	1 C: 422		N: 1305	3 unassigned genera
	1 C or 1 L: 7		Y/N: 12	
			?: 1	1694 species
ssDNA, including a family of satellites	2 L: 1	A, P	N: 1	2 families
	1 C: 72		Sat: 72	73 species
ssDNA(+)	1 C: 53	Ar, B, P	N: 65	4 families
	6 or 8 C: 12			65 species
ssDNA(−)	1 C: 68	A	N: 68	1 family
				68 species
ssDNA(+/ −)	1 L: 62	A, C, F, P	N: 575	4 families
	1 C: 144			1 unassigned genus
	1–2 C: 369			575 species
dsRNA	1 L: 54	A, B, C, F, P, Pr	Y: 1	11 families
	2 L: 70		N: 197	1 unassigned genus
	3 L: 1		NA: 26	
	4 L: 10		?: 1	
	9–12 L: 90			225 species
ssRNA	1 C: 32	P	NA: 32	2 viroid families
				32 species
ssRNA(−), excluding the members of the family *Phenuiviridae* (*Bunyavirales*)	1 L: 212	A, F, P	Y: 344	17 families
	3 L: 113		N: 7	6 unassigned genera
	3 or 4 L: 7			
	4–8 L: 9			
	6–8 L: 9			
	1 C: 1			351 species

(*Continued*)

TABLE 2.2 (Continued)

Genome	Type: (#)	Host(s)	Env: (#)	Species
ssRNA(+), including ssRNA-RT viruses (120 species)	1 L: 723	A, B, C, F, P, Pr	Y: 273	39 families (3 RT)
	2 L: 12		N: 949	15 unassigned genera
	1 L or 2 L: 360		NA: 11	
	3 L: 37		Sat: 7	
	1–3 L: 108		?: 2	
	4 L: 1			
	10 L: 1			1242 species
ssRNA(+/ −)	2 L: 36	A, P	Y: 47	2 families
Excluding *Phenuiviridae* (+/ −)	3 L: 11			47 species
Mixed				
Pleolipoviridae				1 family
dsDNA	1 C: 3	Ar	Y:3	
Pleolipoviridae				
dsDNA; ssDNA	1 C: 5	Ar	Y: 5	8 species
Phenuiviridae				1 family
ssRNA(−)	3 L: 7	A, P	N: 7	
	4–5 C: 7		Y: 7	
Phenuiviridae				
ssRNA(+/ −)	3 L: 10	A	Y: 10	24 species
Summary	Linear: 3209	A, Ar, B, C, F, P, Pr	N: 3174	122 families
DNA: 2483	1 L: 2316		Y: 1066	26 unassigned genera
dsDNA: 1702[a]	2 L: 119		Y/N: 12	
ssDNA: 73	3 L: 179		NA: 69	
ssDNA(+): 65	1 or 2 L: 360		Sat: 79	
ssDNA(−): 68	1–3 L: 108		?: 4	
ssDNA(+/ −): 575	4 L: 11			
	3 or 4 L: 7			

(*Continued*)

TABLE 2.2 (Continued)

Genome	Type: (#)	Host(s)	Env: (#)	Species
RNA: 1921	4–8 L: 9			
dsRNA: 225	6–8 L: 9			
ssRNA: 32	9–12 L: 90			
ssRNA(+): 1242[b]	10 L: 1			
ssRNA(−): 365				
ssRNA(+/ −): 57	Circular: 1188			
	1 C: 800			
	1 or 2 C: 369			
	4–5 C: 7			
	6 or 8 C: 12			
	1 L or C: 7			4404 species

Viral genomes can be linear (L) or circular (C); their hosts are coded here as A (Animalia), Ar (Archaea), B (Bacteria), C (Chromista), F (Fungi), P (Plantae), or Pr (Protozoa).Virions can be enveloped (Y) or not (N); when information is not available a "?" is used. *NA*, not applicable; *Sat*, satellite. Number of species (#) are indicated in each case, when appropriate.
[a]*Including all members of the family* Pleolipoviridae, *as well as all dsDNA-RT virus species.*
[b]*Including all dsRNA-RT virus species.*

of life, the vast majority are limited to species of Bacteria or Archaea (more than 60%), while roughly another third infect members of the kingdom Animalia. Virions of dsDNA viruses, particularly those able to infect Archaea, display extreme variability of forms. About half of these viruses produce virions with heads and tails, icosahedral viruses are also common, as are bacilliform, spherical, prolate ellipsoid as well as droplet-, lemon-, rod-, and spindle-shaped virions. Most dsDNA viruses are in the B form of the nucleic acid, and there is at least one report of an A form in a virus that infects a hyperthermophilic acidophile Archaea (*Sulfolobus islandicus rod-shaped virus 2*, family *Rudiviridae*).

Of note, among the 38 families and three unassigned genera of dsDNA viruses, members of the *Caulimoviridae* and *Hepadnaviridae* families are the only ones that generate new genome copies by retrotranscription (DNA to RNA to DNA) instead of canonical DNA replication (DNA to DNA).

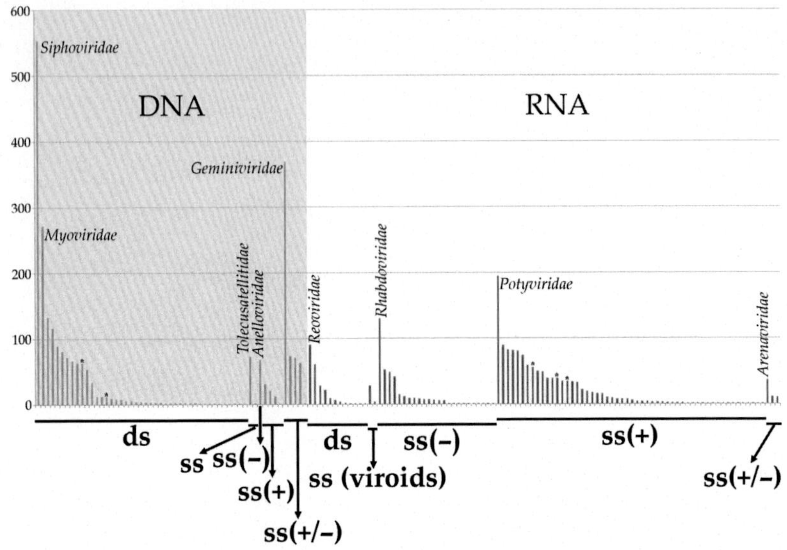

FIGURE 2.2 Schematic representation of virus families and unassigned genera as of April 2017. Number of species is indicated on the left, and all families and genera are presented according to the nature of their genome, DNA (*gray*) or RNA (*white*). In each case, families and genera have been subdivided in terms of the molecular type [ds, ss, ss(+), s(−), and ss(+/ −)] and names of the largest families for each molecular type are indicated. The asterisk represents virus families that use a **r**etro**t**ranscribing (RT) mechanism. The figure has been drawn to scale.

ssDNA viruses—No Polarity or Mixed Polarity

These particular ssDNA viral entities belong to either one of two families: the newly proposed *Tolecusatellitidae* family that groups 72 satellite viruses—which are always associated with geminiviruses, and the monospecific family *Bidnaviridae* whose only member infects silkworms causing silkworm flacherie disease. Members of this class of viruses can have a bipartite, linear genome (bidnavirus), or a monopartite, circular genome (tolecusatellitids). Their virions are nonenveloped icosahedra.

Members of the family *Tolecusatellitidae* are satellites (i.e., they are nonindependent entities) that rely on a helper virus for their spread (a begomovirus or a mastrevirus from the family *Geminiviridae*). On the other hand, the only member of the bidnaviruses, *Bombyx mori bidensovirus*, the bipartite, linear genome is composed of segments VD1 and VD2 that are encapsidated separately. Since equal amounts of positive (+) and negative (−) molecules are encapsidated, four different types of full particles are produced: VD1 + /VD2 + , VD1 + /VD2 − , VD1 − /VD2 + , and VD1 − / VD2 − . This bidnavirus is not related to tolecusatellitids; it is listed here in order to emphasize the ss nature of its genome. ssDNA viruses convert into a double stranded form prior to transcription to mRNAs.

ssDNA(−) viruses

The only family of ssDNA(−) viruses is the family *Anelloviridae* that is comprised by 68 species. Anelloviruses are able to infect diverse groups of vertebrates, and are sometimes asymptomatic in humans. They are associated with various diseases that include hepatitis, lupus, and miopathy, among others. The virus can be transmitted sexually and by human excreta. The genome, represented by a single circular molecule of ssDNA of negative polarity, is protected in nonenveloped icosahedral virions. It is produced through dsDNA intermediates and contains four potential ORFs. Proteins are expressed by alternative splicing of a single premRNA.

ssDNA(+) viruses

Viruses with a ssDNA(+) genome are hosted mainly by bacteria (2 families, *Inoviridae* and *Microviridae*, encompassing 52 species), plants (one family, *Nanoviridae*, with 12 species) and archaea (with one monospecific family, *Spiraviridae*). The genome is circular, but the number of genomic molecules varies from 1 (in 53 species) to 6–8 (only for the nanoviruses). In nanoviruses, each genomic molecule is monocistronic, but in rare cases one gene can carry two overlapping ORFs. Each genomic molecule is encapsidated separately.

ssDNA(+/−) viruses

This group presents a different scenario than that explained for the bidnaviruses in the section on *ssDNA viruses*. In that case an assortment of ssDNA molecules gives rise to genomic combinations of + or − strands. In the case of ssDNA(+/−) viruses the genome is ambisense. That is, in the same, or different molecules, the genome is composed of DNA sequences coding in opposite directions. Most of these viruses are monopartite, but geminiviruses can have a monopartite or bipartite genomes. The genome is circular, except for members of the *Parvoviridae* family of animal viruses, which present a linear genome. Four families and one unassigned genus group all viruses with an ambisense ssDNA genome. This represents slightly more than 13% of all recognized viruses. Virions are icosahedral and nonenveloped. ssDNA(+/−) viruses are hosted only by eukaryotes, with the exception of members of the kingdom Protozoa at the time of this writing.

dsRNA viruses

All viruses with a dsRNA genome possess a linear molecule in variable numbers: 1, 2, 3, 4—or even between 9 and 12 (*Reoviridae*). Two unusual features of endornaviruses, not found in any other RNA viruses, is that their genome codes for an unusually long, single ORF and that the genomic molecule possesses a site-specific nick in the 5′ region of the coding strand. Although all dsRNA viruses are nonenveloped, it is also true that not all seem to produce virions. For example, members of the *Amalgaviridae* capsid-less family of viruses do code for a coat protein. Amazingly, however, the RdRp of amalgaviruses is closely related to the same protein belonging to members of another family of dsRNA viruses (*Partitiviridae*), while the CP is homologous to the nucleocapsid proteins of ssRNA(−) viruses of the genera *Phlebovirus* and *Tenuivirus* (family *Phenuiviridae*). That is, the genome of amalgaviruses is chimeric (different origins), and probably evolved by recombination between a partitivirus and a tenuivirus. Amalgaviruses are transmitted by seeds and cannot be mechanically transmitted between plants. The other dsRNA virus family whose members do not produce virions is *Endornaviridae*. Members of this family infect chromists, fungi, and plants.

The only noneukaryote dsRNA virus is the tripartite *Pseudomonas virus phi6*, the lone species of the family *Cystoviridae*. It is also the only dsRNA virus with enveloped virions consisting of double layered icosahedrons. The genomic dsRNAs of this virus are never exposed in the cytoplasm. Instead they remain enclosed in the inner capsid where they are transcribed by the viral RdRp to create the corresponding mRNA for translation (ssRNA(+) molecules). Generation of new dsRNA occurs within assembled progeny capsids after the three different genomic sRNA(+) segments are translocated from the original capsids.

ssRNA (Viroids)

Members in this group are all viroids classified into two families; *Avsunviroidae* (four species) and *Pospiviroidae* (28 species). All infect only plant hosts. Viroids possess a circular genome that is not encapsidated in a protein coat. Biologically speaking, they behave like viruses, but their genomes do not code for proteins. Viroids depend on proteins of their hosts for their replication (via rolling circle model) and other functions. Curiously, viroid genomes of the family *Avsunviroidae* share some organization and replication characteristics with the human ssRNA(−) *Hepatitis delta virus*, from the unassigned, monospecific genus *Deltavirus*. Presumably, this virus and viroids have a common ancestor. It is not unreasonable to expect the discovery of more of these virus/viroids entities in the future, at least in the kingdoms Animalia and Plantae.

ssRNA(−) viruses

Viruses with a ssRNA(−) genome, classified into 18 families and 6 unassigned genera, include 365 viruses hosted almost exclusively by plants (two families, and all members of the genus *Tenuivirus* from the family *Phenuiviridae*) or animals (13 families, six unassigned genera, and the genera, *Goukovirus* and *Phasivirus* from the family *Phenuiviridae*). Notably the *Rhabdoviridae* is one of the most ecologically diverse families of RNA viruses, with 131 members identified in a range of plants and animals, including mammals, birds, reptiles, and fishes. The only exception to this two kingdom restricted host range is the virus *Sclerotinia sclerotimonavirus* which is able to infect a hypovirulent isolate of the fungus *Sclerotinia sclerotiorum*, and is classified in the monospecific family *Mymonaviridae*.

The vast majority of ssRNA(−) viruses possess a linear genome. Exceptions are found with members of the genus *Tenuivirus* (family *Phenuiviridae*) where there are 4–5 circular genomic molecules, and the only member of the genus *Deltavirus* that has a monopartite circular genome. Those with a linear genome can be monopartite, bipartite, tripartite, while some viruses have their genome proportioned into four, 4–8 or 6–8 molecules. Prior to translation, the genome of ssRNA(−) viruses is copied into the plus strand by the virus-encoded RdRp. Apart from members of the *Ophioviridae* family of plant viruses, ssRNA(−) virions are enveloped.

Negative-sense RNA viruses are the etiological agents of diseases in humans like rabies, influenza, hemorrhagic fever, and encephalitis. It has been hypothesized that arthropods host many (perhaps all) ssRNA(−) viruses that cause disease in plants and animals. This is because similar variation, in terms of number and type of genomic molecules, in negative-sense RNA viruses are also found in arthropods. We might expect that the study of these viruses in invertebrates will shed some light on the origin and radiation of this peculiar and somewhat specialized group of viruses in the future. Arthropods are a major reservoir of viral genetic diversity and it is quite possible that they were central to the evolution ssRNA(−) viruses.

ssRNA(+) viruses

After dsDNA viruses the positive-sense viruses are the most common in the virosphere—and for good reason. ssRNA(+) viruses, after reaching the cytoplasm, can be immediately translated giving rise to the proteins encoded in the viral genome, and serve, as well, as the template for the generation of more copies of the virus genome. From the point of view of economy and minimalism, there is no better option. ssRNA(+) viruses are classified into 39 families and 15 unassigned genera. Among these families we can find the three retrotranscribing families *Metaviridae*, *Pseudoviridae*, and *Retroviridae*—which add a layer of complexity to the realm of simplicity.

Viruses belonging to this group can infect members of virtually all kingdoms of life—except for Archaea. Only four viruses with ssRNA(+) genomes have been reported to infect bacteria, all belonging to the family *Leviviridae*.

Not surprisingly, all positive-sense RNA viruses have a linear genome (as no messenger RNAs in cellular organisms are circular). The most numerous family of this group of viruses is the *Potyviridae*—probably the most important plant virus family after *Geminiviridae*, not only in terms of the number of species but from the point of view of food security and economic losses they incur.

ssRNA(+/−) viruses

A few viruses possess an ambisense genome. That is, the genome has all the genetic information encoded in segments either in sense or antisense orientations. Two families (*Tospoviridae* and *Arenaviridae*) and the genus *Phlebovirus* of the family *Phenuiviridae* possess 57 species with ssRNA (+/−) genomes. All form enveloped, spherical virions that encase bipartite or tripartite linear genomes. Viruses of this group only infect animal or plant species. RdRp is packaged within virions, which makes the generation of plus segments from the minus segments of the genome possible early in the replication cycle. With more than 1000 species susceptible to infection, *Tomato spotted wilt orthotospovirus* (*Tospoviridae*) is reputed to be the virus with one of the widest host ranges of all.

VIRUS TAXONOMY: ORGANIZATION OF THE VAST DIVERSITY AMONG VIRUSES

The early 1990s saw a tremendous surge in the discovery of new viruses, and the need to classify viruses (and maintain a consistent naming system) was recognized as a necessity. Various classification schemes were proposed in an attempt to bring to order the apparent diversity, and to facilitate the study of these entities. Early schemes were based solely on size. Viruses were classified as distinct from cellular organisms by virtue of their ability to pass through unglazed porcelain filters known to retain the smallest of bacteria. But as the numbers of filterable agents increased, viruses were distinguished from each other by more measurable characteristics, namely the disease or symptoms caused in an infected host. Under this scheme, e.g., animal viruses that caused hepatitis or jaundice were grouped together as hepatitis viruses and viruses that induced mottling symptoms in plant hosts were grouped together as mosaic viruses. The advent of new technologies in virus purification, serology, and electron microscopy spurred the use of physical characteristics for distinguishing viruses.

Baltimore Classification

In 1971 the classification of viruses according to the genomic nature of viruses and method by which they replicate was proposed. It represented the easiest, and yet logical, way of grouping viruses. The central theme of the proposal, referred to as the Baltimore system of virus classification, is that all viruses must synthesize positive strand mRNAs from their genomes in order to produce proteins and replicate in their hosts. The proposal gave origin to six groups, and a seventh group was later added in order to accommodate the dsDNA-RT viruses (*Caulimoviridae*, of plant and animal hosts, and *Hepadnaviridae* of only animal hosts). The seven groups are as follows:

Group I. dsDNA viruses, in which mRNAs are produced by direct transcription using a host RNA polymerase. These mRNAs can be produced from the genome of infecting virus (early mRNAs) or from progeny viral dsDNAs (late mRNAs). The dsDNA-RT viruses are not included in this group, but in group VII (see below).

Group II. ssDNA viruses regardless of the genome polarity(ies) (+, −, +/−, and + plus −) are included in this group since the production of the mRNAs involves the step of generating dsDNA first. In eukaryotes, virus replication occurs in the nucleus, most probably by a rolling circle mechanism.

Group III. dsRNA viruses, as explained previously, produce mRNAs and replication templates by means of their own RdRP. This almost never occurs outside of the virus capsid.

Group IV. ssRNA(+) viruses with genomes that are recognized immediately by the host cell machinery as mRNA. Translated proteins subsequently direct the replication of the viral genome. This messenger can be polycistronic, in which case the polyprotein resulting from translation is later cleaved by proteases encoded in the viral genome to give rise to functional viral proteins. In other cases, transcription is more complex and may involve the generation of subgenomic mRNAs, ribosomal (+ or −) frameshifting or proteolytic processing of an initial precursor translation product. Retrotranscribing ssRNA(+) viruses are not included in this group, but in group VI (see later).

Group V. ssRNA(−) viruses that must be copied first into a sense RNA molecule in order to produce viral mRNAs. There are two main types of ssRNA(−) depending on where virus replication occurs: those that replicate in the cytoplasm are monopartite and are transcribed by the RdRp to produce the sense molecule that can be translated and serve also as the template for the generation of the minus strand (genomic molecule). And those with segmented genomes that are replicated in the nucleus, in which case separate mRNAs are produced by the viral RdRp resulting in both messenger and replication template molecules. ssRNA (+/−) viruses are also included in this group.

Group VI. ssRNA(+)-RT viruses that, by means of a virus-encoded **r**everse **t**ranscriptase (RT), generate a dsDNA copy of the genomic molecule. This dsDNA molecule is integrated into the genome of the host (becoming a provirus) where it can be replicated and transcribed in the nucleus using the host cell's machinery to provide viral mRNAs as well as ssRNA(+) genomic molecules.

Group VII. dsDNA-RT viruses that possess a gapped genome (open circular, OC, dsDNA genome) that upon infection is filled to generate a **C**ovalently **C**losed **C**ircular (CCC) form. Transcription of the CCC genome in the nucleus generates viral mRNAs and a subgenomic RNA. Using a virus-encoded RT, the subgenomic RNA is converted to the genomic OC dsDNA molecule.

Altogether, the combination of virion architecture, variations in genome type, number of genomic molecules, and variations of genomic strategies, presence or not of an envelope, host range, and other particularities, makes every virus a unique, evolved construct of nature. The virion provides much information toward the characterization of viruses and the way they interact with cellular organisms, but should not be the main emphasis in distinguishing viruses. For one thing, virion morphology in different groups may be the result of convergent evolution. That is, there is no evolutionary relatedness involved, nor shared ancestry. The demonstration of chimerism in virus genomes also adds to the unreliable use of morphology when trying to analyze virus evolution and classification since we would be comparing features of different origins (replicase and capsid proteins, for example) in the same subject of study. Importantly also is the fact that not all viruses produce virions, as explained. Virion morphology, however, is still a very useful way to characterize (and even identify) viruses since the expression of their molecular features are identifiable and measurable, and derive directly from the information the virus encodes.

The International Committee on Taxonomy of Viruses

The first internationally organized initiative for developing a universal taxonomic scheme for viruses was the formation of the International Committee on Nomenclature of Viruses in 1966, which later became the **I**nternational **C**ommittee on **T**axonomy of **V**iruses (ICTV) in 1973. The system that was developed is essentially based on the familiar systematic taxonomy scheme of Order, Family, Subfamily, Genus, and Species. Levels higher than

Order were not included as such levels imply a common ancestry; multiple independent lineages for viruses now seem the more likely scenario. As of March 2017 ICTV subdivides viruses into 8 orders, 122 families, 35 subfamilies, 735 genera, and 4404 species.

The classification scheme unites viruses using a range of similar attributes. This way viruses are grouped by comparing numerous properties of individual viruses without assigning universal priority to anyone property. Orders and families are typically assigned by virus morphology, genome composition, orientation (+/− sense), segmentation, gene sets, and replication strategies. The further subdivision of families into genera generally segregates viruses into those possessing the same complement of homologous genes. Because of the congruence with evolutionary histories, taxon assignments are often recapitulated in virus nucleotide or amino acid sequence relationships. The lowest level of classification considered by the ICTV is species. According to the ICTV in 1991, "*A virus species is a polythetic class of viruses that constitute a replicating lineage and occupy a particular ecological niche.*" The advantage of defining virus species as polythetic is particularly relevant as viruses undergo continual evolutionary changes and show considerable variability. Of note, a type species is identified for each genus and is usually the virus that necessitated the creation of the genus and best defines or identifies the genus. Some criteria used in the classification of viruses are summarized in Table 2.3.

Taxonomic levels lower than species (e.g., strains, variants) are not officially considered by the ICTV, but are left to specialty groups. It is widely accepted that a strain refers to isolates of the same virus from different geographical locations. Variants refer to viruses with phenotypes that differ from that of the original wild type strain, including serotypes and pseudotypes.

Further guidelines were formulated to facilitate the development of a reasonably uniform nomenclature for all viruses. Table 2.4 highlights a few of the rules. Taxon names are designated with suffixes: Order (*-virales*), Family (*-viridae*), Subfamily (*-virinae*), Genus (*-virus*). In most cases the English common names of viruses have become the species names. Generally the virus name provides information regarding the host that the virus was originally isolated from (e.g., *Escherichia virus Lambda, Siphoviridae*, a virus originally isolated from *Escherichia coli*, that is regarded as the mother of Molecular Biology), a symptom associated with the disease it incites (*Papaya ringspot virus*, *Potyviridae*, first isolated from symptomatic *Carica papaya* plants showing characteristic ring spots on the fruits), or the disease itself (as mentioned previously, or *Yellow fever virus*, *Flaviviridae*, the causal agent of that disease), or the place where it was found for the first time (*Lake Sinai virus 1*, from the genus *Sinaivirus* belonging to as of yet an unassigned virus family, and detected in honey bee samples from an apiary

TABLE 2.3 Some Criteria Used for the Characterization, Description, and/or Classification of Viruses

General Properties	Specific Properties
Physical (Virions)	• Molecular weight, buoyant density, sedimentation values • Physical stability under pH, temperature, ions, irradiation, detergents, organic solvents • Viability outside the host(s) • Number of particles
Chemical (Virions)	• Macromolecular composition: Content of proteins, lipids, carbohydrates, and nucleic acids
Biochemical (Virions)	• Number, size, type and interactions of structural proteins, lipids, and carbohydrates • Presence or not of internal layers and of an external envelope • Type of nucleic acid (RNA or DNA), strandedness (ss or ds), linear or circular, and polarity (positive or negative sense, or ambisense) • Chemical modifications of virions components
Structure (Virions)	• Number and composition of capsomers • Form, shape, and symmetry of virions (virion morphology); ornamentation • Size and dimensions
Genomic, genetic, and biochemical	• Genome length • Number of genomic molecules (genome partition) • Strategies followed to produce mRNAs • Genome sequence and/or gene sequences • Amino acid sequence of proteins (mostly deduced from the genomic/gene nt sequence) • Posttranslational processing and/or protein modifications • G + C content (total and along the virus genome) • Presence or absence and type of capping molecule at the 5′ end (true cap, modified tRNA, genome-linked protein) • Presence or absence, and length, of the genomically encoded polyA tail at the 3′ end • Presence, type, and function (if any) of repetitive sequences in the genome • Genome organization and replication strategies • Number of ORFs and intended product(s) and their corresponding functions • Genome expression strategies and site of expression • Phylogenetical relationships with related viruses and group of viruses • Capability of genomic insertion and heritability

(*Continued*)

TABLE 2.3 (Continued)

General Properties	Specific Properties
Immunological	• Specific reactivity with monoclonal and polyclonal antibodies of virion proteins and genome-encoded proteins • Cross-reactivity with related viruses • Allergenicity (if any)
Cellular and organismal	• Attachment and internalization (specific receptors and pathways) • Site of virion accumulation • Interactions with cell components • Tissue tropism, cytotoxicity, and tissue pathogenicity • Number, type, and localization of inclusion bodies • Site and mode of replication and virion assemblage • Characteristics of cell-to-cell movement and of systemic spread • Symptomatology (if the virus causes a disease) • Latency time (if any)
Others	• Host range • Mode of transmission, both vertical and horizontal • If vectored, type and identity of vectors, and mode of transmission • Geographical distribution • Cooccurrence and interactions with other viruses, satellites, and/or viroids

near Lake Sinai, South Dakota, United States). Additionally, the rules are in line with the way in which other formal taxon species names are written; they are written in italic script and the first word in the taxon name begins with a capital letter.

Viruses can also be referred to by acronyms, sigla (singular: siglum) or abbreviations. An acronym of a virus name is created by using the initial letter in the words of a virus species name: *Cucumber mosaic virus* (*Bromoviridae*) is widely referred as CMV. Sigla, on the other hand, are constructed using letters or other characters taken from words of a compound term, like arbovirus for **ar**thropod-**bo**rne **virus**. Most of the times, if not always, sigla are proposed by a group of experts. Abbreviations, finally, can be constructed by a combination of the aforementioned criteria: PiRV- (from 1 to 4) refers to RNA viruses hosted by the oomycete *Phytophthora infestans* (Chromista): the unrelated PiRV-2 and PiRV-3, and the still unclassified PiRV-4, and PiRV-1 of the family *Narnaviridae*. Acronyms, sigla and abbreviations of viruses or viral groups are never italicized.

TABLE 2.4 Selected Rules for the Classification and Nomenclature of Viral Ranks and Its Orthography

Number[a]	Summary of Rules	Notes and Examples
3.2	Hierarchical levels of classification must include Order, Family, Subfamily, Genus, and Species (In general, most of the times, only the virus specific name is provided in the literature; sometimes, Family and Order). The suffixes used are: • *virales* for Orders (3.34) • *viridae* for Families (-*viroidae* for viroids) (3.32; 3.36) • *virinae* for Subfamilies (3.30) • *virus* for Genera (-*viroid* for viroids) (3.27; 3.36)	Not all virus families have been assigned to a specific order. Similarly, not all families are divided into subfamilies and, in some cases, a virus species does not belong to a defined genus (unassigned) although it does belong to a family. Some genera are also floating (unassigned) since they are not allocated to a specific family. The use of "*Virus name (Order: Family: Subfamily: Genus)*" is not required and it is used here just as an example
	In 2017 it was approved that the rules concerning the classification of viruses should also be applied to viroids and satellite nucleic acids. Accordingly, suffixes for satellite ranks are: • *satellitidae* for Families • *satellitinae* for Subfamilies • *satellite* for Genera	Example 1 (all ranks):
		Human alphaherpesvirus 1 (*Herpesvirales*: *Herpesviridae*: *Alphaherpesvirinae*: *Simplexvirus*)
		Example 2 (unassigned genus in a family):
		Banana mild mosaic virus (*Tymovirales*: *Betaflexiviridae*: *Quinvirinae*)
		Example 3 (accepted genus, but unassigned family):
		Citrus leprosis virus C (*Cilevirus*)
		Example 4 (majority of viruses: no order assigned, no subfamilies defined):
		Chikungunya virus (*Togaviridae*: *Alphavirus*)
3.8	A name is valid if accepted by the ICTV given that (1) it has been published, (2) it is associated with descriptive material, and (3) it conforms to the Rules of the *The International Code of Virus Classification and Nomenclature*	Of course the approval of a name must follow the recommendations and fulfill the requirements established to that effect (i.e., the paperwork related to the application). These are studied by special committees devoted to specific virus groups. Viruses described based on metagenomics studies only, generally lack descriptive material

(*Continued*)

TABLE 2.4 (Continued)

Number[a]	Summary of Rules	Notes and Examples
3.11	Names for taxa should be easy to use and easy to remember (particularly for virus species names)	In general names in use are short and *probably* easy to remember. Nonetheless, cases exist in which the name of the rank is of complex spelling, or too long
		Example 1 (short name, easy to remember):
		Aura virus
		Example 2 (long name):
		Choristoneura fumiferana DEF multiple nucleopolyhedrovirus
		Example 3 (complex spelling, besides example 2):
		The Family *Tolecusatellitidae*, or the virus species *Thottapalayam orthohantavirus*
3.15	Sigla (singular: siglum) may be accepted as part of the name of a taxa	Sigla are constructed using letters or other characters taken from words of a compound term, like arbovirus for **ar**thropod-**bo**rne **virus** (this is not a taxonomic rank)
		Example 1 (from an Order)
		Bunyavirales, sigil for ***Bunya****mwera orthobunyavirus* and the suffix -***virales***
		Example 2 (from a Family):
		Tristromaviridae, from the Greek words ***tri****a* (three) and ***stroma*** (layer), plus the suffix −***viridae***
		Example 3 (from a Genus):
		Anphevirus, sigil for ***Anophe****les* and the suffix −***virus***

(*Continued*)

TABLE 2.4 (Continued)

Number[a]	Summary of Rules	Notes and Examples
3.23	The name of a virus species must contain few words, and never be constructed with only the name of the host and the word *"virus"*	Many virus names include the vernacular or scientific name of the host from which the virus was first described. Sometimes a symptom name is included, while in other instances the name of the location where the virus was found is used instead
		Example 1 (vernacular host name and symptom included):
		Rice grassy stunt tenuivirus
		Example 2 (scientific name and symptom included):
		Echinochloa ragged stunt virus
		Example 4 (generalized host name included):
		Avian coronavirus
		Example 5 (Geographic name included):
		St Croix River virus
		Example 6 (no clue provided, except that is a *certain* virus of a specific genus):
		Rotavirus A
3.39 and 3.40	Names of all accepted ranks should be italicized with the first letter of the first word in capitals. First capital letters can also be used if belonging to a proper noun	If the not yet approved rank (especially virus species) contains a scientific name of the host, this name should not be italicized. For instance, Phytophthora infestans virus 4 is used, although *Phytophthora infestans* is the scientific name of a widely known oomycete. On the other hand, when talking about a specific group of viruses (not species) in general, italics are not required. That is, *"potyviruses comprise the largest group of viruses with a ssRNA(+) genome,"* or *"the newly created order that accommodates bunyaviruses,"* or *"a new coronavirus is described here"*

[a]*From The International Code of Virus Classification and Nomenclature (2002).*

The Goal of Virus Taxonomy

Virus taxonomy can then be defined as the arrangement of viruses into related clusters, identification of the extent of relatedness within and among these clusters, and the assignment of names to the clusters (taxa). In other words the goal of taxonomy is to categorize the multitude of known viruses so as to maximize organization, stability, and predictivity. The first goal, probably the most basic, is easily attainable given that the rules of classification are clear and followed by all. That is, the vast diversity of viruses is organized by grouping all similar viruses that share the same denomination and, for the most part, established evolutionary relationships. If this preliminary classification is based on solid grounds, stability ensues. Nonetheless, taxonomy cannot be a static, never-changing tool for virologists. As our realization of viral diversity widens and new technologies become available, virus taxonomy has to be flexible enough to accommodate occasional revisions and reinterpretation of perceived relationships between viruses. As for predictability, this is more complex and challenging; i.e., being able to speculate on the characteristics of near relatives of known species. Whether this is even possible at this point is questionable, given that the current knowledge of virus biodiversity is both biased and fragmentary, reflecting a focus on culturable or disease causing agents.

However, the traditional approach to virus classification has been challenged by recent technological developments. Rapid advances in metagenomics have revealed a virosphere that is more phylogenetically and genomically diverse than that depicted in the current classification scheme. Indeed a number of studies have generated large sequence datasets of viral metagenomic sequences that cannot readily be classified using conventional criteria since these are based on phenotypic properties as well as genetic relationships. Classification based on sequence comparisons alone represents a pragmatic solution to the problem of classifying viral metagenomic sequences, but is a significant departure from current methods for classifying viruses. Viral metagenomic sequences may have sufficient defining characteristics to enable their classification as additional taxa in existing virus families and may even be used to justify the creation of new virus families. Further information on diversity and clustering may, however, be required to justify the formation of genera and species ranks within the family. Perhaps though, a reevaluation of the concept of viral species, genera, families, and higher levels of classification is needed; particularly a consensus on which diagnostic properties are the most useful for identifying individual members of a virus species. Just as microbiologists discarded dubious morphological traits in favor of more accurate molecular yardsticks of evolutionary change, virologists can gain new insight into viral evolution through the rigorous analyzes afforded by the molecular phylogenetics of viral genes, which are essential for the classification of viruses below the family taxon level. On the other hand, others have gone as far as suggesting a new division of biological entities into two classes of organisms: ribosome-encoding organisms that include all archaea, bacteria and eukarya, and capsid-encoding organisms, the viruses. Doubts have been raised on the suitability of relying only on the capsid structure for resolving phylogenetic uncertainties. Some argue that a more integrated account of all three categories of containment, replication, and reliance on the information processing systems of the host, recognizes the origin and evolution of viruses. The last point is worth emphasizing; evolutionary processes have shaped the genetic structure and diversity of viruses and the viruses themselves have influenced the evolution of cellular organisms in all domains of life.

FURTHER READING

Adams, M.J., Lefkowitz, E.J., King, A.M.Q., Harrach, B., Harrison, R.L., Knowles, N.J., et al., 2017. Changes to taxonomy and the International Code of Virus Classification and Nomenclature ratified by the International Committee on Taxonomy of Viruses (2017). Arch. Virol. 162, 2505–2538. Available from: https://doi.org/10.1007/s00705-017-3358-5.

Koonin, E.V., Dolja, V.V., 2014. Virus world as an evolutionary network of viruses and capsidless selfish elements. Microbiol. Mol. Biol. Rev. 78, 278–303. Available from: https://doi.org/10.1128/MMBR.00049-13.

Mateu, G.M. (Ed.), 2013. Structure and Physics of Viruses. An Integrated Textbook. Springer, Dordrecht, The Netherlands. Available from: https://doi.org/10.1007/978-94-007-6552-8.

Paez-Espino, D., Eloe-Fadrosh, E.A., Pavlopoulos, G.A., Thomas, A.D., Huntemann, M., Mikhailova, N., et al., 2016. Uncovering Earth's virome. Nature 536, 425–430. Available from: https://doi.org/10.1038/nature19094.

Peterson, T., 2014. Defining viral species: making taxonomy useful. Virol. J. 11, 131. Available from: https://doi.org/10.1186/1743-422X-11-131.

Ruggiero, M.A., Gordon, D.P., Orrell, T.M., Bailly, N., Bourgoin, T., Brusca, R.C., et al., 2015. A higher level classification of all living organisms. PLoS ONE 10, e0119248. Available from: https://doi.org/10.1371/journal.pone.0119248.

Simmonds, P., 2015. Methods for virus classification and the challenge of incorporating metagenomic sequence data. J. Gen. Virol. 96, 1193–1206. Available from: https://doi.org/10.1099/jgv.0.000016.

Chapter 3

Replication and Expression Strategies of Viruses

Sephra Rampersad[1] and Paula Tennant[2]
[1]*The University of the West Indies, St. Augustine, Trinidad and Tobago,* [2]*The University of the West Indies, Mona, Jamaica*

Nature's stern discipline enjoins mutual help at least as often as warfare
Theodosius Dobzhansky

One of the key features of viruses is their reliance on living cells for replication and propagation. On their own, viruses lack the complete machinery necessary for many life-sustaining functions. Infection of a host cell and viral propagation are dependent on the transcription of viral mRNA, and in turn, translation of viral proteins as well as genome replication. Specifically, viruses depend on host cells for: (1) energy, mainly in the form of nucleoside triphosphates, for polymerization involved in genome and viral protein synthesis; (2) a protein-synthesizing system for synthesis of viral proteins from viral mRNAs (some viruses also require host enzymes for posttranslational modification of their proteins; e.g. glycosylation); (3) nucleic acid synthesis, for although some viruses code for an enzyme or enzymes involved in the synthesis of their nucleic acids, they do not usually contribute all the polypeptides involved and are reliant on various host factors; and (4) structural components of the cell, in particular lipid membranes, involved in virus replication.

Even though the dependence on host cell functions varies between virus groups and largely relates to genome complexity, nowhere in the biosphere is genome replication accomplished with greater economy and simplicity than among viruses. Viral genomes carry out multiple functions serving as mRNAs for translation in some instances and/or templates for genome transcription and replication. Accordingly, viral genomes contain regulatory RNA elements that promote, regulate, and coordinate these molecular processes. The central role played by viral genomes is frequently executed by

Viruses. DOI: https://doi.org/10.1016/B978-0-12-811257-1.00003-6

viral or host cell proteins that interact with these genomes, but other partners include other RNA molecules (e.g., cellular tRNA primer that initiates reverse transcription of the retroviral RNA genome). Together, protein and RNA factors interact with cellular pathways to allow viruses to successfully hijack and customize the host cell machinery for virus production. As a result, viruses and their hosts have been involved in a long-standing battle of adaptation and counter-adaptation for gene expression and nucleic acid synthesis.

In this chapter, we address genome replication strategies including the diverse strategies that exploit the biology of the hosts, control gene expression, and ensure preferential propagation of the virus. The other phases of virus replication are discussed in Chapter 10, Host–Virus Interactions: Battles Between Viruses and Their Hosts. Work with bacteriophages identified the essential phases of virus replication. The process, beginning with entry of the virus into the host cell to the release of progeny viruses, is referred to as the replication cycle. The replication cycle of all viruses involves three key phases: initiation of infection, genome replication and expression, and finally, egress or release of mature virions from the infected cell.

OVERVIEW OF VIRUS GENOME TRANSCRIPTION AND REPLICATION

From the perspective of the virus, the purpose of viral replication is to allow production and survival of its kind. By generating abundant copies of its genome and packaging these copies into virions, the virus is able to continue infecting new hosts. All viruses must therefore express their genes as functional mRNAs early in infection in order to direct the cellular translational machinery to synthesize viral proteins. The pathways leading from genome to message vary among different viruses (Fig. 3.1). As shown in Chapter 2, Virion Structure, Genome Organization, and Taxonomy of Viruses, viral genomes are quite different from cellular genomes, which consist uniformly of dsDNA. Viral genomes provide examples of almost every structural variation imaginable. There are four main categories of viral genomes: dsDNA, ssDNA, dsRNA, and ssRNA. These categories are further divided on the basis of distinct modes of transcription. For some RNA viruses, the infecting genome acts as mRNA. For other RNA and DNA viruses, viral mRNA is synthesized upon entry into the host cell.

RNA viruses replicate their genomes via one of two unique pathways—either by RNA-dependent RNA synthesis, or among the retroviruses, by RNA-dependent DNA synthesis (reverse transcription) followed by DNA replication and transcription. Both pathways require enzymatic activities that are not usually found in uninfected host cells and as a result, these viruses code for the requisite enzymes, which are either expressed early in infection

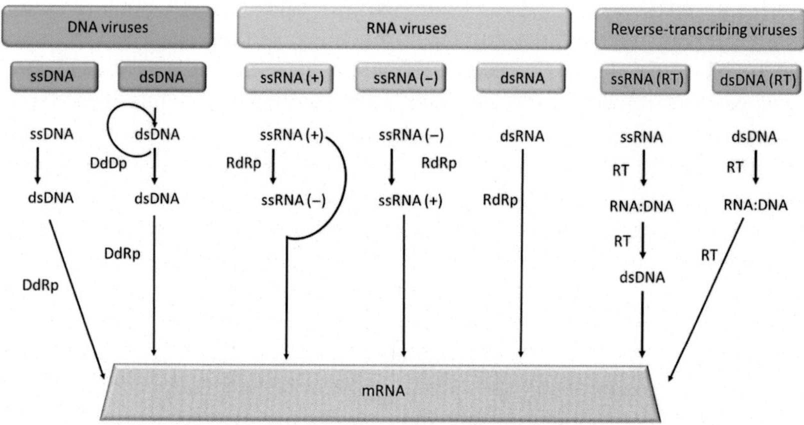

FIGURE 3.1 Summary of replication and transcription modes of different classes of viruses. DdDp, DNA-dependent DNA polymerase; DdRp, DNA-dependent RNA polymerase; RdRP, RNA- dependent RNA polymerase; RT, reverse transcriptase. The ssRNA(+) can serve as the template for translation and does not undergo any modification prior to translation.

or they are copackaged with the viral genome during the assembly of virions in preparation for the next round of infection. Note that the process of synthesizing RNA complementary to an RNA template is often called "*transcription,*" and differs from the conventional definition of transcription where DNA is used as template for mRNA synthesis.

Double-stranded RNA viruses carry their own RNA synthesizing enzymes within their virions. Once inside a cell, the enzymes transcribe one of the RNA strands of the dsRNA genome to ssRNA(+) within the virion, which upon release into the cytoplasm (through channels in the virion) is translated by the host cell machinery. In addition, these ssRNA(+) can act as a template for (−) RNA strand synthesis that is converted back into a dsRNA genome for packaging. Whereas DNA viruses need only to generate mRNA, these RNA viruses without a DNA stage have to synthesize both viral RNA and mRNA. The viral RNA is generated through a replication intermediate, referred to as the "*antigenome*" or "*minus*" (−) strand, which serves as a template for viral RNA synthesis. Single-stranded RNA genomes, however, exist in two forms: either the sense or the "*plus*" (+) strand or the nonsense or "*minus*" (−) strand. ssRNA(+) genomes act as mRNA, are infectious upon entry into host cells, and are immediately translated into protein, including the enzymes required for viral reproduction. On the other hand, (−) RNA, which is not mRNA, requires a virus-encoded RNA-replicating enzyme that is carried within the virion, to copy (−) RNA to monocistronic (+) RNA that is recognized by the host cell translation machinery. The enzyme also synthesizes viral (progeny) genome using the (+) RNA as template. The other class of (+) RNA viruses is the retroviruses. Here, the virion carries an enzyme that converts the ssRNA(+) into dsDNA upon infection. The virus genome integrates into the host genome and can be passed from parent to offspring should integration occur in germline cells. The integrated virus genome, referred to as a provirus, is transcribed as a cellular gene (some may require splicing) and translated by the cellular synthesis machinery on export to the cytoplasm. Some full-length ssRNA(+) transcripts are packaged into new retrovirus virions. It should be noted that in all these examples, the balance between the processes of transcription and genome replication must be properly maintained to allow efficient viral proliferation. It appears that the transition occurs by the (1) action of *trans*-acting proteins that are either absent, or at low levels in virions, but which accumulate over the course of infection; (2) regulatory role of promoter RNA secondary structure along with the action of specific viral (e.g., capsid) and host cellular factors; (3) alteration of the virus RNA synthesizing enzyme complex that changes its role as "*transcriptase*" to a "*replicase*"; or (4) a combination of the above.

DNA viruses need only to generate mRNA and thus replication involves strategies that are familiar in cell biology: transcription of mRNA from dsDNA and replication of dsDNA within the cell nucleus. Many use host enzymes for these processes, while some larger viruses code for their own enzymes. ssDNA viruses, however, first convert their ssDNA to dsDNA intermediates (using host cell DNA enzymes), which are then transcribed into mRNA. Cellular splicing machinery typically generates mature viral mRNAs. The switch from transcription to replication, that is the switch from antigenome production to genomic nucleic acid for packaging, is highly regulated, and unlike RNA viruses, there is the strict demarcation with respect to timing of genomic DNA replication. Early genes, which code for catalytic (e.g., polymerase) and regulatory proteins, are transcribed prior to the initiation of viral DNA replication. Late genes that code for the structural components of the capsid and envelope are transcribed only after viral DNA replication.

POLYMERASES

As indicated earlier, some viruses encode and/or carry the enzymatic repertoire required for genome replication and/or transcription, while others recruit host polymerases. Large DNA viruses, for example, members of *Herpesviridae, Adenoviridae*, and *Poxviridae*, and giant viruses, are among those viruses that encode most of their own proteins for replication. Proteins encoded by these viruses are those involved in recognition of the origin of replication, DNA-binding proteins, helicases and primases, DNA polymerase and accessory proteins, exonucleases, thymidine kinase, and dUTPase. Small DNA viruses, for example, *Papillomaviridae, Polyomaviridae*, and *Parvoviridae*, do not encode the entire repertoire of proteins required for viral replication because of their limited genome size. They do, however, encode proteins that usurp and control cellular activities. Viruses that do not replicate in the nucleus and do not have access to host polymerases, typically encode their own polymerases for replication.

The four main types of polymerases used by viruses depend on their genomic constitution and site of replication, that is, nucleus or cytoplasm, and include: **R**NA-**d**ependent **R**NA **p**olymerases (RdRps), RNA-dependent DNA polymerases, DNA-dependent RNA polymerases, and DNA-dependent RNA polymerases. DNA viruses replicate their genomes using DNA polymerase enzymes and transcribe their mRNA using DNA-dependent RNA polymerase enzymes. Both (+) and (−) ssRNA viruses replicate and transcribe their genomes using RdRp enzymes (Fig. 3.1). **R**everse **t**ranscriptase (RT) is the enzyme used to produce DNA from RNA templates, and viruses that replicate via an RNA intermediate require this enzyme. RT is virus-encoded as the host cell does not require this enzyme for its nuclear metabolism.

Although high fidelity of virus genome replication is crucial for the long-term survival of viruses, some polymerases are less faithful than others when incorporating the correct nucleotide during replication. The rate by which mutations occur is universally determined as the number of nucleotide substitutions per base per generation. DNA viruses experience low mutation rates. Their DNA polymerase has an error rate of approximately 10^{-6} to 10^{-8} mutations per base pair per generation. This is because of the proofreading ability of the polymerase. A 3′ to 5′ proofreading exonuclease domain is intrinsic to most DNA polymerases. It allows the enzyme to verify each inserted nucleotide during DNA synthesis and excise mismatched nucleotides in a 3′ to 5′ direction. With the exception of nidoviruses, the replicative enzymes of RNA viruses (RdRps) lack proofreading ability and these viruses exhibit the highest mutation rates. RNA viruses possess an error rate of approximately 10^{-4} to 10^{-6} mutations per base pair per generation. Not all mutations generated will persist in a virus population, however. Mutations may be neutral or silent (because of genetic code redundancy) and those that interfere with viral replicative mechanisms are eliminated from the viral population. Mutations that do not affect essential viral functions may persist and eventually become fixed within the viral population (see Chapter 4: Origins and Evolution of Viruses).

Initiation, Elongation, and Termination

Unlike cellular DNA and RNA polymerases, which require oligonucleotides to initiate nucleic acid synthesis, viral polymerases initiate genome replication using a variety of mechanisms, that presumably reflect their adaptation to the host cell. Nucleic acid synthesis by polymerases is divided into three phases: initiation, elongation, and termination. Both virus genome transcription and mRNA synthesis occur in three stages.

Initiation

During initiation, the polymerase machinery is recruited to the viral promoter and synthesis begins at or near the 3′ end of the template. Two different start sites are used in the synthesis of mRNA and viral genome RNA in a primer-independent (*de novo*), or a primer-dependent mechanism. Primer-dependent initiation requires either an oligonucleotide primer or a protein primer, to provide the initial 3′-hydroxyl for addition of the first incoming nucleotide. In *de novo* initiation, a nucleoside triphosphate, sometimes referred to as the one-nucleotide primer, provides the 3′-hydroxyl for the addition of the next nucleotide. Structures in the polymerase or conformational changes apparently contribute to the process. *De novo* initiation is used by RNA viruses, including those with genomes of positive, negative, ambisense, and dsRNA. Unlike DNA polymerases, most RNA polymerases do not require primers.

After initiation of viral mRNA transcription, DNA viruses and some dsRNA viruses (e.g., reoviruses) acquire a 5′-terminal cap structure (m7Gppp[5′]N-; where N is the first nucleotides of the nascent RNA) using host enzymes. Most ssRNA(+) viruses (e.g., flaviviruses and nidoviruses) encode their own capping enzymes and some unsegmented ssRNA(−) viruses (including Rabies virus (*Rabies lyssavirus*), measles virus (*Paramyxoviridae*), and Ebola virus (*Filoviridae*)) are capped by their polymerase. Another capping mechanism used by negative-sense RNA viruses (e.g., influenza viruses) is that of "*cap snatching*" from nascent host pre-mRNAs (see later).

Elongation

Once viral genome replication factors and the template are assembled into a complex, the polymerase synthesizes a new complementary strand, without dissociation from the template, and by the repeated addition of a nucleoside monophosphate to the 3′ end of the growing RNA strand. Poliovirus RdRp, for example, adds about 5000 nucleotides and so in a single-binding event it can synthesize the entire genome.

Termination

Termination leads to the release of the newly synthesized RNA strand and the dissociation of the polymerase from the template. Transcription termination involves secondary structure mechanisms or in eukaryotic cells, RNA signals direct polyadenylation and termination. Unlike polyadenylation of host mRNAs, which is carried out by a specific poly(A) polymerase, polyadenylation of viral mRNAs is catalyzed by the viral polymerase. In nonsegmented negative RNA viruses, obligatory sequential transcription dictates that termination of each upstream gene is required for initiation of downstream genes. Therefore, termination is a means of regulating expression of individual genes within the framework of a single transcriptional promoter.

MECHANISMS OF GENOME TRANSCRIPTION AND REPLICATION

The diversity in mechanisms used by polymerases to replicate and/or transcribe viral genomes is summarized in the following section. As will be seen, the mechanisms are dictated by the nature and structure of the viral genomes.

Rolling Circle Replication

Rolling **c**ircle **r**eplication (RCR) is important to circular ssDNA genome replication (e.g., ΦX174 phage and geminiviruses). The DNA polymerase involved must exhibit a high level of processivity and strand displacement characteristics. Upon infection, a complementary strand designated as the (−) strand is generated by host enzymes from the virus' ssDNA genome that is referred to as the (+) strand. A duplex or **r**eplicative **f**orm (RF) results. Subsequently, a viral endonuclease enzyme (Rep protein) produces a nick in the (+) strand of the RF, exposing a free 3′ end and a free 5′ end. DNA synthesis begins by addition of deoxyribonucleotides by DNA polymerase to the free 3′ end of the (+) strand using the (−) strand as template. The (−) strand revolves while serving as template hence the name rolling circle. The addition of nucleotides displaces the free 5′ end outward in the form of a free tail. Once the entire (−) strand template is copied, the (+) strand tail is cut into correct genome lengths by an endonuclease to provide many copies of free, linear DNA molecules. The latter are eventually by joined DNA ligase to form closed, circular ssDNA molecules. The RF can be used to generate mRNA for the manufacture viral proteins. The RCR mechanism is also thought to be important in the replication of dsDNA viruses such as herpesviruses (Fig. 3.2). Here, multiple cycles of continuous copying of a circular template, followed by discontinuous DNA synthesis on the displaced strand template produces linear dsDNA molecules containing multiple copies of the genome (concatemers).

Rolling-Hairpin Replication

Non-circular genomes may also replicate using an RC-like mechanism, that is, a variation of RCR named rolling-hairpin replication. Linear ssDNA **a**deno-**a**ssociated **v**iruses (AAVs) are an example. AAVs encode the Rep78 protein that contains amino acid sequence motifs similar to RCR initiator proteins. These viruses contain secondary structures at the ends of their DNA called **i**nverted **t**erminal **r**epeats (ITRs). ITRs are seen as terminal hairpin structures. These ITR regions interact with the viral-encoded Rep protein

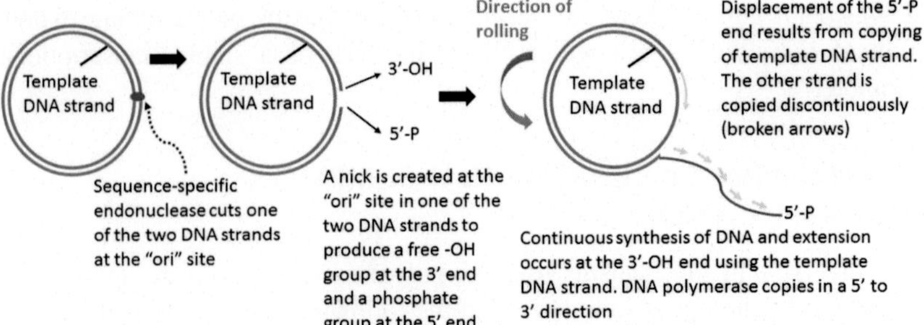

FIGURE 3.2 Rolling circle genome replication. Concatemeric DNA molecules are synthesized from a circular template by a rolling circle mechanism in which nicking of one strand allows the other to be copied continuously multiple times. Discontinuous DNA synthesis on the displaced strand template produces linear dsDNA containing multiple copies of the genome. That is, the dsDNA molecules generated consist of head-to-tail linked genomes. They are eventually cleaved at precise locations to release unit length genomes.

at specific binding sites to initiate replication using the host replication machinery. Rep creates a nick between the hairpin and coding sequences. This creates a free 3′-OH at the 3′ ITR that serves as a primer, and DNA synthesis continues to the end of the genome, duplicating the 5′-terminal ITR structure. Refolding of the termini generates the same secondary structures present in the template DNA. The end result is a fully replicated viral genome with the same secondary structures.

dsDNA Bidirectional Replication

This is the classical mode of replication used by eukaryotes and most nuclear dsDNA viruses, including the majority of phages. DNA replication begins at a specific site in the viral genome, called origin of replication, or "*ori*." Some viruses (e.g., *Human alphaherpesvirus 1*) have multiple "*ori*" sites. The step-wise assembly of replication initiation complexes at these *ori* sites then occurs followed by recruitment of topoisomerases that unwind dsDNA at each *ori*, and prevents supercoiling and torsional stress of the partially unwound template DNA. ssDNA-binding proteins keep the single strands of DNA separate. A replication fork or bubble is produced. A primase synthesizes short RNA primers that are used by the DNA polymerase to begin DNA synthesis. DNA is synthesized in a 5′ to 3′ direction. DNA polymerase and several replication-associated factors copy the leading strand at the fork, starting at the 3′ end of the primer in a continuous manner. Copying of the lagging strand requires discontinuous DNA synthesis that results in production of short DNA (Okazaki) fragments, which must then be ligated after the primers are removed by RNase H degradation.

dsDNA (RT) Transcription and Replication

Pararetroviruses (e.g., members of *Caulimoviridae*) are circular, dsDNA viruses that transition through an RNA intermediate (**pre**genomic **RNA** or pgRNA) in a manner that is reminiscent of retrovirus replication. Replication involves two phases; transcription of the pgRNA from virus DNA in the nucleus followed by reverse transcription in the cytoplasm. In contrast to retroviruses, virus DNA remains episomal and does not integrate into the host genome. Covalently closed virus dsDNA serves as a template for host polymerase transcription and the generation of viral pgRNA. Upon transportation to the cytoplasm, capped and polyadenylated pgRNA is translated to viral proteins including the RT and is also used as template for subsequent reverse transcription catalyzed by virus RT. The resulting dsDNA is either packaged into a new virion or targeted to the nucleus for another round of transcription.

ssRNA (RT) Replication

This mechanism pertains to all members of the family *Retroviridae*. The process takes place in the cytoplasm, after viral entry. Here, the genomic RNA is reverse transcribed into dsDNA. A RNA:DNA duplex is produced, which is then resolved by degradation of the RNA by the virus-encoded RNase H. Only a small stretch of polypurines is resistant to degradation and this serves as a primer to initiate the synthesis of the cDNA. The cDNA, integrated into the host cell's genome, is now referred to as a provirus and undergoes cellular transcription and translational processes to express viral genes. Integration is a key event in the replicative process of all retroviruses. Integration is signaled after the RNA genome of these viruses is reverse transcribed into ds linear, viral DNA. Linear viral DNA contains termini **l**ong **t**erminal **r**epeats (LTRs) sequences at the 3′ and 5′ ends, which are specifically recognized by the viral **in**tegrase (IN) enzyme. In some retroviruses, nuclear localization signals facilitate migration to the nucleus. Depending upon the retrovirus, preintegration complexes either enter the nuclei of nondividing cells through the **n**uclear **p**ore **c**omplex (NPC) (e.g., HIV) or enter when the nuclear membrane dissolves during cell division [e.g. Moloney murine sarcoma virus, (*Murine leukemia virus*)]. Once inside the nucleus and after association with host chromosomes, viral IN catalyzes insertion of viral sequences into the host DNA (Fig. 3.3). The LTR sequences at the 3′ and 5′ ends of the linear viral DNA are joined to the host's DNA in two steps called

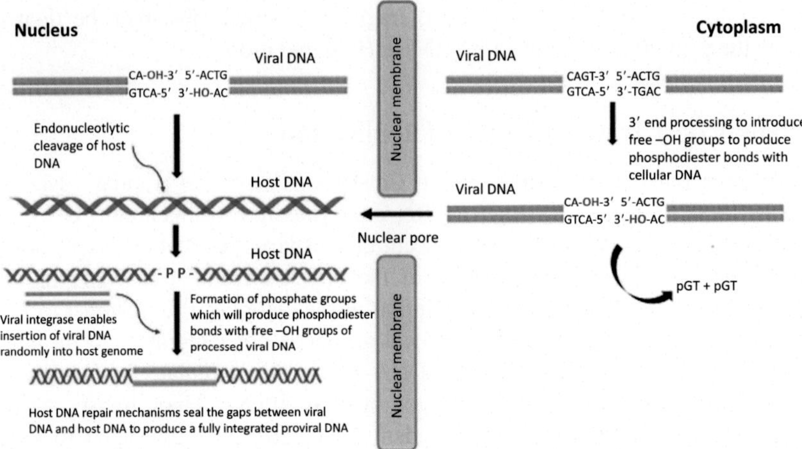

FIGURE 3.3 Integration of viral DNA (e.g., retroviruses) into host's genome. The LTRs of linear viral DNA are joined to the host's DNA in two steps called end-processing and end-joining.

end-processing and end-joining. After successful integration of the viral DNA into the host DNA, cellular repair proteins fill the gap and ligate the DNA ends. Autointegration is prevented by cellular proteins (HMG family of proteins for **H**igh **M**obility **G**roup proteins, and BAF for **B**arrier to **A**utointegration **F**actor). Retroviruses have different target sites of integration; for example, lentivirus DNA insertion occurs preferentially in active **t**ranscription **u**nits (TUs), whereas *Murine leukemia virus* integrates at or close to promoter regions located at the 5′ terminus of CpG and TUs islands.

Positive-Strand RNA Virus Replication

This mode of replication occurs in the cell host cytoplasm for all ssRNA(+) viruses. ssRNA(+) molecules serve as templates for replication and transcription. The 5′ of the genome may be naked, capped, or covalently linked to a viral protein. The 3′ end may be either naked or polyadenylated. The ssRNA(+) strand is translated after viral entry into the host cell. A viral polyprotein is typically produced, which encodes the proteins required for replication. The replication process results in the formation of a dsRNA intermediate that is detected by the immune system. Replication, therefore, occurs in membrane-associated replication factories in an effort to avoid the host's immune defense response. This dsRNA is transcribed into genome length ssRNA(+) and serves as the template for either replication or translation. Many ssRNA(+) viruses produce smaller sections of the original transcribed template strand or **s**ub**g**enomic **RNAs** (sgRNAs) for the translation of structural or movement proteins. Depending on the virus, sgRNAs may be generated during internal initiation on a minus-strand RNA template and require an internal promoter or there is the generation of a prematurely terminated minus-strand RNA that is used as template to make the sgRNAs. The third mechanism uses discontinuous RNA synthesis while making the minus-strand RNA templates. This involves template switching of the RNA polymerase and the production of chimeric RNAs consisting of a 5′ common leader sequence derived from the 5′ terminus of the genomic RNA fused to the "*body*" of the transcript (i.e., the 3′ terminal end). The resulting chimeric sg minus-strand RNA can in turn function as a template for the production of subgenomic positive-strand RNAs. Although most sgRNAs are translated into proteins, other sgRNAs regulate the transition between translation and replication, function as riboregulators of replication or translation, or support RNA–RNA recombination.

Double-Strand RNA Virus Transcription and Replication

The 5′ end of the dsRNA viral genome may be naked, capped, or covalently linked to a viral protein. Genomic dsRNA is transcribed into viral mRNA that serves as a template for both translation and genome replication. Translation of this mRNA generates proteins required for replication and viral encapsidation. However, dsRNA is typically detected and cleared by the host's immune response. As such, many dsRNA viruses undergo replication within their icosahedral capsids. The replicating RNA polymerases are located within the capsid and produce mRNA strands that are extruded from the particle. In this way, the viral dsRNA does not enter the cytoplasm and evades the hosts' immune system.

Negative-Strand RNA Virus Transcription and Replication

This mode of replication is employed by all ssRNA(−) viruses genomes, except for deltaviruses. Replication occurs in the cytoplasm. The virus codes for its own RdRp, which converts the (−) stranded RNA into (+) RNA template strands. The (+) RNA serves two roles: (1) as the viral mRNA, which is then translated into viral gene products, and (2) as template to produce more (−) RNA strands. The viral RdRP complex is assumed to be the same for both replication and transcription and can switch off functions as required. The newly synthesized (−) RNA is later encapsidated. Of note, two genome subgroups can be distinguished in this group: nonsegmented and segmented. Viruses with segmented genomes replicate in the nucleus, and the RdRp produces one monocistronic mRNA strand from each genome segment.

dsDNA Template Transcription and Replication

This type of genome template for replication/transcription is observed in all dsDNA viruses that replicate in the nucleus or in the cytoplasm. The mode of transcription is similar to eukaryotic transcriptional events in which the process is divided into three steps: (1) the initiation step, when a transcription initiation complex is assembled at the promoter region located upstream of the transcriptional start site, allowing for the recruitment of the RNA polymerase, (2) the elongation step, in which, the polymerase is recruited to template DNA, is activated by phosphorylation of the **c**arboxy-**t**erminal **d**omain (CTD), and proceeds to transcribe the template DNA to RNA, and (3) the termination step, which involves the recognition of specific signals, including the polyadenylation signal.

VIRAL GENOME EXPRESSION

With few exceptions translation in the host cell begins at the 5′ initiation codon, the ribosome ratchets along the mRNA template, incorporating each new amino acid, translocating from one codon to the next up until a termination codon. For productive infection, viruses must then utilize this machinery, and remain both stable and undetected in the cell. As alluded to in the previous sections, many viral transcripts have marked structural differences from cellular mRNAs that preclude translation initiation, such as the absence of a 5′ cap structure or the presence of highly structured 5′ untranslatable leader regions containing replication and/or packaging signals. Furthermore, while the great majority of cellular mRNAs are monocistronic, viruses must often express multiple proteins from their mRNAs. As a result, viruses have evolved a number of mechanisms to allow translation to be customized to their specific needs.

Straightforward exploitation of the cellular capping machinery is typical of DNA viruses that replicate in the nucleus. Other strategies used by viruses include internal initiation of translation of uncapped RNAs in picornaviruses and their relatives, snatching of capped oligonucleotides from host pre-mRNAs to initiate viral transcription in segmented negative-strand RNA viruses, and recruitment of genes for the conventional, eukaryotic-type capping enzymes that apparently occurred independently in diverse groups of viruses (flaviviruses, reoviridae, poxviruses, asfarviruses, some iridoviruses, phycodnaviruses, mimiviruses, baculoviruses, nudiviruses).

Poly(A) tails, at the 3′ end, are associated with poly(A)-binding proteins that stabilize the mRNA in the cytoplasm by protecting the 3′ end of the newly synthesized mRNA against exoribonucleolytic degradation. Many ssRNA(+) viruses lack a poly(A) tail, but are still efficiently translated. For instance, flaviviruses (e.g., *Dengue virus*, *West Nile virus*) have a capped RNA genome that contains conserved sequences at the 5′ and 3′ ends, allowing for circularization and efficient translation. Other examples that follow the same strategy include rotaviruses, barley yellow dwarf viruses, and possibly *Hepacivirus C* (HCV).

Since eukaryotic cells are not equipped to translate polycistronic mRNA into several individual proteins, DNA viruses overcome this limitation by using the cellular mechanism of splicing of their polycistronic mRNA to monocistronic mRNA. RNA viruses, on the other hand, that mostly replicate in the cytoplasm, do not have access to these host mechanisms and consequently produce monocistronic sgRNAs (e.g., coronaviruses and closteroviruses), use segmented genomes where most segments are monocistronic (e.g., reoviruses and orthomyxoviruses) or translate their polycistronic mRNA into a single large protein (polyprotein) that is subsequently proteolytically cleaved (by viral or host enzymes) into functional individual proteins (e.g., picornaviruses and flaviviruses). However, the use of these mechanisms is not without consequences: (1) some viral proteins may be expressed from sgRNAs but the components of the replication complex that are needed early in infection must still be translated from the genomic RNA, (2) viruses with segmented genomes have to ensure the correct packaging of the different segments, and (3) polyprotein expression represents an inefficient use of host cell resources as all virus proteins are produced in equal amounts, even though catalytic proteins are often required in much smaller quantities than the structural proteins. Alternative and more efficient mechanisms of expressing multiple proteins from a single viral mRNA involve internal ribosome entry, leaky scanning, ribosome shunting, reinitiation, ribosomal frameshifting, and stop codon read-through. In addition, several viruses have evolved proteins and/or RNA structural elements that further enhance translation of the viral mRNAs. Viral gene expression is facilitated by the possession of regulatory signals within viral mRNAs that are recognizable by the host cell. These signals ultimately enable the virus to shut off host gene expression to ensure preferential viral gene expression. The strategies are reviewed in the section that follows.

Disruption of Transcription Initiation Complex Assembly

Transcription can be viewed as a highly regulated 3-phase process: initiation, elongation, and termination. Initiation of transcription requires the recruitment and assembly of a large multiprotein DNA-binding transcription initiation complex. During the course of evolution, several viruses have developed strategies that affect the loading of host transcription initiation factors into transcription complexes, which ultimately shuts down host protein synthesis (Fig. 3.4). Viral mRNA transcripts compete against cellular mRNAs and preferentially gain access to the cellular gene expression machinery.

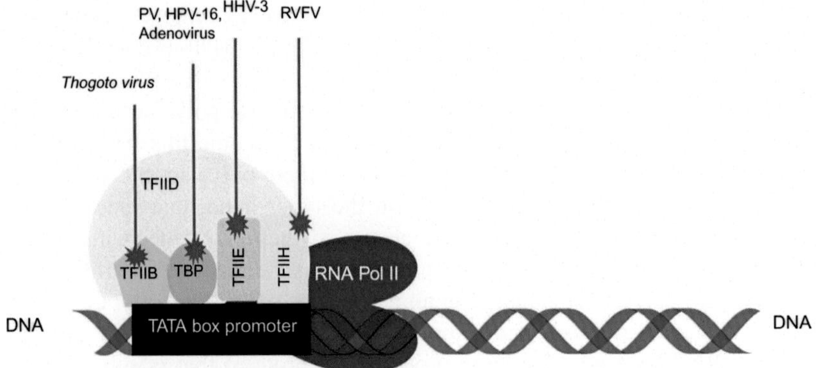

FIGURE 3.4 Different strategies used by viruses to down regulate host transcription. The **TATA b**ox-binding **p**rotein (TBP) is inhibited by HPV-16 E7 protein or by adenovirus E1A protein, and it is cleaved by the poliovirus 3C protease; *Thogoto virus* ML protein interacts with host TFIIB and strongly inhibits IRF3 and NF-kappa-beta-regulated promoters; HHV-3 IE63 targets TFIIE, while *Rift Valley fever phlebovirus* (RVFV) targets TFIIH complex. Viruses: PV, Poliovirus; HPV-16, *Gammapapillomavirus 16*; HHV-3, *Human alphaherpesvirus 3*.

Termini Maturation and Modification

Elongation and termination of transcription are coupled to end-processing of the mRNA where the 5′ cap (added co- or posttranscriptionally) and 3′ poly (A) tail are generated on the ends of the mRNA strand. The 5′ cap refers to the 7-methylguanosine (m7G) that is added onto the 5′ end of mRNA transcript. The processed 5′ end is important because the 5′ cap is the site for the assembly of the translation initiation complex, and it protects and stabilizes the mRNA strand from 5′-3′ exonucleolytic degradation when it is exported out the nucleus and into the cytoplasm for translation. The scanning process to locate the start codon for initiation of translation begins at the 5′ end, and the 5′ cap sequence allows immune recognition of foreign RNAs (including viral transcripts) as "*nonself*." Cap snatching refers to a mechanism used by ssRNA(−) viruses that are incapable of synthesizing their own 5′ cap. Cap snatching involves cleavage of a short nucleotide sequence, 10–20 nts in length, from the 5′ end of cellular mRNAs (Fig. 3.5). The capping apparatus can be either host- or virus-derived. If virus-encoded, cleavage is carried out by the endonuclease activity of the viral RdRp. Sequence complementarity shared between the nucleotides within the cleavage site of the donor mRNA and the viral RNA facilitates successful cap snatching. Members of the *Arenaviridae* and the *Orthomyxoviridae* families, along with most if not all of those of the order *Bunyavirales*, steal the 5′ capped ends of host mRNAs to incorporate this cis-acting stability element into their own transcripts. Recent discoveries indicate that 2′-O-methylation of cap structures is recognized by innate immune interferons to differentiate host versus virus transcripts. The cap-stealing mechanisms used by segmented RNA viruses to generate their mRNAs circumvent this innate detection system.

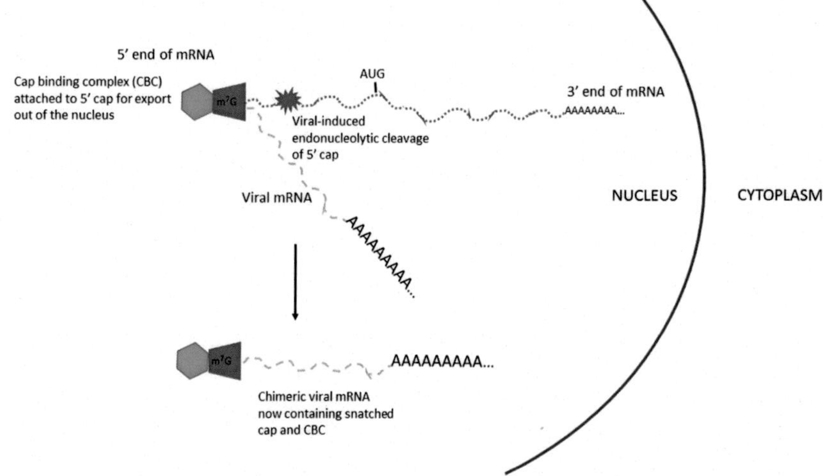

FIGURE 3.5 Cap snatching of cellular mRNA. The mechanism involves cleavage of short fragments from the 5′ end of cellular mRNAs and use of these capped fragments for the synthesis of viral mRNAs.

Internal Ribosome Entry

Viral initiation of translation in the absence of a 5′cap requires activation of noncanonical translation mechanisms. An RNA domain, called the **I**nternal **R**ibosome **E**ntry **S**ite (IRES), enables cap-independent initiation of translation, and can allow initiation of translation at a site not specific to the 5′cap. IRES elements are important to viruses without a 5′ cap as they allow ribosome recruitment under conditions where cap-dependent protein synthesis is severely repressed. *Cripavirus* IRES also allows translation initiation on an alanine or glutamine tRNA and not necessarily the methionine tRNA. Downstream hairpin loops are RNA structures that facilitate initiation of cap-dependent translation in the absence of eIF2 translation initiation factors.

In addition, the physiological state of the infected cell dictates whether host mRNA transcripts undergo cap-dependent translation or cap-independent translation. When the cell exhibits normal housekeeping functions, translation of cellular mRNAs is carried out by a cap-dependent mechanism; however, under stressful conditions, such as heat shock, viral infection, hypoxia, and irradiation, the translation mechanism switches from cap dependency to IRES-driven mechanisms. Infection by a range of viruses induces the activation of the ER stress response, resulting in the stimulation of IRES-dependent translation. This switch is widely observed in picornaviruses since their viral mRNA transcripts do not contain the m7G cap at their 5′ ends. As such, viruses containing IRES are able to efficiently benefit from the host cells ER stress response for their own multiplication. Inhibition of cap-dependent translation of cellular mRNAs by either viral infections or stress factors is achieved by (1) specific cleavage of cellular translation initiation factors by HIV and picornaviral proteases, or cellular caspases, (2) active phosphorylation of factors and cofactors of translation, such as eIF2α, (3) excessive production of the cap-binding protein eIF4E, which interacts with eIF4G and results in the impairment of the eIF4 complex, and (iv) restriction of eIF4E function by activation of microRNAs which, again, disrupts the assembly of the eIF4 complex.

Poly(A) Tailing

Addition of the 3′ poly(A) tail is another end-processing mechanism required to protect the mRNA transcript from nucleolytic degradation in the cytoplasm and enable mRNA stability. This is the site of binding of poly(A) binding protein in the cytoplasm. Cellular mRNA transcripts undergo polyadenylation through cleavage at the signal sequence AAUAAA by the CPSF (**c**leavage and **p**olyadenylation **s**pecificity **f**actor) and CSTF (cleavage stimulation factor). Viral mRNAs are synthesized without this signal sequence. Stuttering occurs at a site containing a slippery sequence (mononucleotide repeats) and involves 1-base repeated frameshifts on the mRNA strand (Fig. 3.6). It is a mechanism used by various negative RNA(−) viruses of the families *Bornaviridae*, *Filoviridae*, *Paramyxoviridae*, *Rhabdoviridae*, and

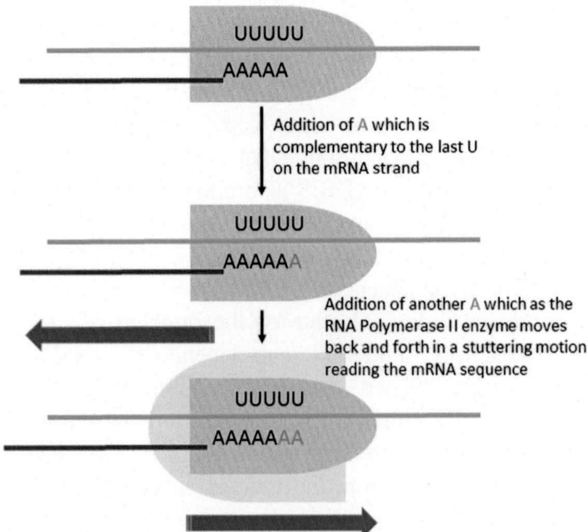

FIGURE 3.6 Stuttering mechanism. Stuttering involves the slippage of RNA polymerase over a stop codon sequence on the cellular mRNA sequence that results in addition of A residues on the viral mRNA strand effectively introducing a poly(A) tail on the viral mRNA. The addition of the 3′ poly(A) tail is another end-processing mechanism that protects the mRNA transcript from nucleolytic degradation in the cytoplasm and enables mRNA stability.

Orthomyxoviridae to polyadenylate their mRNAs. The addition of the poly(A) tail to the 3′ terminus of mRNA transcripts carries a stretch of five to seven uridine residues located in close proximity to the 5′ terminus of the viral RNA template. To achieve this, viral-encoded RdRp remains attached to the 5′ terminus of the viral RNA template, resulting in steric hindrance at this site. Upon recognition of the polyadenylation signal, RdRp moves back and forth over this stretch of U residues, reiteratively copies these residues, and produces a stretch of adenines effectively, a poly(A) tail at the 3′ end of the viral mRNA.

RNA Editing

This mechanism, also observed in some eukaryotes, allows RNA viruses (except dsRNA viruses) to produce multiple proteins from a single gene. In these viruses, the RNA polymerase reads the same template base more than once, creating insertions or deletions in the mRNA sequence, thereby generating different mRNAs that encode different proteins. There are two kinds of mRNA editing: (1) cotranscriptional editing through polymerase slippage and (2) posttranscriptional editing. RNA editing in members of the *Ebolavirus* genus increases their genome coding capacity by producing multiple transcripts encoding variants of structural and nonstructural glycoproteins from a single gene, ultimately increasing its ability for host adaptation.

Alternative Splicing

Also observed in many cellular organisms, alternative splicing allows production of transcripts having the potential to encode different proteins with different functions from the same gene (Fig. 3.7). The sequence of the mRNA is not changed as with RNA editing; rather the coding capacity is changed as a result of alternative splice sites. Alternative splicing is regulated by cellular and viral proteins that modulate the activity of the splicing factors U1 and U2, both of which are components of the spliceosome. The spliceosome is made up of the snRNAs (small **n**uclear **RNAs**) U1, U2, U4, U5, and U6, together with various regulatory factors. Activation of the spliceosome is facilitated by cis-acting signals in the mRNA sequence. Some of these signals include donor splice sites (5′ terminus), acceptor splice sites (3′ terminus), polypyrimidine tracts, and branch point sites. Serine/arginine-rich proteins, as well as heterogeneous nuclear ribonucleoproteins, play a key role in splice site recognition. Alternative splicing (1) increases the virus' ability to encode several proteins in a given transcript (e.g., adenoviruses and retroviruses can encode ~12 different peptides from one pre-mRNA), (2) is a mechanism to regulate early and late expression for viruses (e.g., papillomaviruses), and (3) splicing is coupled to export of mRNA out of the nucleus. While only mature, spliced mRNA transcripts are exported out of

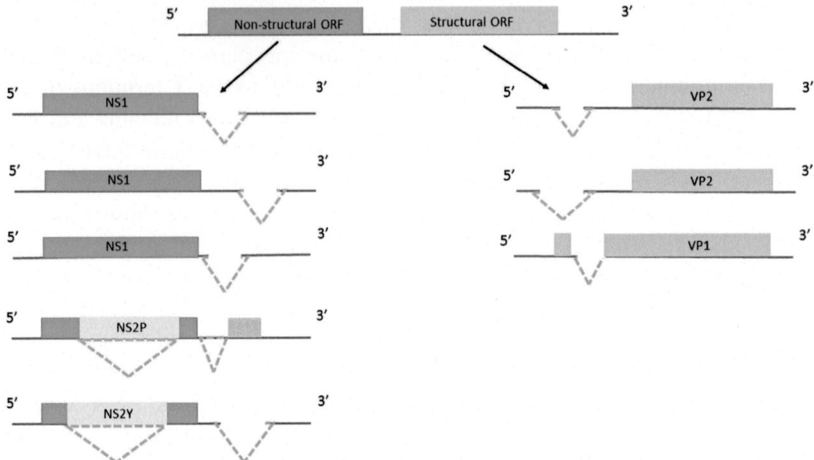

FIGURE 3.7 Alternative splicing. Alternative splicing is common in parvovirus pre-mRNA transcript processing and allows for the generation of different proteins from a specific nucleotide sequence on the viral mRNA strand. *Dotted lines* indicate alternative splice sites.

the nucleus, hepadnaviruses and retroviruses are able to export nonspliced mRNA transcripts out of the nucleus for translation. On the other hand, the NS1 protein (**n**onstructural **p**rotein **1**) of influenza viruses can interact with multiple host cellular factors via its effector- and RNA-binding domains. It is capable of associating with numerous cellular spliceosome subunits, such as U1 and U6 snRNAs, and can inhibit cellular gene expression by blocking the spliceosome component recruitment and its transition to the active state.

Both conservation and evolution of viral splice site sequences allow for improved adaptation to the host, and ensure recognition by the host's splicing machinery. Therefore, viruses can induce preferential induction of viral mRNA splicing by the cellular splicing machinery. Knowledge concerning the coordination between cellular and viral genome splicing comes from adenoviruses and retroviruses, but only limited data are available for other viruses, for example, influenza viruses.

Suppression of Termination

This is also referred to as stop codon read-through, and is a programmed cellular and viral-mediated mechanism used to produce C-terminally extended polypeptides, and in viruses, it is often used to express replicases. Termination of translation occurs when one of three stop codons enters the A-site of the small 40S ribosomal subunit. Stop codons are recognized by release factors (eRF1 and eRF3), which promote hydrolysis of the peptidyl-tRNA bond in the peptidyl transferase center (P-site) of the large ribosomal subunit. Termination is a very efficient mechanism that is tightly controlled by the type of stop codon encountered (UAA, UAG, or UGA). Some termination codons are referred to as "*leaky*" depending on the nature of the base positioned after the stop codon (e.g., UGAC) where they allow "*read through*" at frequencies ranging from 0.3% to 5%. Read-through occurs when this leaky stop codon is misread as a sense codon with translation continuing to the next termination codon. Read-through signals and mechanisms of prokaryotic, plant, and mammalian viruses are variable and are still poorly understood.

Programmed Ribosomal Frameshifting

Programmed ribosomal frameshifting is a tightly controlled, programmed strategy used by some viruses to produce different proteins encoded by two or more overlapping open reading frames (Fig. 3.8). Ordinarily, ribosomes function to maintain the reading frame of the mRNA sequence being translated. However, some viral mRNAs carry specific sequence information and structural elements in their mRNA molecules that cause ribosomes to slip, and then readjust the reading frame. The frameshift results from a change in the reading frame by one or more bases in either the 5′ (−1) or 3′ (+1) directions during translation. This ribosomal frameshift enables viruses to encode more proteins in spite of their small size.

Leaky Scanning and Translation Reinitiation

In the world of viruses, this extended mechanism of "*modified*" translation occurs when the start codon is bypassed by the translation initiation complex during translation, but continuous scanning allows locating another AUG start codon at a downstream site. This occurs because the initiation codon can be part of a weak Kozak consensus sequence. As a result, there can be

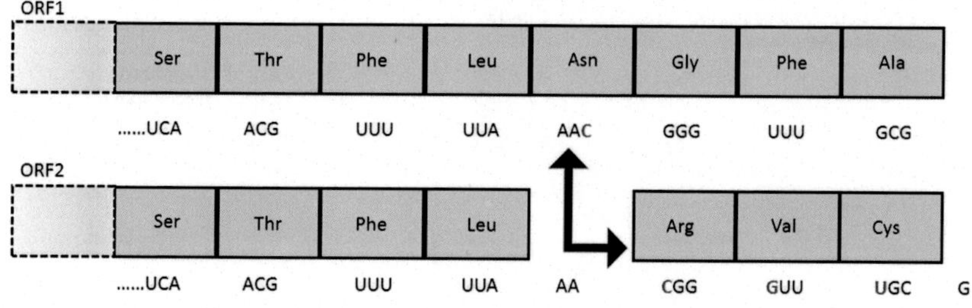

FIGURE 3.8 Ribosomal frameshifting. −1 ribosomal frameshift, common to parvoviruses, changes the amino acids encoded in the mRNA strand by moving the reading frame 1-base down (−1).

the production of several different proteins if the AUG codon is not in frame, or proteins with different N-termini if the AUGs are in the same frame. A number of viruses engage in leaky scanning, including members of the families *Herpesviridae*, *Orthomyxoviridae*, and *Reoviridae*.

Ribosomal Shunting

Ribosomal shunting occurs when the ribosomal initiation complex is loaded onto an mRNA at the 5′-cap but the process of scanning for the start codon progresses for only a short distance, bypasses a large internal leader region, and initiates translation at another start codon located downstream of the leader sequence (Fig. 3.9). It is, therefore, referred to as cap-dependent discontinuous scanning. The mechanism of ribosome shunting has not been described in molecular detail. Shunting expands the coding capacity of mRNAs of viruses such as caulimoviruses.

"2A" Oligopeptides and "*Stop-Carry On*" Recoding

Production of viral proteins often requires noncanonical decoding events (or "*recoding*") on certain codons during translation due to the restricted coding capacity of a small genome size. "2A" oligopeptides coded for by viruses (e.g., *Foot-and-mouth disease virus*, FMDV) are important to "*stop-carry on*" recoding. 2A oligopeptides interact with the ribosomal exit tunnel to initiate a stop codon-independent termination of translation at the final proline

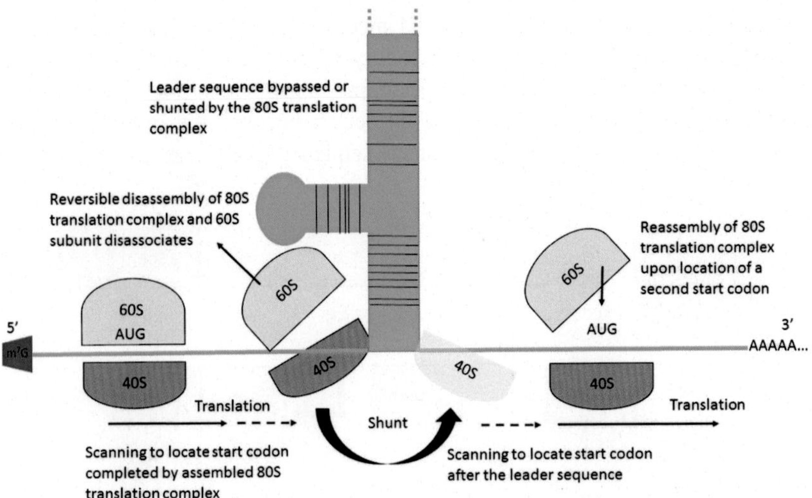

FIGURE 3.9 Ribosomal shunting. Ribosomal shunting by the 80S ribosome involves bypassing the AUG start codon closest to the 5′ end of the mRNA sequence to locate a second AUG start codon downstream.

codon of 2A. Ribosomes, therefore, skip the synthesis of the glycyl-prolyl peptide bond at the C-terminus of a 2A peptide (cleavage of the peptide bond between a 2A peptide and its immediate downstream peptide). Translation is then reinitiated on the same codon, which leads to production of two individual proteins from one open reading frame.

SUBVERSION OF HOST GENE EXPRESSION

Viruses not only employ strategies that maximize the coding capacity of their small genomes, disguise their mRNA with the same structural elements found in host mRNA, regulate their genome expression in a time- and space-dependent manner, but they have also evolved ways of subverting host cell functions in order to favor their own replication and translation.

Inhibition of Cellular RNA Polymerase

The CTD of host RNA polymerase contains 52 heptapeptide repeats (YSPTSPS) and is phosphorylated primarily at serine amino acids multiple times during the transcription process (Fig. 3.10). These phosphorylation events serve to activate or deactivate the enzyme. Some viruses (herpesviruses, bunyaviruses) counteract this phosphorylation at serine amino acids

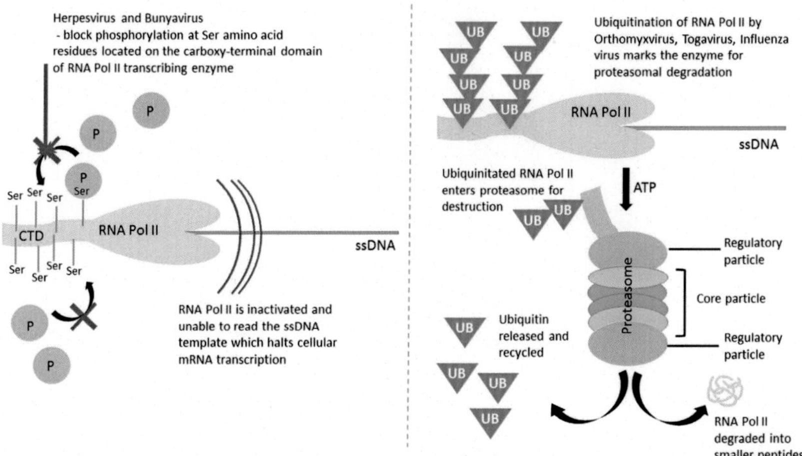

FIGURE 3.10 Examples of mechanisms used by viruses to block the activity of **RNA p**olymerase (RNA Pol) from transcribing cellular mRNA. Phosphorylation of serine residues located on the CTD of the enzymes is blocked by some viruses. Other viruses arrest RNA Pol activity by signaling ubiquitination of the transcribing enzyme, which is subsequently degraded by the proteasome. Virus-encoded protein, ICP-22, of human herpesviruses blocks CTD Ser-2 phosphorylation of host RNA Pol; virus-encoded polymerases, PB1, PB2, PA, cause host RNA Pol ubiquitination and proteasomal degradation; virus-encoded protein, nsp2, is produced by togaviruses and functions in ubiquitination of Rpb1 subunit of host RNA Pol and proteasomal degradation.

to inactivate RNA polymerase, while other viruses (orthomyxviruses, togaviruses) disrupt cellular RNA polymerase function by signaling ubiquitination of the enzyme and its subsequent degradation by proteasomal action.

Disruption of Cellular mRNA Export Pathways

Viruses can engage in targeted disruption of cellular mRNA export pathways to promote preferential viral gene expression (Fig. 3.11). All DNA viruses replicate within the nucleus except poxviruses, asfarviruses, and phycodnaviruses. Few RNA viruses, including bornaviruses, orthomyxoviruses, and retroviruses, replicate in the nucleus. Trafficking between the nucleus and cytoplasm is usually unidirectional for large macromolecules like the mRNA transcript, and occurs through the **n**uclear **p**ore **c**omplex (NPC). Viruses that replicate in the nucleus must out-compete cellular mRNAs to export viral mRNAs out of the nucleus for translation into virus gene products in the cytoplasm. Several viruses can inhibit nuclear export of cellular mRNAs by

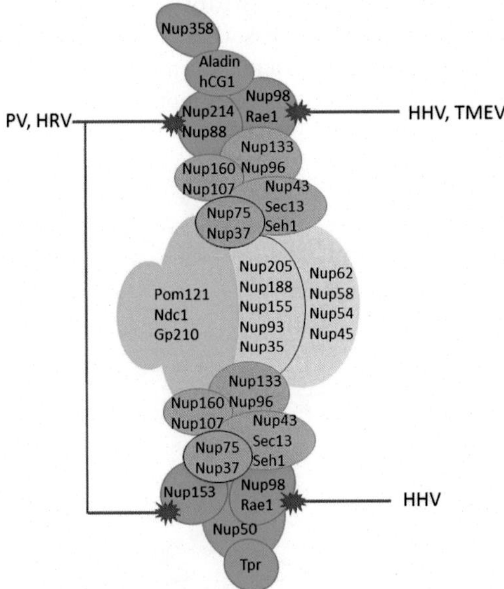

FIGURE 3.11 Inhibition of cellular mRNA export out of the nucleus by targeted disruption of the structure of the NPC. One half of the NPC is shown in the diagram. Vesiculoviruses are negative-stranded (−) RNA viruses that prevent proper cellular mRNA export by interfering with the action of the VSV matrix (M) protein. This effect decreases competition of viral mRNAs with cellular mRNAs for use of the translational machinery. Many DNA viruses (e.g., adenoviruses) also selectively inhibit host mRNA export, while ensuring that viral mRNAs are efficiently exported after transcription (Yarbrough et al., 2014). Viruses: PV, Poliovirus; HRV, Human rhinovirus; HHV, Human herpesvirus; TMEV, Theiler's murine encephalomyelitis virus.

disrupting nuclear export receptors (exportin1 and TIP-associated protein) and nucleoporins that comprise the NPC to compromise their function in nucleocytoplasmic trafficking of cellular mRNA.

Decay of Host mRNAs by Viruses

Viruses have developed different strategies to effectively degrade host mRNAs and to allow preferential translation of their own mRNA (Fig. 3.12). Most viruses produce an endonuclease that cleaves host mRNAs, which are then degraded by host exonucleases (e.g., Xrn1 in mammals). The mRNA fragments are then degraded by mRNA decay pathways (nonsense-mediated decay), and exosomal degradation. *Human gammaherpesvirus 8*, *Human alphaherpesvirus 1*, and *Influenza A virus* all induce destruction of cellular mRNAs. Betacoronaviruses, influenza viruses, vaccinia viruses, and herpesviruses can produce viral endonucleotyic products to an extent that saturates cellular RNA decay-related quality control mechanisms and limit their function. Invariably this mechanism helps viral RNAs escape detection by the cellular RNA decay machinery.

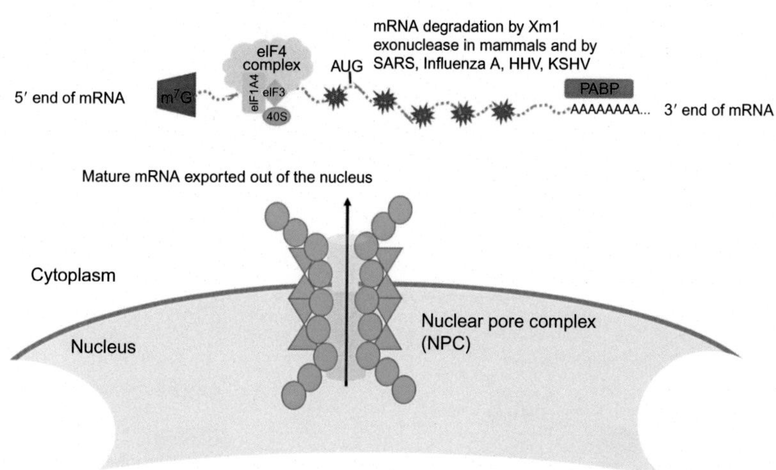

FIGURE 3.12 Degradation of cellular mRNA by viral-encoded endonuclease enzymes allows for preferential translation of viral mRNA. M^7G—5′ cap, eIF—eukaryotic initiation factor, AUG—start codon, PABP—poly(A) binding protein. Viruses: SARS, SARS-coronavirus; Influenza A, *Influenza A virus*; HHV, Human herpesvirus; KSHV, *Human gammaherpesvirus 8* (herpesvirus associated with Kaposi's sarcoma in humans).

Circumvention of Cellular RNA Decay Machinery

Eukaryotic cells utilize a surveillance mechanism to constantly monitor transcripts for aberrant RNA molecules including misfolded, "*mis*"-translated (e.g., mRNAs with a premature termination codon), and mispackaged mRNAs and these transcripts are quickly degraded in the cytoplasm. Transcripts of cytoplasmic viruses must circumvent the cellular mRNA decay machinery to enable virion production. Several cytoplasmic viruses repress key aspects of the cellular RNA decay machinery to escape RNA degradation. Picornaviruses are able to suppress cellular RNA decay factors, and polioviruses and human rhinoviruses produce viral proteases that degrade Xrn1, Dcp1, Dcp2, Pan3 (a deadenylase), and AUF1decay factors. HCV RNAs have similarly been shown to bind LSm1-7 and knockdown RNA decay factors.

Shutoff of Cellular mRNA Translation

Initiation of translation of cellular mRNA occurs through the recruitment and assembly of a multisubunit translation initiation complex at the 5′ end of the mRNA strand (Fig. 3.13). Viruses capable of inducing the shutdown of

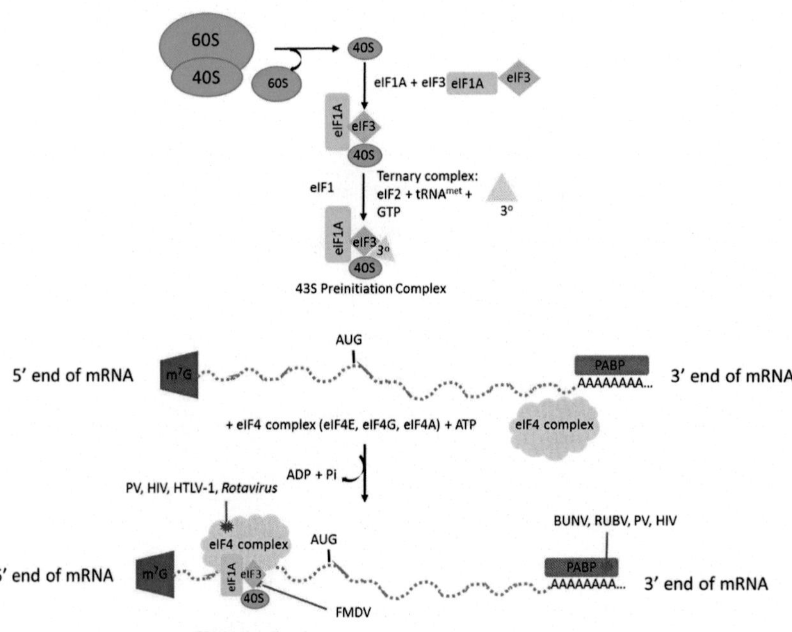

FIGURE 3.13 Shutoff of host translation machinery by viral interference with specific eukaryotic translation initiation factors and poly(A) binding protein (PABP). Viruses: PV, Poliovirus; HIV, *Human immunodeficiency virus 1*; HTLV-1, Human T-lymphotropic virus-1; FMDV, *Foot-and-mouth disease virus*; BUNV, *Bunyamwera orthobunyavirus*; RUBV, *Rubella virus.*

cellular mRNA translation are able to continue to translate at least part of their mRNAs using noncanonical translation mechanisms, for example, cap-independent translation, ribosome shunting, and leaking scanning (e.g., adenoviruses, picornaviruses, reoviruses, and rhabdoviruses).

Recruitment of Cellular Hsp70 Chaperones for Viral Protein Folding

Most viruses interact with cellular chaperones in order to ensure correct folding of viral proteins. Viral proteins often consist of multiple domains or are produced as polyprotein precursors, which must be processed before they can be functional. The coat protein or capsid is a meta-stable structure that must be specifically assembled in a preordered arrangement without reaching minimum free energy; yet must be disassembled upon entry of the host cell. Some cellular chaperones, for example, Hsp70, are used to accelerate the maturation of viral proteins and are involved in regulating the viral biological cycle. The high rate of mutation in RNA viruses may mean an increased dependency on chaperones for the gene products of these viruses. Hsp70 can refold denatured proteins, which negates some of the destabilizing alterations in structural proteins as a result of mutated genes. This ensures that a high proportion of viral proteins is accurately configured to function in virus multiplication.

Compromising Cellular Lipid Metabolism

Viruses can manipulate the cellular metabolism to provide an increased pool of molecules, for example, nucleotides and amino acids, which are required for viral gene expression and virion assembly. Some viruses need to create a lipid-rich intracellular environment favorable for their replication, morphogenesis, and egress. Replication of HCV occurs on specific lipid raft domains, whereas assembly occurs in lipid droplets. As such, in order for HCV to create replication compartments and increase sites of assembly, the RNA virus requires both the synthesis of fatty acids, for example, cholesterol, sphingolipids, phosphatidylcholine, and phosphatidylethanolamine, and formation of lipid droplets. Lipids are especially required for assembly of virions of enveloped viruses as these molecules are a major component of membranes. Cellular lipid metabolism is affected at three levels: enhanced lipogenesis, impaired degradation, and disruption of export, which is subsequently manifested in the host as HCV-associated pathogenesis. HCV infection is also associated with reduced serum cholesterol and β-lipoprotein levels.

CELL CYCLE DISRUPTION FOR PREFERENTIAL VIRAL REPLICATION

Transition between the phases of the cell cycle is driven by activities of **c**yclin/**c**yclin-**d**ependent **k**inase (Cdk) complexes. Cyclins are a diverse family of proteins whose structure includes a conserved region known as the "*cyclin box,*" which is necessary for Cdk binding and activation. The cyclins are classified according to the cell cycle phase they regulate: Cyclin D proteins are G1 phase cyclins; Cyclin E and Cyclin A proteins promote cell cycle progression through G1/S phases; Cyclin B proteins are associated with the M-phase (Fig. 3.14).

Viral interference of the host cell cycle can result in the dysregulation of cell cycle checkpoint control mechanisms to promote viral replication and to facilitate efficient virion assembly. Both DNA and RNA viruses specifically encode proteins responsible for targeting and arresting essential cell cycle regulators to create intracellular conditions that are favorable for viral replication and propagation. Retroviruses and other RNA viruses also interfere with the host cell cycle. Viral-mediation of the cell cycle can increase the efficiency of viral gene expression and virion assembly. Cell cycle arrest may delay apoptosis of infected cells. In addition, a specific G2/M-phase cell cycle arrest as induced by *Human immunodeficiency virus 1* prevents new cell production, which aids the virus in immune evasion.

Viruses can initiate the cell cycle and activate TFIIF transcription factors by stimulating cyclin D activity or dissociation of the **r**etino**b**lastoma (Rb)/TFIIF complex in a cyclin-D-independent manner. Many viruses encode a cyclin-D homolog protein (v-cyclin) that associates with Cdk6 to phosphorylate Rb, which regulates G1 phase. The EBV-EBNA2 and EBV-LP,

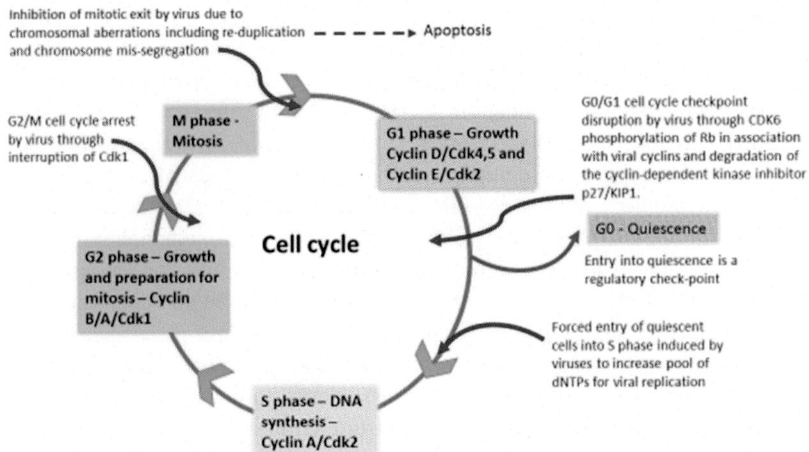

FIGURE 3.14 Cell cycle dysregulation by viruses.

HTLV-1-Tax, and HBV-HBx proteins upregulate cyclin D, which leads to Rb hyperphosphorylation and TFIIF-mediated transcriptional activation. Distinct viral families are able to target the Rb/TFIIF association and promote the premature or unscheduled transition of the infected host cell into the S phase. Other viruses utilize mechanisms that result in G2/M arrest through either inactivation of Cdk1 at the G2/M checkpoint and/or at the interference with mitotic progression. Viruses can also dysregulate mitosis through the activity of viral oncoproteins such as HTLV-1-Tax, Ad-E1A, and HPV-E6/E7, which serve to induce chromosomal aberrations and chromosome mis-segregation that ultimately can result in apoptosis of the cell. Various DNA viruses primarily infect quiescent or differentiated cells, which contain low levels of deoxynucleotides (dNTPs) as these cells do not undergo active cell division. As such, a restricted pool of dNTPs will not provide an ideal environment for viral replication. It has been proposed that such viruses can induce quiescent cells to enter the cell cycle, specifically the S phase, in order to create an environment that generates factors, such as nucleotides, that are required for viral replication. Large DNA viruses, for example, herpesviruses, can cause cell cycle arrest as a mean of competing for cellular DNA replication resources.

The viral-mediated modifications of host cell cycles, which may be detrimental to cellular physiology, significantly contributes to associated pathologies, such as cancer progression and cell transformation. Viral infections may account for approximately 20% of all human cancers worldwide.

A summary of most of the strategies developed by viruses to ensure viral replication and gene expression is provided in Fig. 3.15.

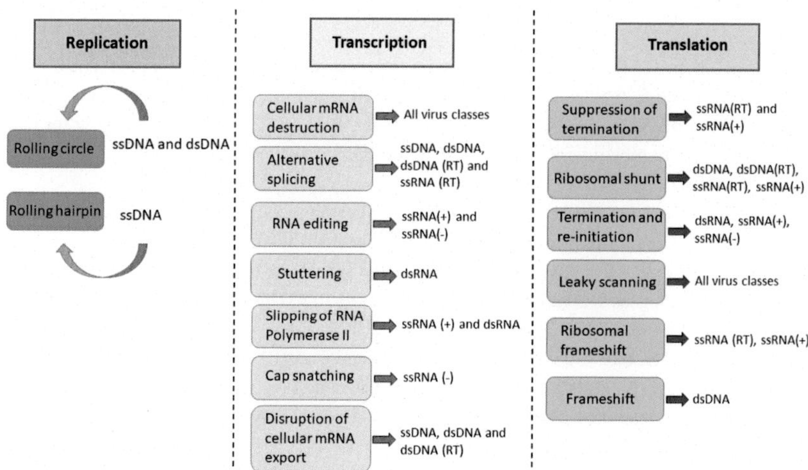

FIGURE 3.15 Summary of strategies developed by viruses to ensure viral replication and gene expression.

VIRUS GENOME REPLICATION COMPLEXES

Finally, viruses have developed a number of targeted strategies to manipulate cellular activities, which enable specific recruitment of macromolecules required for viral replication and gene expression at specific locations in the host cell. Typically, these macromolecules are recruited and concentrated into specific cytoplasmic or nuclear compartments. Formation of such specialized cellular microenvironments, also termed viroplasms, virus factories, virus replication centers, complexes, or compartments, requires the coordinated control of cellular biosynthesis in addition to (1) alterations in the dynamics and distribution of the cytoskeleton and associated motor proteins, (2) relocalization of cellular organelles, and (3) reorganization and redeployment of cellular membranes associated with membrane-bound organelles, for example, the endoplasmic reticulum, mitochondria, chloroplasts, Golgi apparatus, endosomes, and lysosomes in eukaryotes. Formation of viral factories often requires sequestration of mitochondria (and/or chloroplasts) and chaperones to perinuclear sites. In addition, DNA viruses that replicate in the nucleus induce nuclear reorganization and redistribution of chromatin and nuclear domain components such as the nucleolus, interchromatin granules, and Cajal bodies. These compartments provide a scaffold for efficient viral gene expression, while simultaneously concealing viral genomes (refer to Chapter 10: Host–Virus Interactions: Battles Between Viruses and Their Hosts) and their products from immunological detection.

FURTHER READING

Cavallari, I., Rende, F., D'Agostino, D.M., Ciminale, V., 2011. Converging strategies in expression of human complex retroviruses. Viruses 3, 1395–1414. Available from: https://doi.org/10.3390/v3081395.

Dubois, J., Terrier, O., Rosa-Calatrava, M., 2014. Influenza viruses and RNA splicing: doing more with less. mBio 5. Available from: https://doi.org/10.1128/mBio.00070-143e00070-14.

Gaglia, M.M., Covarrubias, S., Wong, W., Glaunsinger, B.A., 2012. A common strategy for host RNA degradation by divergent viruses. J. Virol. 86, 9527–9530. Available from: https://doi.org/10.1128/JVI.01230-12.

Hoeben, R.C., Uil, T.G., 2013. Adenovirus DNA replication. Cold Spring Harb. Perspect. Biol. 5. Available from: https://doi.org/10.1101/cshperspect.a013003a013003.

Moon, S.L., Wilusz, J., 2013. Cytoplasmic viruses: rage against the (cellular RNA decay) machine. PLoS One Pathogens 9. Available from: https://doi.org/10.1371/journal.ppat.1003762e1003762.

Nascimento, R., Costa, H., Parkhouse, R.M., 2012. Virus manipulation of cell cycle. Protoplasma 249, 519–528. Available from: https://doi.org/10.1007/s00709-011-0327-9.

Swanton, C., Jones, N., 2001. Strategies in subversion: de-regulation of the mammalian cell cycle by viral gene products. Int. J. Exp. Pathol. 82, 3–13. Available from: https://doi.org/10.1046/j.1365-2613.2001.00165.x.

Vreede, F.T., Fodor, E., 2010. The role of the influenza virus RNA polymerase in host shut-off. Virulence 1, 436–439. Available from: https://doi.org/10.4161/viru.1.5.12967.

Yarbrough, M.L., Mata, M.A., Sakthivel, R., Fontoura, B.M.A., 2014. Viral subversion of nucleocytoplasmic trafficking. Traffic 15, 127–140. Available from: https://doi.org/10.1111/tra.12137.

Chapter 4

Origins and Evolution of Viruses

Jerome E. Foster[1] and Gustavo Fermin[2]
[1]The University of the West Indies, St. Augustine, Trinidad, [2]Universidad de Los Andes, Mérida, Venezuela

---

When considering the virosphere, extremely unlikely events become probabilistic certainties.

Forest Rohwer and Katie Barott

Almost daily we hear expressions related to the purported emergence of new viruses. It is true that, thanks to the dedicated efforts of virologists and the advancement of new and powerful technologies, viruses unheard of before are being *discovered* almost on a daily basis. And they catch our attention for a good reason: they are fascinating. For one thing, they are challenging our previous paradigms regarding their nature and origin; additionally, they exert varying effects on all living organisms, communities, and the biosphere as well. No one can claim for sure that they have discovered an "*emerging*" virus if we accept the idea that all living beings are in continuous change (biological evolution) and thus continuously "*emerging*" with renovated faces. All viruses have ancestors that somehow, in some time in history, appeared and later changed to the forms (and with the capabilities) we now know. How this happened is probably one of the most crucial questions of biology due to the likely impact on our comprehension of the phenomenon we call life. So far we have presented on the structure and composition of viruses, and we will present later on the species they interact with, and the many consequences these relationships have on the hosts, the viruses themselves and the environment where both reside. In this chapter the focus is on the origin and evolution of viruses.

In studying the evolutionary history of viruses, we cannot rely on the discovery and analysis of material that have helped us greatly in the study of some cellular organisms: fossils and/or sequences of shared genes (for instance, ribosomal genes), or a gene common to all members of a particular group (chloroplast genes in plants, for example). With viruses this is not

Viruses. DOI: https://doi.org/10.1016/B978-0-12-811257-1.00004-8

possible. There is not a single gene shared among all viruses. They are very "*simple*" informational entities, which have many structural and functional features in common, but they do not seem to have a single piece of information that might lead to the postulation of the (inexistent entelechy) **L**ast **U**niversal **C**ommon **A**ncestor (LUCA) of Viruses. Similarly, and consequently, a Tree of Viruses does not exist.

Life has a chemical, abiotic origin that cannot be refuted. The same can be expected of viruses but we have to resort to a plethora of different methods of inference (some of them not used with cellular organisms) when trying to uncover their evolution. It is quite curious that despite the fact that our early ancestors were probably molecules very similar to a "*virus*," we are unable to quite understand the present day entities. It is very likely that viruses have multiple origins (i.e., they are polyphyletic) and the emergence of an exquisite multitude of cellular species contributed to their radiation (and recreation?). What follows is a brief account of the events that presumably shaped the route by which viruses managed to emerge and thrive in our environment.

ARE VIRUSES LIVING ORGANISMS?

Despite biology being the study of life, a definition of what constitutes life is far from clear-cut. The Merriam-Webster dictionary defines life as "*an organismic state characterized by capacity for metabolism, growth, reaction to stimuli, and reproduction.*" Another definition is that of the theory of autopoesis, by Maturana and Varela, in which life is envisioned as "*the ability of a system to maintain and reproduce itself.*" As explained later a virus-infected cell is essentially a cell that has surrendered to the infecting agent that takes control of all of its capacities: in a way, a virus becomes the cell it infects—with a viral (informational) twist. From this point of view a virus is a living entity *only* when inside its host. As for the second definition, is a virus not the epitome of maintaining and reproducing itself? As discussed in Chapter 2, Virion Structure, Genome Organization, and Taxonomy of Viruses, we are probably running in circles trying to apply a definition to viruses that should not be used with viruses in the first place.

Classic Virus Concept

Lwoff (1957) gave us one of the earliest definitions of viruses based on his studies of bacteriophages—viruses that infect bacteria. He described viruses as strictly intracellular infectious entities of potential pathogenicity, with only one type of genetic material, which are unable to grow, divide (by binary fission) and lack the ability to produce energy. A virus was not considered a living entity as it differed from cells in several aspects: (1) it did not reproduce in its entirety but only from its genetic material, (2) there was no binary fission involved in its reproduction, (3) it had only one type of genetic material (RNA or DNA) rather than both, (4) it lacked an enzymatic system capable of producing energy from food sources—what was then called a "*Lipmann system,*" and (5) they lacked ribosomes and relied on their host cells' translation machinery.

Viruses were envisioned as entirely dependent on host cell metabolism, transforming the infected cells into virus factories, and therefore thought of as metabolically inert infectious particles, rather than living organisms. They relied on the cell's machinery to replicate and propagate themselves and were therefore obligate parasites. As a result, they have traditionally been excluded from the **T**ree **o**f **L**ife (ToL), which contains only cellular **r**ibosome-**e**ncoding **o**rganisms (REOs).

The Virocell Concept

It was first proposed in 1983 by Bandeau that viruses do in fact display all the hallmarks of what scientists define as a living organism if one considers the infected cell, which the virus controls, as a complete entity. That infected cell, termed a virocell, would display organized restructuring (viral factories), metabolism, growth, and reproduction, i.e., would conduct all the processes mentioned previously. Similarly, in a lysogenic situation, we can consider the infected cell a ribovirocell as it has two functions—encoding both ribosomes and capsids. Virions themselves would then not represent more than the spores or reproductive cells of this virocell, much as plants reproduce through their seeds. It was therefore put forward that the classic definition of viruses as particles was a misrepresentation of their true nature and that they should be identified by their intracellular form. One may consider life as the ability to maintain and reproduce oneself as suggested by Maturana and Varela in their theory of autopoesis (1973). Viruses, with their virus factories and the virocell concept, fit such a definition of life. If we decide to exclude viruses from the biological definition of life because they must reproduce through a host then we risk excluding other organisms such as obligate intracellular parasites. These also rely on a host to reproduce, good examples being the bacteria belonging to the genera *Chlamydia* and *Rickettsia*.

Virocell Versus Ribocell: Are Giant Viruses the Missing Link?

The ribocell (ribosomes-encoding cell) model puts forward a hypothetical minimal cell based on a self-replicating minimal genome coupled with a self-reproducing lipid vesicle compartment. Current ribocells have, in addition, complex machinery that allows for the synthesis of proteins by means of dedicated organelles called ribosomes. By contrast, there is the virocell model—where biological entities depend on a coat protein to define the physical object they represent (i.e., virions), and rely on the ribosomes of the cell they infect to complete their life cycles. At times, it is not easy to draw a separating line between the two. In the case of lysogeny, e.g., we have a REO and a capsid-encoding replicative entity inhabiting the same cellular environment. The discovery of giant viruses blurred the boundary between cellular parasites and viruses even further; it became very clear that different viruses show different levels of dependence toward their host.

The first member of the genus *Mimivirus*, a **n**ucleo**c**ytoplasmic **l**arge **DNA** **v**irus (NCLDV) was discovered in 2006, and with it, the discussion on virus origins was reopened. The virus was named *Acanthamoeba polyphaga mimivirus* (APMV; *Mimiviridae*), and later a second member of the genus, and of the family, was discovered (*Cafeteria roenbergensis virus*), both of which possess a dsDNA genome. APMV had several characteristics that suggested it could be the theoretical missing link between viruses and cellular organisms. After being sequenced and compared through phylogenetics, only about 1% of the APMV genome was homologous to the genome of its amoeba host DNA (*A. polyphaga*). Only 25% of that mimivirus genome is homologous to that of other NCLDVs, while up to 70% of its genes have no homologs in any other known viral or cellular genomes, with such genes being termed as ORFans. Mimiviruses, along with other megaviruses, also have been found to contain partial translation apparatus, including several functional virus-encoded aminoacyl-tRNAs. Thus although they are translationally incompetent, these viruses have some components of the translation machinery. Another characteristic unique to this group of viruses is that the mimivirus virus factory can be infected (parasitized) by another virus (*Mimivirus-dependent virus Sputnik*, a virophage of the family *Lavidaviridae*), which gives even more credence to the concept of these viruses as possessing some of the characteristics that define living entities. Nonetheless despite these characteristics, none of the giant viruses actually code for ribosomal proteins or structural RNAs, and so cannot be described as *bona fide* ribocells. But the giant viruses demonstrate that viruses can be more complex than some eukaryotic cells. The fact that these giant viruses have almost no equivalent among the three recognized domains of cellular life points to a greater variety of precellular forms than those conventionally considered. We now turn our attention to the fundamental mechanisms of viral evolution as these shed light on the ancestry of viruses.

SOURCES OF VARIATION IN VIRUS GENOMES: THE MANY WAYS BY WHICH VIRUSES CHANGE

Genomic molecules of viruses are subject to the same sources of variation seen with genomic molecules of cellular organisms. It could not be different since viruses replicate and express their informational content in the cytoplasm and/or nucleus of their hosts. Although virus genome replication and expression involve the use of their own encoded proteins (replicases and transcriptases), virus proteins experience the same biological, chemical, and physical constraints imposed on the host's proteins. One notable difference, however, is that in cellular organisms, if RNAs (mRNAs, tRNAs, microRNAs, small nuclear and nucleolar, and other small RNAs) mutate, but their coding DNA sequence does not, then these changes are not heritable. In viruses, particularly RNA viruses, sources of variation include those derived from changes directly experienced by RNA molecules. For example, through the mechanism known as copy-choice, replicases that produce viral RNAs can switch from one virus RNA template to another one, thus creating a recombinant RNA molecule that, most of the times, is viable and not identical to any of the two involved in the "*recombination*" event.

The rapid and sustained generation of virus variants has led to the formulation of the virus quasi-species concept. That is, viruses that circulate in a host are comprised of a population of mutants (variants) generated during error-prone replication of RNA viruses (and some DNA viruses). These individual molecules (or virus ensemble) compete and are subject to selection, thus creating a changing landscape in terms of the distribution of variants over time. Members of this varied population of molecules interact among themselves within the host, possibly causing positive effects (complementation) or negative effects (interference), thus rendering selection at the highest possible communal level. This in turn affects virus fitness in such a way that the "*virus*" (understood as a varied population that acts as a unit of selection) experiences continuous pressure for its maintenance as a defined entity—that, by force, *has* to change over time. A spectrum of variants, then, and not individual genomes are the target of evolutionary events. Thus the way viruses reproduce in their host cells makes them particularly susceptible to the genetic changes that help to drive their evolution. Most of the biologically relevant variation observed is the result of genetic variation and competitive selection, together with random events acting on multiple replicative units. Virus evolution is decisively influenced by mutation rates during viral replication, and in some cases, also by recombination, genome segment reassortment, and **h**orizontal **g**ene **t**ransfer (HGT).

Mutation and High Viral Replication Error Rates

In the classical genetic sense a mutation is any change in the linear arrangement of nucleotides making up a particular DNA sequence that is stably inherited (i.e., a mutation must to be fixed for it to be considered as such). Most DNA mutations, if not all, are fixed in the second round of replication after the mutagenic event has taken place. Mutations at the level of viral RNA molecules are said to be fixed if the variant is capable of dissemination (i.e., the mutation did not impair the virus' ability to replicate, assemble, and be transmitted). Either way, another interesting fact about mutations in viruses is that their nucleic acids can be subjected to modifications within their harboring virions (e.g., by UV radiation). Once in the infected cell, viral nucleic acids can be repaired leading to the restoration of their original sequence; if they are not, the mutation (insertion or deletion) probably will persist. Mutations in virus genomes occur firstly by nonrepaired errors incurred during replication; rates of mistakes are an intrinsic property of polymerases and of their possession or not of proofreading activities. Mutations can also occur by the action of mutagens. Finally, mutations may appear as a consequence of the action of host repairing mechanisms, particularly in SOS-derived responses in bacteria and their analogs in eukaryotes. In general mutation rates in viruses far exceed those observed in cellular organisms. Mutation rates in RNA viruses have been estimated to be in the range of 10^{-3} to 10^{-5} mutations per nucleotide, while in cellular organisms, although variable, are at least several orders of magnitude lower.

Evolutionary studies have shown there is a strong inverse relationship between mutation rate per genome replication and genome size. The greater the error rates in replication, the higher the chance of deleterious mutations. If we assume life started in the form of simple RNA replicating elements, then a reduction in such error rates, or some buffer against the effects of such deleterious mutations during replication likely occurred. Extant ssRNA and dsRNA viruses exhibit the characteristics of small genome size and large replication error rates. This suggests that RNA viruses may have evolved from primitive RNA replication elements that never gained the error-correcting ability observed in modern cellular polymerases. This would be more plausible than imagining that elements of a cellular organism with high-fidelity polymerases would regress to an error-prone replicating parasite (something that would have been negatively selected for during evolution). Only dsDNA viruses have been shown to have reduced error rates of any significance, implying that replication stability, and thus greater complexity, may have depended on the advent of dsDNA in cellular organisms.

Recombination

Recombination occurs when two or more related viruses infect the same host cell and exchange sequences. The net result is a new variant derived from two parental genomes. Such exchanges are important sources of genetic variability in viruses, particularly RNA viruses. RNA viruses can also recombine with hosts' RNAs, with surprising consequences. It has been observed, e.g., that the insertion of the host cell 28S RNA into the hemagglutinin gene of an influenza virus (*Orthomyxoviridae*) led to increased pathogenicity of the hybrid virus variant. Recombination thus also allows viruses to create novel genomes and to modify and expand their sequence space by interacting with molecules of their own sequence space (the same viral species) or different sequence spaces (other viral species). When such interaction occurs with sequences outside the sequence space of either virus (e.g., their host sequence space) the virus experiences an extreme lateral acquisition (HGT from the host).

Even more, evidence suggests that small DNA viruses might have emerged as a consequence of recombination events (and selection) between RNA viruses and DNA plasmids. Not only do RNA molecules recombine with each other, as DNA molecules do, but mechanisms exist that allow the exchange of sequences *between* RNA and DNA molecules—two related, and yet different, molecules.

Reassortment

Coinfection of a host cell with multiple viruses comprised of two or more separate nucleic acids (i.e., the virus genome is segmented) may result in the exchange of genome segments to generate progeny viruses with novel genome combinations. The process, referred to as reassortment, is also observed with multipartite viruses (or multicomponent viruses) in which the genome is not just segmented, but in some instances the genome components are encapsidated in separate virus particles. Reassortment can be of importance in creating new variants of the virus following mixed infections with distinct but related viruses, or with variants (strains) of the same virus species. One can envision reassortment as a kind of recombination. In an evolutionary framework, this might have helped create variants that eventually became species (upon further, and possibly, extensive changes) and contributed to an accelerated rate of acquisition of genetic markers that can overcome adaptive host barriers faster than the slower process of incremental increase due to mutation alone. But the emergence of these viruses is puzzling. There are some advantages in having a segmented or multipartite genome composed of small, but not independent fractions: namely, (1) viruses with this kind of genome organization may better tolerate high mutation rates, (2) the individual genomic components can replicate faster, (3) genetic exchange, by means of reassortment of the individual genomic components, is facilitated, and (4) a higher stability of their virions has been observed (except for some segmented viruses). These benefits come with a cost: a reduced chance of infection of new cells or organisms given that all functional components are required to establish infection (and to be efficiently transmitted).

Horizontal Gene Transfer

The exchange of genetic material between viruses and cellular organisms, in either direction (or between viruses), is termed **h**orizontal or **l**ateral **g**ene transfer (HGT). Given the large number of viral genes that lack homologs in cellular organisms versus the high number of viral retro-elements found in the cellular genome, it stands to reason that such HGT has been mostly from viruses to cellular life forms, rather than vice versa. The large number of viruses in relation to hosts, as revealed by metagenomics studies, also supports this theory, of a higher chance of the many donating to the few.

It is broadly acknowledged that low fidelity of replication confines the evolution of many viruses within a narrow genome size range. This is especially true for RNA viruses. It is then less likely that cellular genomes transferred many genes to viruses successfully without decreasing the fitness of the virus. Arguably, there is a high probability that many of the novel genetic features that may have arisen in viruses through selection would have been laterally transferred to cellular genomes at some stage. The evidence suggests that viruses have probably been driving cellular evolution from the beginning of the three domains of life. Specifically, rather than escaping from cells, modern viruses might have coevolved with them, at a constant selective pressure.

Nonetheless, recent evidence indicates that the opposite is also true. That is, some viruses seem to have been able to recruit diverse protein-coding genes from their hosts (HGT from hosts to viruses), that in some instances eventually ended up representing the coat protein genes of some viruses. As we understand it now, viruses followed different routes of speciation and radiation. While some groups of viruses might be descendants of primordial genetic selfish elements, others might have evolved on many occasions via the recruitment of host cell protein-coding genes.

WHERE DO VIRUSES COME FROM?

Hypotheses on Virus Origin

It is theorized that cellular organisms have a common ancestor due to shared genetic, biochemical, and structural traits, most notably the ribosomes and its associated translation activity—for which genes encoding their protein and structural RNAs are similar. This population of organisms is referred to as the **L**ast **U**niversal **C**ommon **A**ncestor (LUCA, or more recently as the **L**ast **U**niversal **C**ell **A**ncestor) and represents the point at which cellular organisms began their divergence into the three extant domains of Archaea, Bacteria, and Eukarya. It is not clear whether viruses predate the LUCA or are descendants of LUCA.

Precellular Theory (Virus-First Hypothesis)

Much evidence has shown that viruses (or virus-like elements) are not only very ancient, but possibly predated cells. From this came the "*virus-first*" hypothesis which states that viruses predated cells, existing as self-replicating units, and played a key role in the appearance of modern cells we now know. Furthermore it has been suggested that viral proteins without cellular homologues originated in an ancestral virus world that predated the cellular world of organisms. These then evolved as parasites of LUCA and the eventual three domains of life. One major flaw in this hypothesis is that all known viruses need host cells to replicate and survive.

Endogenous Hypothesis (Escape Hypothesis)

The first hypothesis on virus origins was the endogenous hypothesis (also known as the escape hypothesis). This hypothesis describes viruses as components of host cells that escaped control. Virus genomes originated from fragments of cellular prokaryotic or eukaryotic genomes (or both) and became autonomous and infectious, presumably through the acquisition of capsid genes. These later evolved by pickpocketing genes via HGT. Support for this hypothesis was provided by the discovery of lysogeny in bacteriophages. In this process a bacteriophage's genome integrates into that of the infected bacterium and is then replicated through the bacterial replication cycle as a prophage. The ability of the prophage to later emerge as a virion through a lytic cycle was a model for the mechanism of the escape theory. The fact that viruses shared genetic material with cells (RNA and DNA) and could interact with the nucleus and other cellular organelles was further evidence for proponents of this theory.

One of the major flaws of the escape model is this: how did structures such as complex capsids and nucleic acid injection mechanisms evolve from cellular structures, since there are no known cellular homologs of these viral components? There are other examples of viral proteins that contradict the escape hypothesis. The Topo II enzyme encoded by the bacteriophage T4, for instance, is markedly different in structure and activity from that found in its host (bacterial Topo II DNA gyrase), suggesting that these two enzymes are of different origins.

Viral-Oncogene Hypothesis

This hypothesis was proposed in 1969 by Todaro and Huebner, and came about through the discovery of proto-oncogenes, the precursors of oncogenic viruses (e.g., human papillomaviruses, Epstein-Barr virus (*Human gammaherpesvirus 4*), *Hepatitis B virus*). It states that somatic animal cells have all the genetic information needed to create oncogenes and oncogenic viruses, which are stably inherited. These genes are normally suppressed by our cells, and become switched on by agents such as chemicals or radiation.

The protovirus concept, put forth in 1970 by Temin is an extension of the prophage and viral-oncogene hypothesis applied to animal viruses and host cells. It is based on the mechanism of transcription; that genetic information is transcribed back and forth between DNA and RNA regularly. During this process, mutations, recombination, and other events led to the *de novo* creation of oncogenic virus elements from cellular genetic material.

Reductive Theory

Forterre suggested in 1991 that virus-specific proteins and genes arose from ancient now-extinct cellular lineages that predated LUCA of the three domains of life. This alternative reduction hypothesis postulates that viruses are reduced forms of parasitic organisms. It postulates that life began originally in the form of RNA cells (which only had RNA genomes). Within this RNA cellular world, a small RNA cell became an endosymbiont of a larger RNA cell, and at some point lost its ability to encode most proteins, but could still replicate autonomously and eventually became infectious.

Support for this theory has come in the form of the discovery of giant or NCLDVs such as mimiviruses (described earlier). These viruses have features which overlap physically and genomically with parasitic bacteria that behave in the same endosymbiotic manner. *Acanthamoeba polyphaga mimivirus*, in particular, has been studied extensively for the past decade since its discovery. This virus has a morphology that resembles that of cells including the presence of more than one form of nucleic acid within the capsid (viral genomic DNA and prepackaged viral-encoded mRNA), some functional components of the translation machinery (including tRNA synthetases and translation factors), and genomes larger than most other viruses and some bacteria. Even more, mimiviruses can be parasitized by other viruses. The existence of these so-called giant viruses gives credence to the reductive theory that viruses may have originally been RNA cells that lost their ability to independently perform the functions they are now obliged to garner from host cells.

Evidence for Viruses Being Ancient

The concept of an ancient virus lineage is gaining support. There is a growing consensus that life on earth began in the form of RNA—a world inhabited by cells with RNA genomes based on the circumstantial evidence: (1) DNA is essentially a modified form of RNA and probably derived from some prototype of genetic material, (2) RNA is less chemically stable (more reactive) than DNA due to the presence of a hydroxyl group on Carbon-2 of the ribose sugar backbone in RNA (it is thought that evolution into more complex cellular life arose through the emergence of the more stable DNA as a genetic repository), (3) extant bacteria that use the RNA base uracil rather than thymine in their DNA exist—possible proof of the transition from RNA to DNA, but incomplete in these instances, (4) viruses can have either RNA or DNA genomes—proof that RNA by itself can serve as a basis for life, and (5) experiments have shown that *in vitro* RNA can arise from a prebiotic environment.

All these combined give some support to the idea of an ancient RNA world. But proving that the first living world was based on RNA does not directly imply that the first forms of life were viruses or virus-like entities. Most probably both cellular organisms and viruses evolved from primal selfish genetic elements *and* from the interactions later established between them.

Extreme Chimerism and Modularity

Viruses are well known as vehicles of transmission of genetic information between hosts (e.g., transduction). But another interesting fact regarding viral genome molecules is that, in evolutionary times, they may have emerged as a consequence of recombination events in which a virus hijacked a gene from a different virus or from their hosts, as mentioned earlier. In other cases, HGT can result in extreme chimerism and newly described viruses. For example, it has been demonstrated that some ssDNA viruses encode a capsid protein gene originally "*inherited*" from a positive sense RNA virus. Replicative proteins of these chimeric viruses have distinct evolutionary histories that differ from that of the actual viral genome carrier, *and* the replicative genes are themselves chimeras of functional domains inherited from viruses belonging to different *contemporary* virus families. This results in chimeric virus particles in which the genome of an ssDNA virus is encapsidated by proteins derived from a RNA virus, and replicated by proteins of differing origins.

This extreme form of chimerism creates complications in deciphering the evolutionary history of such viruses. On the other hand, discoveries like this reinforce the idea that viral capsids are restrictive: that is, regardless of the genome molecule (DNA or RNA) a virus possesses, capsid proteins between the two kinds of viruses (DNA viruses and RNA viruses) seem to be related and there are not many structural variants. Furthermore, when resorting to the strategy of using structural comparisons to infer phylogenetic relationships and the use of nucleotide sequences related by homology, we might end up assuming that viruses belong to the same evolutionary lineage—when this is not in fact the case—and only because their capsid proteins are structurally similar.

Finally, it has also been postulated that the genomes of dsDNA viruses (encompassing members that infect species from the three domains of cell life) derive from a robust hierarchical modularity, thus pointing to a potential direct evolutionary relationship among these viruses. Briefly, it has been found that the large-scale structure of the dsDNA virosphere has been shaped by (1) vertical inheritance of gene ensembles, that helped define cohesive group of viruses, (2) hallmark, and other connector genes, from compartmentalized groups that allowed connectivity- and modularity-sharing among the members of the dsDNA virosphere network, (3) HGT among members of this very network, yielding thus increased modularity-sharing, and (4) host genes capturing. Viruses with other kinds of genome molecules might have taken a different evolutionary route, but it would be not surprising to find similar features as those observed when analyzing dsDNA viruses only.

What Are Metagenomics Studies Telling Us?

In the past few years metagenomics studies, in which genetic data have been generated using mass parallel sequencing techniques, show a vast number of viruses in existence in comparison to cellular organisms. This recent accumulation of sequence data from both viruses and other organisms has led to the realization that there are numerous viral genes in existence for which there are either no homologs (so-called ORFan sequences), or homologs are only distantly related. The only exceptions have been proviruses, where viral genomes are integrated into cellular genomes, such as endogenous retroviruses in humans.

Metagenomics is helping to uncover the previously unknown catalog of proteins (and genes) that viruses have used for their successful colonization of the planet. The data should eventually assist in establishing connections between viruses, and between viruses and their hosts, and contribute to an integrated and coherent hypothesis on virus origin and evolution.

Virus Hallmark Genes

No single gene discovered to date is shared by all known viruses, which is in contrast to that seen in the cellular world (e.g., the genes for ribosomal RNAs). However, there are virus hallmark genes, which as a subset, provide certain functions key to all virus replication strategies. These hallmark genes are shared by a wide range of viruses despite these viruses utilizing different replication strategies and infecting a diverse range of hosts. They include genes encoding polymerases (DNA and RNA polymerases), helicases, primases, capsid-encoding genes, terminases, and in the case of retroviruses, integrases. These virus hallmark genes have no close homologs in cellular organisms, and distant homologs observed are actually viral retro-elements integrated into host genomes. This implies that viral hallmark genes separated from cellular genomes early in evolutionary history.

Viral Capsid Proteins and Their Relatedness

It is thought that the emergence of viruses as we know them today probably coincided with the acquisition of capsid-encoding genes. The ability of viruses to encapsulate their genetic materials within capsids distinguishes them from other mobile genetic elements, like transposons and retrotransposons. Hence the capsid is regarded as one of the hallmark features of viruses as mentioned previously. No analog or homolog structures have been observed in cellular REOs. Viruses, due to this unique feature, are described as **c**apsid-**e**ncoding **o**rganisms (CEOs). This includes viruses of all three cellular domains and viruses of viruses—all require REOs as hosts for the production of proteins and energy.

There is also theoretical evidence to show that the capsid was a possible preadaptation necessary for early evolution as it may have allowed for horizontal transfer of genetic material between cellular compartments. The ability of a self-replicating genetic element to survive (and thus maintain its genes) depends on a sufficient degree of genetic exchange. When there is too little exchange, then parasitic selfish elements that arise tend to overuse the system's resources leading to a population collapse—a tragedy of commons. When there is too much exchange, the selfish elements spread too quickly to previously parasite-free compartments. Models indicate early CEOs may have facilitated an intermediate level of HGT between protocells, giving rise to ancient virions.

Since very different viruses infecting very different hosts can share the same capsid topology and structure, despite having little similarity in their primary genetic sequences, it appears that RNA and DNA viruses share a common origin that probably predates cellular organisms. The jelly-roll capsid structure, e.g., is found in most virions with an icosahedral structure including some NCLDVs (*Mimiviridae*, *Iridoviridae*, *Asfarviridae*, *Phycodnaviridae*) along with *Adenoviridae* (dsDNA), *Birnaviridae* (dsRNA), *Microviridae* (ssDNA), and *Picornaviridae* (ssRNA), to name a few. Ancestral viruses with jelly-roll capsids most likely existed before their hosts diversified from the LUCA. As cellular domains emerged and evolved from the LUCA, their viruses would also have diversified into the extant viruses we know of, keeping the positively selected jelly-roll capsid.

The counterargument to this is that viruses underwent convergent evolution in their capsids (in which selective pressures force survivors to adopt similar strategies). However, there is evidence against this counterargument:

1. These similar structures are present across a broad range of virus species and taxa meaning that there would need to be numerous instances of convergence (which reduces the likelihood of it actually having happened this way).
2. Selective pressures would be expected to drive evolution toward a single structure that is most favored, rather than the many different structures that currently exist. In the case of the viral capsid protein, this is not so.

The jelly-roll is not the only capsid fold variation in extant species as some viruses, such as members of the dsDNA virus orders *Caudovirales* and *Herpesvirales*, display what is known as the HK97-fold.

Although several recent studies provide evidence for HGT the mechanism is complicated and still poorly understood. Take the example of *Thermoproteus tenax virus 1*. It was demonstrated that the nucleocapsid protein (TP1) of *Thermoproteus tenax virus 1* (*Tristromaviridae*) is a derivative of the Cas4 nuclease of its archaeal host. This case of protein exaptation is the first explanation as to the origin of a viral capsid protein. Here it seems that the archaeal Cas4 gene fragmented in two, that the N-terminal encoding portion of the gene was retained to later become TP1, and that loss of catalytic amino acid residues led to the removal of nuclease activity. We are still far from elucidating all the events that resulted in the present day viruses.

Viroids and Mavericks

Viroids are small infectious short circular strands of RNA without a protein coat that replicate autonomously when introduced into host cells. At present, there are 30 known viroid species ranging from 170 to 450 kb, all of which only infect plants, with some causing diseases while others are benign. Viroid genomes do not encode proteins, and, unlike viruses which parasitize host translation, viroids are parasites of the cellular transcriptional process, using host cellular RNA polymerases for their replication. Their existence is thought to be evidence of relics of the theoretical ancient RNA world; their ancestors protoviroids—self-replicating elements existing before the emergence of the capsid protein.

Extant viroids have several characteristics that support this hypothesis: (1) they are restricted to small genome sizes due to error-prone replicases, (2) they possess a high G + C content that increases thermal stability and replication fidelity, (3) their circularized genomes allow for complete replication without genomic tags, (4) they lack protein-coding ability, and (5) replication in some species is mediated by ribozyme—catalytic RNA molecules analogous to enzymes. All would be applicable to an ancient RNA world devoid of ribosomes, proteins, or DNA. DNA and proteins would then have been later developments that eventually relegated RNA to the role of messenger and other intermediate genetic functions in other organisms. It is thought that modern viroids are descendants from such ancient RNA genetic elements and, in evolving into plant parasites, lost their self-replicating ability.

In contrast, however, other selfish genetic elements tell a different story. Polintons, also known as mavericks, are large DNA transposons widespread in the genome of eukaryotes (except for members of the group Archaeplastida that includes land plants as well as glaucophytes, and red and green algae). Very recently polintons were shown to possess two coat protein genes and the capacity of forming virions under certain conditions. More interestingly, however, it seems that they evolved directly from bacteriophages, and later became the center from which, most of the large dsDNA viruses of eukaryotes evolved, along with several groups of plasmids and transposons. It appears then that viruses and virus-like elements, provided informational blocks for the emergence of new variants of themselves and of other selfish genetic elements, and at the same time host protein exaptation widened their spectrum of capabilities.

EVOLUTION OF VIRUSES WITH THEIR HOSTS

Any review on virus evolution is incomplete without consideration of the role played by the host(s) on which viruses depend. Various lines of evidence suggest a role for viruses in cellular evolution and that viruses have helped to drive the evolution of their hosts. We briefly present on those aspects of virus–host interactions that are important to the understanding of virus evolution.

Arms Race Between Hosts and Parasites, and Unplanned Mutualism

Research has shown that the history of life as we know it has been a constant evolutionary battle between viruses (and other "*parasitic*" entities) and their cellular hosts, in what has been described as a genetic arms race. Viruses infect hosts, and as a counter, hosts have evolved various defense mechanisms to combat viral infection (such as RNA silencing or RNA interference), and disrupt viral replication and local and/or systemic movement (Chapter 10: Host–Virus Interactions: Battles Between Viruses and Their Hosts). Viruses that are successfully propagated are those that acquired counter measures to such defenses. An example being the use of RNAi suppressors seen in some plant RNA viruses. This more or less occurs with vectors of viruses and the same kind of mechanisms provide additional obstacles to replication in the vector and transmission. This constant arms race between viruses and their hosts does not explain viruses' origin, but it surely has shaped their evolution.

Evidence has also shown that cellular organisms benefit from viral genes and vice versa—as seen in virtually every studied virus–host system. In any such system, theoretically, unstructured interaction would eventually lead to an imbalance ultimately resulting in the random extinction of the virus, the host or both. In the case of hosts and their viruses that have coevolved successfully, there exists a fine balance between the aforementioned arms race *and* cooperation. In prokaryotes we find, e.g., that temperate phages integrate genetic data (sometimes beneficial) in bacterial and archaeal chromosomes via a lysogenic cycle, and that prophage domestication allows for the appropriation of phage genes by their prokaryote hosts (e.g., **g**ene **t**ransfer **a**gents, GTAs) for cellular processes. In eukaryotes, similar outcomes of cooperation have been observed. Introns, e.g., originated from prokaryote group II self-splicing introns, themselves virus-like reverse transcribing selfish elements. Moreover the telomerase-RT needed for chromosome end replication originated from bacterial group II introns. Clearly, antagonism and cooperation, by means of an inevitable coexistence between viruses and cellular organisms, is the driving force of evolution of all biological entities. If we recall, ecological interactions are evolvable and inheritable, and so separating the evolution of viruses from that of cellular organisms is a mistake we cannot afford to make.

VIRUSES AND THE TREE OF LIFE: TREE OF LIFE OR TREE OF VIRUSES?

The ToL is a schematic representation of the history of life. These tree-like portrayals of biological relationships—specific evolutionary models—were proposed in the mid-19th century, but it was in the late 1970s that comparative studies of rRNA sequences proved the universal relatedness of all life and the first outlines of a universal sequence-based ToL generated. Known life was seen to fall into one of three phylogenetic domains: Archaea, Bacteria, and Eukarya. The ToL as we have known it for decades though, is presently challenged by (1) links between viruses infecting the three domains of life that point to a very ancient origin of viruses, and (2) the realization of HGT as a powerful evolutionary force throughout the prokaryotic and eukaryotic domains (and the lack of methods to account for HGT).

Several arguments still exist as to whether viruses belong in such a tree. If we consider the virocell concept to be true, then the traditional ToL is really a tree of cellular organisms and is not universal. Irrespective of whether viruses are living or nonliving, there is concrete evidence that they have driven the evolution of cellular organisms, and vice versa, and continue to do so. Given the strong evidence of HGT between host cells and their viruses, we might consider the evolution of life as a tree of genes (of any origin) or a composite.

In summarizing the current thinking, including the drivers of virus origin and evolution:

- Evolution of life started with virus-like organisms in the primordial gene pool.
- The primordial pool probably consisted of positive-strand RNA, retro-transcribing, ssDNA, and dsDNA virus-like elements.
- The next life forms to emerge would have been dsRNA and negative-strand RNA virus-like elements.

- Positive-strand RNA viruses are likely direct descendants of the primordial RNA–protein gene pool.
- Three hypotheses coexist; namely, that (1) modern cellular organisms are a later evolutionary event arising from virus-like forms, (2) that extant viruses emerged from primordial cell forms (escape theory), and (3) both primordial viruses and cellular organisms coevolved.
- Retro-transcribing elements may have been the first step toward DNA-based organisms.
- It is now widely believed that RNA preceded DNA. RNA is more labile and less stable. Only DNA has the characteristics necessary to serve as stable genetic repository for cellular (and multicellular) organisms.
- Many viruses have RNA or DNA genomes.
- DNA is essentially a modified form of RNA.
- Some bacteria actually possess uracil rather than thymine in their DNA genome—possibly a relic from the transition of life from RNA to DNA.
- Based on rRNA analyses the cellular tree includes the three traditional branches of life: Archaea, Bacteria, and Eukarya.
- There are three separate branches of viruses (related to their hosts) evolving in tandem.
- These two classes of life forms (cellular and viral) are connected via a network of HGT in either direction.
- The evolution of cellular and virus/virus-like elements was possibly driven by virus–host coevolution.

Lastly, the reasons why it is so difficult to analyze virus evolution in a way that would eventually allow for their inclusion in a ToL (by themselves or as part of with the tree of cellular organisms):

- Viruses have different types of genomic molecules, as opposed to cellular organisms that only rely on DNA. Rates of mutation between RNA virus genomes and DNA virus genomes, as compared to the DNA genomes of cellular organisms, are also much higher.
- Virus fossils are not available for analysis. Even if some viruses are found in well preserved conditions, and later "*revived*," they are still too close to our times to count on them as witnesses of the far past.
- It difficult to unify all viral families using sequence-based phylogenetic analysis. There are no nucleotide sequences (genes) common to all viruses. Interestingly, however, genes in the sequence space of viral groups (species, genera, families, and the few taxonomic orders of viruses that have been proposed) allow for their grouping together.
- As opposed to the nucleotide sequence space of viruses, there are very few protein domains. The limited forms of blocks useful to the construction of a virus are of multiple genetic origins despite the protein product being more or less similar.

- Chimerism (i.e., multihybrid nature) in the world of virus "*construction*" adds an additional layer of complexity to the origin and evolution of viruses and is the reason phylogenetic studies are biased.
- It appears that viruses have changed considerably along with their hosts. In addition, switching of hosts during their evolution and the interactions established with other viruses complicate any analysis of the virus–host coevolution—unless host switching is considered in the analysis (and in any tree).
- Metagenomics studies have revealed that (1) we only know a very small fraction of all viruses, and (2) that the recent discoveries show no significant similarity to known viruses and extant cellular organisms. We rely on very few known viruses to make inferences regarding the many.
- No less important is the fact that, in order to position viruses in a ToL, we should first come up with a working definition that allows some consideration of cellular organisms and viruses as equal or as equivalent biological entities relying on the creation, change, and transfer of information.

In summary, for now we only have a collection of virus trees reflecting membership to a defined sequence space, but not a single tree including all. In this form, it would not be useful to include viruses in a ToL until we solve the phylogenetic relationships among all viruses (if such is possible and attainable), and between viruses and their hosts. For instance, what we call a tree of species is most of the time a tree of genes, though in a few cases it may be a tree of genomes. Such trees corroborate conclusions derived from other kinds of analyses (e.g.morphological, ecological), but they are still far from being robust. Ultimately a phylogenetic tree is simply a hypothesis rather than a metaphor—but this is not a reason to stop looking for an answer.

FURTHER READING

Andino, R., Domingo, E., 2015. Viral quasispecies. Virology 479-480, 46–51. Available from: https://doi.org/10.1016/j.virol.2015.03.022.

Forterre, P., 2006. Three RNA cells for ribosomal lineages and three DNA viruses to replicate their genomes: a hypothesis for the origin of cellular domain. Proc. Natl. Acad. Sci. USA 103 (10), 3669–3674. Available from: https://doi.org/10.1073/pnas.0510333103.

Holmes, E.C., 2011. What does virus evolution tell us about virus origins? J. Virol. 85 (11), 5247–5251. Available from: https://doi.org/10.1128/JVI.02203-10.

Iranzo, J., Krupovic, M., Koonin, E.V., 2016. The double-stranded DNA virosphere as a modular hierarchical network of gene sharing. MBio 7. Available from: https://doi.org/10.1128/mBio.00978-16e00978-16.

Krupovic, M., Koonin, E.V., 2015. Polintons: a hotbed of eukaryotic virus, transposon and plasmid evolution. Nat. Rev. Microbiol. 13, 105–115. Available from: https://doi.org/10.1038/nrmicro3389.

Raoult, D., Forterre, P., 2008. Redefining viruses: lessons from Mimivirus. Nat. Rev. Microbiol. 6 (4), 315–319. Available from: https://doi.org/10.1038/nrmicro1858.

Sicard, A., Michalakis, Y., Gutiérrez, S., Blanc, S., 2016. The strange lifestyle of multipartite viruses. PLoS Pathog. 12, e1005819. Available from: https://doi.org/10.1371/journal.ppat.1005819.

Chapter 5

Host Range, Host–Virus Interactions, and Virus Transmission

Gustavo Fermin
Universidad de Los Andes, Mérida, Venezuela

Of living things, my son, some are made friends with fire, and some with water, some with air, and some with earth, and some with two or three of these, and some with all.

Hermes Trismegistus

Where there is life, there are viruses. For more than a decade, it is clear from ecological studies, and more recently from metagenomic studies, that viruses represent the major part of the modern biosphere; and there is no reason to assume that the same situation did not exist before the **l**ast **u**niversal **c**ommon **a**ncestor (LUCA) in the RNA world. Many of the viruses today are entirely dependent upon their host's cells for their survival and replication. It is erroneous, or extremely naïve, to believe that viruses depend solely on chance for their replication and dissemination. They have developed the ability not only to replicate in their host, but also to move between host cells and ultimately between hosts in order to persist. Indeed, viruses have the remarkable ability to spread from one host to another host, hosts that belong to the three cellular domains of life—Archaea, Bacteria, and Eukarya. This constitutes their host range. Viruses known to infect one or a limited number of species are said to be specialists; whereas those that infect a wide range of hosts, are referred to as generalists. While water and air are very good vehicles for the transportation and spread of many viruses, those viruses that infect eukaryotes often depend on different types of organisms to work as vectors and to facilitate host-to-host spread. Either way, viruses establish

Viruses. DOI: https://doi.org/10.1016/B978-0-12-811257-1.00005-X

relationships with their hosts that encompass all symbiotic associations along the mutualism–parasitism continuum. What we presently know appears to be just the tip of the iceberg. Very recently, it was reported that mammals carry cytoplasmic replicons that might represent an antediluvian virus-derived symbiont. Circular DNA molecules associated with TSE (**t**ransmissible **s**pongiform **e**ncephalopathies) preparations and normal mammalian tissues were found capable of replicating symbiotically in mammalian brain cells. The circular DNA molecules, designated as SPHINX sequences using an acronym given for **S**low **P**rogressive **H**idden **IN**fections of variable (X) latency, show similarities to an *Acinetobacter* phage and possibly represent remnant phage DNAs of commensal *Acinetobacter*. *Acinetobacter* sp. are adapted to a variety of habitats, including soil, and are commonly found in association with humans, animals, and insects. They easily gain entry into animals while grazing or drinking water. These findings lead to a number of questions of how this type of association between a mammalian host and phage virus originated, how it is maintained and whether SPHINX was incorporated during mammalian evolution. More importantly, the findings suggest that mammals share and exchange a larger world of prokaryotic viruses than previously realized.

This chapter provides an overview of the concept of virus host range including the various interactions a virus can initiate and maintain with different hosts. We focus on the strategies adopted by viruses in order to achieve new targets of replication, along with their abiotic and biotic vehicles of dissemination.

DEFINING THE HOST RANGE

The host range, a key property of viruses, reflects the diversity of species that viruses can naturally infect. To be a member of the exclusive "*host range*" club of a virus, the host should support the replication or life cycle of the virus; i.e., the virus should be able to successfully enter a host cell and complete a series of tasks, including unencapsidation and replication in the initial cell and movement to adjacent cells and throughout the host. Nonetheless, it is often difficult to define the host range of a virus as a number of other factors need to be incorporated, such as the host susceptibility to infection, as well as the ability of the virus to undergo sustained transmission from host to host. Moreover, the actual breadth of the host range can be reduced by barriers that prevent contact between vectors and hosts, and the unsynchronized seasonal timing between (1) available infected hosts in a viremic stage and feeding activity of vectors and (2) available uninfected species and infectious vectors in a given environment. And it can be expanded. Depending on the virus, the range may expand to secondary hosts and/or vectors, as a result of selection pressure on the virus generated either in vectors or in hosts as well as of the degree of promiscuity of host seeking behavior of the vectors involved. Moreover, other viruses, such as rhabdoviruses, replicate in their insect vectors. Thus the majority of known rhabdovirus species have two natural hosts, either insects and plants, or insects and vertebrates. The virus host range is also expanded when there are "*spillover*" infections into alternative hosts. In some instances, viruses gain the ability to spread efficiently to a new host that was not previously exposed or susceptible (see the section on zoonosis). These transfers involve either increased exposure or the acquisition of genetic variations that allow the virus to overcome barriers to infection of the new host. Phylogenetic studies suggest that these host shifts are frequent in the evolution of most pathogens, but why viruses successfully jump between some host species but not others is only just becoming clear.

For these reasons, the host range is highly variable among viruses. Some, such as dengue and mumps viruses, whose only known mammalian host are humans, are referred to as specialist viruses. These viruses have evolved to become specialized in infecting one or very few host species. In contrast, generalist viruses successfully infect hosts from different species and even hosts from a higher taxonomical rank. Examples of generalist viruses include *Cucumber mosaic virus* (CMV; *Bromoviridae*), which infects more than 1000 plant species, and *Influenza A virus* (*Orthomyxoviridae*), which infects birds and several different species of mammals. Opinions have been divided regarding the evolutionary significance of host range variation. Some argue that host–virus relationships with a narrow, specific host range are more advanced, while others posit the opposite. The advantages of generalism are more obvious: a generalist virus would be able to exploit multiple hosts and thus enhance its fitness. Since generalist viruses are not the norm, it is generally assumed that generalism comes with a cost. It has also been suggested that evolution should favor specialists because evolution proceeds faster with narrower niches. The answers may lie with genome sequencing. Through genome sequencing, it is now apparent that a variety of organisms carry genes that reflect past infection events by viruses. By using data from multiple potential host species, it may be possible to determine whether extant viruses characterized as presently having a broad host range have been resident in the genomes of their hosts for longer times than current viruses with a narrow host range. Such evidence may suggest that families of viruses with broad host ranges are more evolutionarily ancient, and may have benefited from a greater ability to avoid extinction. Also, it is increasing clear that understanding the evolution and biology of a species cannot be achieved without examining the interactions between the members of the holobiont; i.e., the prokaryotic symbionts, the eukaryotic symbionts, the viruses, and the host. Metagenomics, coupled with biological studies, promise further characterization of these host–virus interactions.

INTERACTIONS BETWEEN A VIRUS AND ITS HOST(S) MAY HAVE DIFFERENT OUTCOMES

Though studies into the entire spectrum of virus–pathogen interactions are still in their infancy, there has been a recent spurt in studies exploring the possibility of a number of symbiotic interactions, not limited to parasitism and disease. More and more evidence is accumulating on the existence of viruses that are essential for the survival of their hosts; viruses that are beneficial and their net effects on the host are positive. The interaction of a virus and its host is being redefined as a two-way biological relationship, which in most cases, if not all, is modulated by the prevailing conditions. In other words, these biological interactions are described in terms of the effects they have on both partners under particular conditions, and the impact on either species can be neutral, positive, or negative. The relationships are not always static, they vary under different conditions, and the magnitude of the effects are continuous instead of discrete. Outside the field of virology, categorization of biological interactions is still controversial; in virology, it is even more so.

It is widely accepted that many of the viruses we know are host-dependent parasites that can cause disease (see Chapters 6–9: Viruses as Pathogens: Plant Viruses, Viruses as Pathogens: Animal Viruses, With Emphasis on Human Viruses, Viruses as Pathogens: Animal Viruses Affecting Wild and Domesticated Species, and Viruses of Prokaryotes, Protozoa, Fungi, and Chromista). By hijacking the metabolism of the infected host in order to fulfill their needs of replication, local and systemic movement and transmission from host to host, a virus causes severe impairments in the physiology of the host; in some cases, leads to the death of the cell or the organism. However, virus–host interactions do not always yield hallmark phenotypic symptoms of viral infections and can influence hosts in which they reside. Case in point a recent study reported direct virus-metazoan symbiosis limiting pathogenic bacterial growth on mucosal surfaces, a phenomenon apparently conserved from cnidarians to humans. Additionally, integration or domestication of viral genetic elements that benefits the host are well illustrated by examples of prophages in bacteria. These interactions are further described in Chapter 11, Beneficial Interactions With Viruses.

In other instances it is very difficult, if not impossible, to categorize host/virus interactions in only one manner. For example, CMV (*Bromoviridae*) is an important plant pathogen with the widest host-range of all known plant viruses, and worldwide distribution. The virus is transmitted by no less than 60 different species of aphids, or via seeds or dodder. Although the virus can dramatically affect the physiology of the infected host ("*intended*" target) where it acts like a true pathogen, its effect on the aphid vector ("*nonintended*" target) is quite different. Depending on the titer of the virus, elicitation of plant defense responses by CMV leads to a reduction of aphid performance; but additionally, some wingless aphids become winged so contributing to virus dissemination. That is, the virus negatively affects its plant host, but at the same time affects its animal host (the biological vector of its dissemination) in a way that we cannot deem *detrimental*. In other kind of interactions, the outcome is very different. Endoparasitoid insects spend part of their development in other invertebrates, e.g., some lepidopteran wasp species. In order to avoid the defense responses of the parasitized organism, endoparasitoid wasps use various mechanisms to ensure the successful development of their larvae. One of such mechanism depends on the presence of viruses that are injected during wasp oviposition. These viruses hosted by the wasps neutralize or decrease the lepidopteran wasp-host defense systems thus enabling the development of eggs and larvae of the endoparasitoid. Interestingly, however, *Diadromus pulchellus toursvirus* (*Ascoviridae*) behaves differently in other wasp species. That is, like a pathogen (negative effect) in *Itoplectis tunetana*, a commensal (neutral effect) in various species of *Eupelmus* and *Dinarmus*, or like a mutualist (positive effect, as mentioned earlier) in *Diadromus pulchellus* and *D. collaris*. Finally, it has been proposed that a shift in the human gut bacteriophage community composition can contribute to a shift from health to disease. Although not a host of bacteriophages, the human receives benefit from the existence of free phages that help in modulating its bacteriome. If *Escherichia virus Lambda* (*Siphoviridae*), e.g., lyses its host (say *Escherichia coli*), and "*free*" virions adsorb to the human gut, they might lyse unwanted bacteria that are part of their host-range. We are not hosting the virus, but we receive some benefit derived from its presence.

THE MANY WAYS OF VIRUS TRANSMISSION

Undoubtedly, one of the critical challenges to initiating the viral replication cycle in a host lies with the virus' ability to overcome multiple barriers as it moves from cell-to-cell to tissue-to-tissue within the host organism and from organism-to-organism, and even across species. In this section, we examine the characteristics and fundamental differences between these types of transmission.

Cell-to-Cell Transmission

Viruses are able to translocate their virions or genomes between neighboring cells via intercellular connections. In plants and algae, cell-to-cell transmission of viruses (local movement) is accomplished by means of plasmodesmata—narrow, intercellular cytoplasmic bridges that connect and enable communication between adjacent cells. In order to overcome the physical constraints imposed by the exclusion size of plasmodesmata, viruses encode a protein, called the movement protein, that dilates plasmodesmata openings to enable the passage of viral nucleic-acid–protein complexes (*Cucumovirus*, *Potyvirus*) or whole virions (*Comovirus*, *Closterovirus*). In some plant viruses, the viral coat protein and replicase also play a role in virus movement through plasmodesmata. Many viruses subsequently utilize the phloem transport system to systemically infect their hosts. *Tobacco mosaic virus* (TMV), e.g., enters minor, major, and transport veins from nonvascular cells in source tissue and exit from veins into sink tissue. In other cases, like *Turnip mosaic virus* (*Potyviridae*) and *Potato virus X* (*Alphaflexiviridae*), systemic infection of the plant is attained by the virus moving through xylem vessels.

Cell-to-cell transmission, and systemic spread, of animal viruses is achieved in two ways: diffusion through the extracellular space (cell-free transmission) or by direct cell-to-cell contact. The first route, cell-free transmission, requires the dissociation of progeny particles from the infected cell either by lysis or exocytosis. This leads to viremia and enables passive dissemination of virus through blood fluids to distant tissues. Lymphatic vessels are one of the principal routes of virus passage from exposed surfaces (skin, respiratory mucosa, and digestive tract) into the interior of the animal body, but blood vessels constitute the main route of dissemination, and represent the source from which hematophagous arthropods obtain the virus for transmission host-to-host. Although this route can allow spread across long distances within the host and permits an easier spread to a new host, the second route involving cell–cell transmission is an effective way of avoiding the various physical and immunological barriers within the organism. In cell-to-cell transmission, virus can use preexisting cell interactions (e.g., neurological or immunological synapses), or viruses can deliberately establish transient cell–cell contact between infected cells and uninfected target cells. In the latter instance, a **v**irological **s**ynapse (VS), i.e., a tight cleft between an infected cell and a target cell is formed as a result of firmly adhering plasma membranes of the two opposing cells. In HIV-1, e.g., assembly of the HIV-1 T-cell VS requires engagement of the HIV-1 Env surface subunit gp120, expressed on the infected cell, with its cellular receptors CD4 and CXCR4 on the target cell, followed by further recruitment of receptors and HIV-1 proteins to the conjugate interface by a cytoskeleton-dependent process in both target and infected T cells. Subsequent clustering of adhesion molecules (such as the integrin leukocyte function-associated antigen 1) presumably contributes to the formation of a stable adhesive junction. HIV-1 and other viruses (Human T-cell lymphotropic virus type 1) utilize both methods of cell–cell transmission. Others, use either existing cell interactions (herpesviruses and rhabdoviruses) or form a VS. Another means of cell-to-cell spread involves the use of actin-rich cellular structures that propel virions from infected cells directly into uninfected cells (poxviruses) and breaking down intercellular barriers by inducing limited membrane fusion between infected and uninfected cells (paramyxoviruses).

Virus Survival in the Environment

Viruses are often shed into the environment from infected hosts or carrier hosts. Bacteriophages, an example of the most abundant of all viruses, can be found not only associated with bacterial host cells, humans, and other animals, but also in deserts, hot springs, warm and cold seas, ground and surface water, soil, food, sewage, and sludge. They are carried by urine, feces, blood, serum, and saliva as well as in the air. Other nonphage viruses, enveloped and nonenveloped (representing various taxonomic groups), are likely present in similar environmental media. It is not an exaggeration to say that viruses are everywhere. While outside of their cellular hosts and in the environment, viruses have the potential to survive, persist, and be transported by various routes to other susceptible hosts. Not all viruses released into the environment are, however, successful in maintaining an infectious status and reaching new susceptible hosts; persistence or survival can vary greatly with virus type and environmental conditions.

In order to survive in the environment, virions must cope with a number of physical, chemical, and biological factors. UV radiation from the sun is the main virucide in the environment. Loss of infectious capacity or inactivation of DNA and RNA viruses primarily occurs in the long-wave (UV-A) range (320–400 nm). This damage is mostly due to chemical changes in the nucleic acids and the formation of pyrimidine dimers; DNA is more susceptible than RNA. Other factors affecting the survival of viruses in the environment include temperature, acidity, salinity, and ions. Of the viruses, phages are most resistant to a range of conditions. Some members of the families *Myoviridae*, *Podoviridae*, and *Siphoviridae* are resistant to highly dry environments and can endure a range of temperature fluctuations and remain infectious after several years. Others are stable under acidic (*Fuselloviridae*, *Tectiviridae*, and *Lipothrixviridae*) or alkaline conditions (*Leviviridae*). For other viruses, it is generally accepted that enveloped viruses are more sensitive to degradation and persist poorly in the environment (as desiccation of viral lipid envelopes typically reduces infectivity) than naked viruses. However, a study investigating enveloped surrogates of coronaviruses (transmissible gastroenteritis and mouse hepatitis) demonstrated that these enveloped viruses remained infectious in wastewater for weeks. Additionally, enveloped influenza viruses have been detected in sewage. Perhaps more environmental persistence studies are warranted rather than a general assumption of negligible persistence.

The survival of free viruses in the environment (an estimated half-life of 48 h in most ecosystems) is of the outmost importance from an epidemiological point of view. A virus found in the air, water, soil, the remains of an infected living being, food, and inanimate surfaces (fomites) can invariably find a way of entry to a susceptible host. Even more, sometimes a nonhost, nonvector animal (e.g., cockroaches) can carry viruses that contribute to infections of true hosts. From the virus' point of view, the window of opportunity for transmission depends crucially on the longevity of infectious stages in the environment. Thus viruses have adapted their life cycles and developed sophisticated strategies to optimize their transmission to new susceptible hosts. Some viruses are transmitted vertically to host offspring and others are transmitted by contact between hosts (e.g., by wind, water, or physical contact), but most viruses rely on vectors for rapid dissemination within host populations.

Horizontal Transmission

At its simplest, transmission is defined as the means by which an infectious agent is passed from an infected host to a susceptible host. It is a function of both the host and the pathogen and consists of pathogen presentation by the host, movement between infected and healthy hosts, and entry into the new host. Transmission dynamics may involve varying degrees of complexity, from single-host species (measles- or rubella viruses) to contrasting multiple host species (*Rift Valley fever phlebovirus*, *Phenuiviridae*). Further, viruses are able to use, simultaneously or sequentially, multiple modes of transmission, including but not exclusive to vertical or horizontal transmission. In vertical transmission, viruses are passed vertically from mother to offspring. In horizontal transmission, viruses are transmitted among individuals of the same generation which encompasses both direct and indirect modes. Horizontal transmission can be further classified as direct or indirect. Horizontal transmission by a direct route includes airborne infection, food-borne infection, and venereal (sexual) infection, whereas transmission by an indirect route involves an intermediate or an inanimate object (fomite) or a biological host, like a mosquito vector, which acquires and transmits virus from one host to another. Table 5.1 summarizes the known modes of transmission of viruses. Viruses that use biological vectors are given in Table 5.2.

Airborne Viruses

Like cellular microorganisms, viruses are ubiquitously found in the air. Viruses can become airborne only if conditions for aerosolization are met (mass, size, shape, and density, among others). An aerosol is a particle suspension in a gaseous medium, e.g., the air. A virus can be naturally aerosolized primarily by sneezing, or secondarily, when an infected surface serves as the source of air transportation by means of other mechanical processes, like splashing, bubbling, sprinkling or even toilet flushing. Human (seasonal) behavior (personal hygiene, closed environments, densely populated areas, transportation hubs, and pollution) is regarded as an important contributor to the spread of viruses by aerosols. Additionally, viruses can be aerosolized by coughing flying animals (bats and birds, for instance); droppings can also serve as a primary source of viruses than can later be secondarily transmitted by air. In the case of *Foot-and-mouth disease virus* (*Picornaviridae*), e.g., computer simulations have estimated that "*in a worst case scenario*" cattle could be infected as far as 20–300 km far from the infectious source.

TABLE 5.1 Transmission Modes of Viruses

Viral Family	Hosts[a]	Transmission	Vector[b]
Adenoviridae	A	Respiratory droplets; orofecal route	N
Alloherpesviridae	A	Passive diffusion	N
Alphaflexiviridae	F–P	Mechanical; vector; grafting; others unknown	Y/N
Alphatetraviridae	A	Oral route. Vertical transmission also possible	N
Amalgaviridae	P	Vertically through seeds; horizontally undocumented	?
Ampullaviridae	Ar	Passive diffusion	N
Anelloviridae	A	Sexual, blood, saliva; possibly also orofecal route and maternal transmission	N
Arenaviridae	A	Zoonosis (saliva, urine, nasal secretions of rodents); fomites; aerosol. Vertically: transuterine, transovarian. Milk, saliva, urine	N
Arteriviridae	A	Genital and respiratory tract secretions; transplacental; urine, semen; probably contact; aerosol	N
Ascoviridae	A	Mechanical; vector. Vertically during oviposition	Y
Asfarviridae	A	Mechanically by biting flies. Contact, fomites, ingestion	Y
Astroviridae	A	Orofecal route	N
Avsunviroidae	P	Mechanical. Seeds; vegetative propagation	N
Baculoviridae	A	Orofecal route; contamination of egg surface. Vertically from infected male or female parent to the egg	N
Barnaviridae	F	Transmission is horizontal via mycelium and possibly basidiospores	N
Benyviridae	P	Mechanical; vector	Y

(*Continued*)

TABLE 5.1 (Continued)

Viral Family	Hosts[a]	Transmission	Vector[b]
Betaflexiviridae	F–P	Mechanical; vector; seeds; grafting, propagating material	Y/N
Bicaudaviridae	Ar	Passive diffusion; vertically by lysogeny	N
Bidnaviridae	A	Oral route?	?
Birnaviridae	A	Horizontally by contact. Vertical transmission also in salmonids	N
Bornaviridae	A	Fomites; direct contact with salivary, conjunctival and nasal secretions, urine and feces	N
Bromoviridae	P-F	Mechanical; vector. Pollen (carried by thrips)	Y
Caliciviridae	A	Direct contact with infected individual; feces, vomitus or respiratory secretions; contaminated food, water, and fomites	N
Carmotetraviridae	A	Oral route	N
Caulimoviridae	A–P	Mechanical; vector; grafting, wounds, seeds, vegetative propagation, dodders	Y
Chrysoviridae	F	Intracellularly during cell division; sporogenesis and cell fusion	N
Circoviridae	A	Orofecal route. Vertical transmission also reported	N
Clavaviridae	Ar	Passive diffusion	N
Closteroviridae	P	Some by mechanical inoculation; propagative material, dodder, grafting. Others by vector	Y/N
Coronaviridae	A	Respiratory droplets, orofecal route, oronasal route, fomites	N
Corticoviridae	B	Passive diffusion	N
Cystoviridae	B	Pilus adsorption	N
Dicistroviridae	A	Contaminated food. Vertically in aphids; horizontally by means of plants (passive reservoir for Cripaviruses). Vector	Y/N
Endornaviridae	C–F–P	Transmission during mitosis, and via pollen and ova in plants	N

Filoviridae	A	Zoonosis; contact with body fluids, blood or injured skin	N
Fimoviridae	P	Grafting; vector	Y
Flaviviridae	A	Sex; blood, semen. Zoonosis (rodents, bats); arthropod bite. Unknown in some. Vertical: transplacental. Transplantation, nonpasteurized milk, aerosols. Indirect contact: urine or nasal secretions, feces, contaminated food	Y/N
Fuselloviridae	Ar	Passive diffusion; vertically by lysogeny	N
Geminiviridae	P	Vector	Y
Genomoviridae	A–F–P	Some vectored; others unknown	Y
Globuloviridae	Ar	Passive diffusion	N
Guttaviridae	Ar	Passive diffusion	N
Hantaviridae	A	Zoonosis (rodents: urine, saliva. Bites); fomites; contact	N
Hepadnaviridae	A	Parental (perinatal transmission), sexual, blood, open skin breaks or mucous membranes. Infection in ovo (birds)	N
Hepeviridae	A	Zoonosis (pigs and others); fomites. Orofecal route; contaminated water	N
Herpesviridae	A	Contact with lesions and body fluids (urine, saliva); sexual; infection at birth by a genitally-infected mother; respiratory route; transplacentary, transplantation, blood transfusion	N
Hypoviridae	F	Cell-to-cell; cytoplasmic exchange; hyphal anastomosis. Conidia	N
Hytrosaviridae	A	Horizontal; food contamination. Vertical: mother to offspring	N
Inoviridae	B	Pilus adsorption	N
Iridoviridae	A	Contact; cannibalism in invertebrates; vector. Cohabitation, feeding or wounding	Y/N
Lavidaviridae	Pr	Passive diffusion?	N

(*Continued*)

TABLE 5.1 (Continued)

Viral Family	Hosts[a]	Transmission	Vector[b]
Leviviridae	B	Pilus adsorption	N
Lipothrixviridae	Ar	Passive diffusion; some pili dependent. Vertically by lysogeny	N
Luteoviridae	P	Mechanical, vector	Y
Malacoherpesviridae	A	Passive diffusion; contact while larvae; experimentally by intramuscular injection	N
Marnaviridae	C	Passive diffusion	N
Marseilleviridae	Pr–A	Passive diffusion. Some water- and soilborne	N
Megabirnaviridae	F	Through hyphal anastomosis; cytoplasmic exchange	N
Metaviridae	A–F–P	Retrotransposons. Vertical transmission only, or cell-to-cell	N
Microviridae	B	Pilus adsorption	N
Mimiviridae	Pr	Passive diffusion	N
Myoviridae	Ar–B	Passive diffusion; vertically by lysogeny	N
Nairoviridae	A	Zoonosis; vector. Some by fomites	Y/N
Nanoviridae	P	Vector	Y
Narnaviridae	C–F–Pr	Horizontally through mating or vertically from mother to daughter cells (no extracellular stage of the virus). Conidia and ascospores	N
Nimaviridae	A	Contact. By predation or cannibalism on diseased animals or via water through the gills. Vertically from nonviable infected eggs or from supporting cell in ovarian tissue	N
Nodaviridae	A	Passive diffusion; direct contact. Vector. Vertical transmission in fishes	Y/N
Nudiviridae	A	Feeding and/or mating route	N
Nyamiviridae	A	Undocumented. May be vectored. In wasps vertical transmission to the progeny	Y

Ophioviridae	P	Mechanical; vegetative propagation. Some are transmitted by zoospores of *Olpidium brassicae*, a root-infecting fungus. *Citrus psorosis virus* rather seems to be propagated by an aerial vector, or grafting	Y/N
Orthomyxoviridae	A	Aerosols (coughing); birds droppings. Zoonosis. Contact with saliva, nasal secretions, feces or blood. In birds: orofecal route. Waterborne (e.g., *Isavirus*). Interspecies infection possible	Y/N
Papillomaviridae	A	Close contact, including sex	N
Paramyxoviridae	A	Contact: feces, secretions from mouth, nose, eyes. Aerosols. Zoonosis; animal bites?	N
Partitiviridae	F–P	Cell division (but not cell-to-cell); sporogenesis; cytoplasmic exchange; hyphal anastomosis. Seeds (*Cryptovirus*)	N
Parvoviridae	A	Respiratory, oral droplets or orofecal route	N
Peribunyaviridae	A	Zoonosis; arthropod bites	Y
Permutotetraviridae	A	Oral route	N
Phenuiviridae: Phasivirus	A	Zoonosis	N
Phenuiviridae: Phlebovirus	A	Zoonosis; arthropod bites	Y
Phenuiviridae: Tenuivirus	P	Mechanical inoculation, usually by an insect	Y
Phycodnaviridae	A–C–P	Passive diffusion. Phaeoviruses are transmitted both horizontally and vertically	N
Picobirnaviridae	A	Orofecal route	N
Picornaviridae	A	Orofecal and respiratory (aerosol) routes; contact; saliva; blood; respiratory secretions. Zoonosis; fomites. Ingestion of virus-contaminated material (*Cardiovirus*)	N
Plasmaviridae	B	Undocumented. Vertically by lysogeny	N

(*Continued*)

TABLE 5.1 (Continued)

Viral Family	Hosts[a]	Transmission	Vector[b]
Pleolipoviridae	Ar	Passive diffusion	N
Pneumoviridae	A	Respiratory secretions	N
Podoviridae	B	Passive diffusion; vertically by lysogeny; some pili dependent	N
Polydnaviridae	A	Vertically transmitted to the offspring as provirus. Horizontally by oviposition in the host larvae	N
Polyomaviridae	A	Orofecal route, contaminated feces and aerosolized dust, hand to mouth. Egg transmission in birds. Human transplantation	N
Pospiviroidae	P	Mechanical. Vegetative propagation. Vector	Y/N
Potyviridae	P	Mechanical, grafting, vector	Y
Poxviridae	A	Direct contact, aerosol or fomites. Some mechanical by arthropod bites	Y/N
Pseudoviridae	A–F–P–Pr	Almost exclusively vertically. Yeast: by conjugation	N
Quadriviridae	F	Cytoplasmic exchange; hyphal anastomosis; sporogenesis. Cell division	N
Reoviridae	A–F–P	Ingestion and surface of eggs (*Cypovirus*); enteric or respiratory routes (*Orthoreovirus*); passive diffusion (*Aquareovirus*); delphacid planthoppers (*Oryzavirus* and *Fijivirus*-also with vegetative propagation); ticks, blood transfusion (*Coltivirus*). Orbivirus: ticks, gnats, phlebotomines, mosquitoes; *in utero* in some vertebrates. *Rotavirus*: orofecal route. *Seadornavirus*: mosquitoes. *Phytoreovirus*: cicadellid leafhoppers	Y/N
Retroviridae	A	Cell-to-cell, airborne; fluids (including blood, milk, and saliva), developing embryo; perinatal routes. Vertically: endogenous provirus (up to 10% of genomic DNA in some vertebrates), but no virions produced	N
Rhabdoviridae	A–P	Zoonosis; animal bites, midges, sandflies, mosquitoes. Passive diffusion; contact. Waterborne. Mechanical inoculation: aphid, leafhopper, planthopper, fungi. Transovarial transmission in insects	Y/N

Roniviridae	A	Passive diffusion. By ingestion of infected material or water. Vertical	N
Rudiviridae	Ar	Passive diffusion	N
Secoviridae	P	Mechanical inoculation; vector: insects or nematodes; some seed or pollen	Y
Siphoviridae	Ar–B	Passive diffusion; vertically by lysogeny	N
Sphaerolipoviridae	Ar–B	Passive diffusion	N
Spiraviridae	Ar	Passive diffusion	N
Tectiviridae	B	Passive diffusion	N
Togaviridae	A	Respiratory (aerosol). Zoonosis; arthropod bite	Y/N
Tolecusatellitidae	P	By vectors of begomoviruses (their helper viruses)	Y
Tombusviridae	P	Mechanical, plant contact, seed, propagating material. Water- and soilborne (stable and infectious). Vector	Y/N
Tospoviridae	P	Arthropod bites; vector	Y
Totiviridae	F–Pr	Passive diffusion, sporogenesis, cell division an fusion	N
Turriviridae	Ar	Passive diffusion	N
Tymoviridae	A–P	Mechanical; vector; poorly by seeds; propagating material	Y
Unassigned: *Cilevirus*	P	Vector	Y
Unassigned: *Deltavirus*	A	Sexual contact, blood, maternal-neonatal	N
Unassigned: *Idaeovirus*	P	Pollination (i.e., pollen-associated); seeds. Mechanical	N

(*Continued*)

TABLE 5.1 (Continued)

Viral Family	Hosts[a]	Transmission	Vector[b]
Unassigned: *Ourmiavirus*	P	Mechanical; undocumented vector	N
Unassigned: *Polemovirus*	P	Grafting; vegetative propagation. Maybe by soil. Vector undocumented	N
Unassigned: *Rhizidiovirus*	C–F	Passive diffusion. Vertically by zoospores of the host *Rhizidiomyces* sp.	N
Unassigned: *Salterprovirus*	Ar	Passive diffusion	N
Unassigned: *Sobemovirus*	P	Mechanical. Vector. Seedborne	Y/N
Virgaviridae	P	Mechanical. Vector. Seeds, pollen. Leaf contact	Y/N

[a]Hosts are defined here as species of the seven kingdoms of life whose members can be infected and support replication of viruses belonging to the families listed in the first column: A (Animalia), Ar (Archaea), B (Bacteria), C (Chromista), F (Fungi), P (Plantae), and Pr (Protozoa).
[b]Vectors are living organisms that carry virions from one infected host to the other (Y); if no vector is known for any member of a family the letter "N" is used; in cases where a virus can be transmitted as an alternative mode of transmission, or some members of the same family are vectored by a living organism while others are not, "Y/N" is used. Most common vectors are listed in Table 5.2. No information was available for the virus families *Alvernaviridae, Feraviridae, Gammaflexiviridae, Iflaviridae, Jonviridae, Phasmaviridae, Sarthroviridae, Solinviviridae, Sunviridae and Tristromaviridae the genus Goukovirus of the family Phenuiviridae, and the unassigned genera Albetovirus, Anphevirus, Arlivirus, Aumaivirus, Bacilladnavirus, Bacillarnavirus, Blunervirus, Botybirnavirus, Chengtivirus, Crustavirus, Dinodnavirus, Higrevirus, Labyrnavirus, Mesoniviridae, Mymonaviridae, Papanivirus, Sinaivirus, Tilapinevirus, Virtovirus and Wastrivirus.*

TABLE 5.2 Examples of Viruses and Their Vectors

Virus Family (Genus)	Hosts[a]	Type of Vector (Some Examples)
Alphaflexiviridae (*Allexivirus*)	P	Arachnids: Mites (*Aceria tulipae*)
Alphaflexiviridae (*Potexvirus*)	P	Insects: Aphids (if a potyvirus provides a helper protein), or buff-tailed bumblebees (*Bombus terrestris*) in greenhouse
Ascoviridae (*Ascovirus*)	A	Insects: Endoparasitic wasps (e.g., *Campoletis sonorensis, Diadegma semiclausum, Microplitis similis, M. croceipes, Toxoneuron nigriceps*)
Ascoviridae (*Toursvirus*)	A	Insects: Endoparasitic wasps (e.g., *Diadromus pulchellus*)
Asfarviridae (*Asfivirus*)	A	Arachnids: Argasid ticks (reservoir) of the genus *Ornithodoros*
		Insects: flies and bugs (genera *Simulium, Stomoxys,* and *Triatoma*)
Benyviridae (*Benyvirus*)	P	Plasmodiophorids: *Polymyxa betae* and *P. graminis*
Betaflexiviridae (*Carlavirus*)	P	Insects: Aphids or whiteflies (*Bemisia tabaci*)
Betaflexiviridae (*Trichovirus*)	P	Arachnids: Mites (*Colomerus vitis, Eriophyes inequalis, E. insidiosus*)
Betaflexiviridae (*Vitivirus*)	P	Insects: Pseudococcid mealybugs (genera *Pseudococcus* and *Planococcus*), scale insects (*Neopulvinaria innumerabilis*), and aphids
Bromoviridae (*Alfamovirus*)	P	Insects: Aphids (*Myzus persicae* and at least 13 more species belonging to the family Aphididae)
Bromoviridae (*Bromovirus*)	P	Insects: Beetles (although with low efficiency). For example, *Diabrotica undecimpunctata howardi* for *Cowpea chlorotic mottle virus*
Bromoviridae (*Cucumovirus*)	P	Insects: Aphids (genera *Aphis, Myzus*) Fungi: *Cucumber mosaic virus* in *Rhizoctonia solani*
Caulimoviridae (*Badnavirus*)	P	Insects: Aphids, mealybugs (*Planococcus citri*), and lacebugs

(*Continued*)

TABLE 5.2 (Continued)

Virus Family (Genus)	Hosts[a]	Type of Vector (Some Examples)
Caulimoviridae (*Caulimovirus*)	P	Insects: Aphids (e.g., *Chaetosiphon fragaefolii*)
Caulimoviridae (*Tungrovirus*)	P	Insects: Leafhoppers of the genera *Nephotettix* and *Recilia* (if a waikavirus provides a helper protein)
Closteroviridae (*Ampelovirus*)	P	Insects: Pseudococcid mealybugs and soft scale insects (*Ceroplastes, Coccus, Dysmicoccus, Heliococcus, Pulvinaria, Neopulvinaria, Parasaissetia, Parthenolecanium, Phenacoccus, Planococcus, Pseudococcus*, and *Saissetia*)
Closteroviridae (*Closterovirus*)	P	Insects: Many different species of aphids (e.g., *Myzus persicae, Aphis citricidus, A. fabae*, and *A. gossypii*)
Closteroviridae (*Crinivirus*)	P	Insects: Whiteflies of the genera *Bemisisa* and *Trialeurodes*
Dicistroviridae (*Aparavirus*)	A	Arachnids: Mites (*Varroa destructor*). Probably fire ants too
Fimoviridae (*Emaravirus*)	P	Arachnids: Possibly the mite *Eriophyes pyri*
Flaviviridae (*Flavivirus*)	A	Insects: Mosquitoes (genera *Aedes, Anopheles, Armigeres, Coquillettidia, Culex, Culicoides, Culiseta, Eretmapodites, Haemagogus, Mansonia, Mimomya*, and *Toxorhynchites*), flies (genera *Musca* and *Phlebotomus*), and bugs (*Cimex*)
		Arachnids: Ticks (genera *Amblyomma, Argas, Dermacentor, Haemaphysalis, Hyalomma, Ixodes*, and *Rhipicephalus*)
Geminiviridae (*Begomovirus*)	P	Insects: Whiteflies (mainly *Bemisisa tabaci*; also reported: *Aleurotrachelus socialis*)
Geminiviridae (*Capulavirus*)	P	Insects: Aphids (*Aphis craccivora*)
Geminiviridae (*Curtovirus*)	P	Insects: Leafhoppers (*Circulifer tenellus*)
Geminiviridae (*Grablovirus*)	P	Insects: Treehoppers (*Spissistilus festinus*)

Geminiviridae (*Mastrevirus*)	P	Insects: Leafhoppers (mostly of the genus *Cicadulina*). Also *Nesoclutha declivata* has been proposed for *Digitaria streak virus*
Geminiviridae (*Topocuvirus*)	P	Insects: Treehoppers (*Micrutalis malleifera*)
Geminiviridae (*Turncurtovirus*)	P	Insects: Leafhoppers (*Circulifer haematoceps*)
Genomoviridae (*Gemycircularvirus*)	F	Insects: Flies (*Lycoriella ingenua*). This virus genus has members that mostly infect animals and plants
Iridoviridae (*Ranavirus*)	A	Insects: parasitic wasps; maybe the mosquitoe *Aedes* sp. as vector to terrestrial turtles, and *Culex territans* or *Lasiohelea* in bullfrogs; parasitic nematodes (e.g. *Thaumamermis cosgrovei*)
Iridoviridae (*Chloriridovirus*)	A	Nematodes: *Strelkovimermis spiculatus*, as vector of *Invertebrate iridescent virus 3* to *Culex pipiens*
Luteoviridae (*Enamovirus*)	P	Insects: Aphids (genera *Acyrthosiphon, Myzus*)
Luteoviridae (*Luteovirus*)	P	Insects: Aphids (genera *Myzus, Schizaphis*)
Luteoviridae (*Polerovirus*)	P	Insects: Aphids (genera *Aphis, Macrosiphum, Myzus, Schizaphis*)
Nairoviridae (*Orthonairovirus*)	A	Arachnids: Argasid and ixodid ticks
Nanoviridae (*Babuvirus*)	P	Insects: Aphids (*Pentalonia nigronervosa, Micromyzus kalimpongensis*)
Nanoviridae (*Nanovirus*)	P	Insects: Aphids (*Aphis craccivora, A. fabae, A. gossypii, Acyrthosiphon pisum, Megoura viciae*)
Nodaviridae (*Alphanodavirus*)	A	Insects: *Aedes aegypti* (to suckling mice)
Nyamiviridae (*Nyavirus*)	A	Arachnids: Ticks (e.g., *Ornithodoros coriaceus*)
Ophioviridae (*Ophiovirus*)	P	Fungi: Zoospores of *Olpidium brassicae*
Orthomixoviridae (*Quaranjavirus*)	A	Arachnids: Ticks (genera *Ornithodoros* and *Argas*)

(*Continued*)

TABLE 5.2 (Continued)

Virus Family (Genus)	Hosts[a]	Type of Vector (Some Examples)
Orthomyxoviridae (*Thogotovirus*)	A	Arachnids: Ticks (genera *Ornithodoros, Hyalomma, Amblyomma,* and *Ripicephalus*)
		Insects: Mosquitoes (genera *Aedes* and *Culex*)
Peribunyaviridae (*Orthobunyavirus*)	A	Arachnids: Ticks, and Insects: mosquitoes, culicoid flies (*Culicoides*), phlebotomines
Phenuiviridae (*Phlebovirus*)	A	Insects: Mosquitoes (genera, *Aedes, Anopheles, Coquillettidia, Culex, Culicoides, Eretmapodites, Mansonia,* and *Toxorhynchites*), flies (*Lutzomia* and *Stomoxys*)
		Arachnids: Ticks (genera *Boophilus, Hyalomma* and *Rhipicephalus*)
Phenuiviridae (*Tenuivirus*)	P	Insects: Planthoppers (genera *Caenodelphax, Javesella, Laodelphax, Nilaparvata, Peregrinus, Sogatella, Tagosodes,* and *Unkanodes*)
Pospiviroidae (*Pospiviroid*)	P	Insects: Aphids (*Myzus persicae* for *Potato spindle tuber viroid,* only if *Potato leafroll virus* (*Luteoviridae*) is present, and *Tomato planta macho viroid*)
Potyviridae (*Brambyvirus*)	P	Unknown: Presumably transmitted by an aerial vector yet to be identified
Potyviridae (*Bymovirus*)	P	Plasmodiophorids: For example, *Polymyxa graminis*
Potyviridae (*Ipomovirus*)	P	Insects: Whiteflies (*Bemisia tabaci*)
Potyviridae (*Macluravirus*)	P	Insects: Aphids (e.g., *Rhopalosiphum maidis* and *Myzus persicae*)
Potyviridae (*Poacevirus*)	P	Arachnids: Eriophyid mites (e.g., *Aceria tosichella*)
Potyviridae (*Potyvirus*)	P	Insects: Aphids, more than 200 species including those of the genera *Acyrthosiphon, Aphis, Aulacarthum, Brachycaudus, Brevicoryne, Hysteroneura, Hyalopterus, Hyperomyzus, Macrosiphum, Metopolophium, Myzus, Phorodon, Rhopalomyzus, Rhopalosiphum, Schizaphis, Sitobion, Therioaphis,* and *Uroleucon.*

Potyviridae (*Rymovirus*)	P	Arachnids: Eriophyid mites (e.g., *Abacarus hystrix*)
Potyviridae (*Tritimovirus*)	P	Arachnids: Eriophyid mites (e.g., *Aceria tosichella*)
Poxviridae (*Avipoxvirus*)	A	Insects: At least 10 different species of mosquitoes (genera *Aedes, Anopheles, Culex, Culiseta, Echydnophaga,* and *Stomoxys*); beetles (*Alphitobius*)
		Arachnids: Mites (*Dermanyssus*)
Reoviridae (*Coltivirus*)	A	Insects: Mosquitoes
		Arachnids: Ticks (genera *Dermacentor, Haemaphysalis, Otobius,* and *Ixodes*)
Reoviridae (*Fijivirus*)	P	Insects: Planthoppers (genera *Delphacodes, Dicranotropis, Javesella, Laodelphax, Perkinsiella, Ribautodelphax, Sogatella, Toya,* and *Unkanodes*)
Reoviridae (*Orbivirus*)	A	Insects: Mosquitoes (genera *Aedes, Culex,* and *Culicoides*), gnats and phlebotomines
		Arachnids: Ticks (genera *Hyalomma* and *Ornithodoros*)
Reoviridae (*Oryzavirus*)	P	Insects: Planthoppers (some members of the genera *Nilaparvata* and *Sogatella*)
Reoviridae (*Phytoreovirus*)	P	Insects: Leafhoppers (genera *Agallia, Agalliopsis, Nephotettix,* and *Recilia*)
Reoviridae (*Seadornavirus*)	A	Insects: Mosquitoes (genera *Aedes, Anopheles,* and *Culex*)
Rhabdoviridae (*Cytorhabdovirus*)	P	Insects: Aphids (*Acyrtoshiphon, Hyperomyzus, Macrosiphum, Megoura,* and *Myzus*), leafhoppers (e.g., *Endria inimica, Recilia dorsalis*), and planthoppers
Rhabdoviridae (*Dichorhavirus*)	P	Arachnids: False spider mites (v. g., *Brevipalpus* spp.)
Rhabdoviridae (*Ephemerovirus*)	A	Insects: Mosquitoes (e.g., genus *Mansonia*) and various *Culicoides* species
Rhabdoviridae (*Ledantevirus*)	A	Insects: Mosquitoes (genera *Aedes, Eretmapodites,* and *Culex*) and wingless bat flies (Nycteribiidae family)
		Arachnids: Ticks (*Amblyomma*)

(*Continued*)

TABLE 5.2 (Continued)

Virus Family (Genus)	Hosts[a]	Type of Vector (Some Examples)
Rhabdoviridae (*Lyssavirus*)	A	Animal bites: Mainly from bats of the subfamily Desmodontinae (except *Mokola Lyssavirus*), *Canis lupus familiaris, Felis catus, Vulpes vulpes*
Rhabdoviridae (*Nucleorhabdovirus*)	P	Insects: Aphids (genera *Aphis, Hyperomyzus*), planthoppers (like *Peregrinus maidis* and *Ribautodelphax notabilis*), leafhoppers (e.g., genera *Agallia, Graminella*, and *Nesoclutha*)
		Arachnids: Mites
Rhabdoviridae (*Tibrovirus*)	A	Insects: Midges (genus *Culicoides brevitarsis* and *C. insignis*)
Rhabdoviridae (*Tupavirus*)	A	Arachnids: Ticks (genus *Ambylomma*)
Rhabdoviridae (*Varicosavirus*)	P	Fungi: Zoospores of *Olpidium brassicae* and *O. virulentus*
Rhabdoviridae (*Vesiculovirus*)	A	Insects: Mostly sandflies (e.g., *Psathyromyia shannonii*), blackflies and cullicoids, but also lice
		Hirudineans: Leeches
Secoviridae (*Comovirus*)	P	Insects: Beetles (especially members of the family Chrysomelidae; e.g., *Cerotoma trifurcata*)
Secoviridae (*Cheravirus*)	P	Nematodes: For example, *Xiphinema americanum*
		Insects: Aphids and thrips
Secoviridae (*Fabavirus*)	P	Insects: Aphids (e.g., genus *Myzus*)
Secoviridae (*Sequivirus*)	P	Insects: Aphid (*Cavariella aegopodii, C. pastinacae*), leafhoppers
Secoviridae (*Nepovirus*)	P	Nematodes: *Xiphinema, Longidorus* or *Paralongidorus* spp.
		Insects: Aphids and thrips
Secoviridae (*Torradovirus*)	P	Insects: Whiteflies (e.g., *Trialeurodes vaporariorum*)

Secoviridae (*Sadwavirus*)	P	Nematodes: (*Xiphinema diversicaudatum*)
		Insects: Aphids and thrips
Secoviridae (*Waikavirus*)	P	Insects: Aphids and leafhoppers (*Graminella nigrifrons, Nephotettix virescens*)
Togaviridae (*Alphavirus*)	A	Insects: Mosquitoes (genera *Aphis, Haemagogus*)
Tolecusatellitidae (all recognized genera)	P	Insects: *Bemisia tabaci* whiteflies. They are satellites of begomoviruses
Tombusviridae (*Auresusvirus*)	P	Fungi: *Olpidium bornovanus* for *Cucumber leaf spot virus,* and *Polymyxa graminis* for *Maize white line mosaic virus*
Tombusviridae (*Avenavirus*)	P	Fungi: Probably zoosporic fungi (not proven yet)
Tombusviridae (*Carmovirus*)	P	Fungi: *Olpidium bornovanus* for various carmoviruses
Tombusviridae (*Dianthovirus*)	P	Fungi: *Red clover necrotic mosaic virus* may be transmitted by the chytrid fungus *Olpidium* sp.
Tombusviridae (*Machlomovirus*)	P	Insects: Thrips (*Frankliniella williamsi*) and chrysomelid beetles (experimentally)
Tombusviridae (*Necrovirus*)	P	Fungi: *Olpidium brassicae*
Tombusviridae (*Tombusvirus*)	P	Fungi: *Cucumber necrosis virus* is transmitted by the chytrid fungus *Olpidium bornovanus*
Tombusviridae (*Umbravirus*)	P	Insects: Aphids (e.g., *Acyrtoshiphon* sp.). *Note*: Umbraviruses depends on one particular polerovirus or enamovirus for virion assembly and thus transmission by aphids
Tospoviridae (*Orthotospovirus*)	P	Insects: More than 10 different species of thrips (like *Frankliniella occidentalis, F. schultzei, Scirtothrips dorsalis,* and *Thrips tabaci*)
Tymoviridae (*Marafivirus*)	P	Insects: Leafhoppers (genera *Aconurella, Dalbulus,* and *Macrosteles*)
Tymoviridae (*Tymovirus*)	P	Insects: Beetles (some members of the families Chrysomelidae and Curculionidae)
Virgaviridae (*Furovirus*)	P	Plasmodiophorids: For example, *Polymyxa graminis*

(*Continued*)

TABLE 5.2 (Continued)

Virus Family (Genus)	Hosts[a]	Type of Vector (Some Examples)
Virgaviridae (*Pecluvirus*)	P	Plasmodiophorids: For example, *Polymyxa graminis*
Virgaviridae (Pomovirus)	P	Plasmodiophorids: *Spongospora subterranea* and *Polymyxa betae*
Virgaviridae (*Tobravirus*)	P	Nematodes: Members of the genera *Trichodorus* and *Paratrichodorus*
Unassigned: *Cilevirus*	P	Arachnids: Mites (*Brevipalpus* spp.)
Unassigned: *Sobemovirus*	P	Insects: Beetles, aphids (*Illinoia pepperi, Myzus persicae, Rhopalosiphum padi*), garden flea-hoppers (*Halticus citri*), leafminers (*Liriomyza langei*), moths (*Diaphaulaca aulica*), leafhoppers (*Circulifer tenellus*), thrips (*Thrips tabaci*), and mirids (*Cyrtopeltis nicotianae*)

[a]Hosts: A (Animalia), F (Fungi), and P (Plantae).

In no-simulation studies, that distance was demonstrated to reach up to 70 km. Metagenomics studies have detected viruses in the atmosphere, including Pseudomonas phage f10, *Pseudomonas virus F116*, and *Escherichia virus P1* from bacteria, Human adenovirus C, along with circoviruses, nanoviruses, microphage-related, and geminivirus-related (plant) viruses. Of note, more contaminated air (smog) harbors an increased number of viral particles. In near-surface atmosphere of different locations (residential, forest, and industrial settings), airborne viruses that infect animals and plants can also be detected, particularly during winter.

Waterborne Viruses

Water is an excellent medium for the transportation and dissemination of viruses. The oceans are a particularly important habitat of bacterial and archaeal viruses—along with those that infect marine plants, animals, fungi, protozoa, and chromista. Water transmitted animal viral pathogens include adeno-, astro-, rota-, noro-, calici-, and polioviruses, as well as hepatitis viruses; urine secreted viruses that can reach water, like polyoma- and cytomegaloviruses, can also be included in the list of water spread viruses. Evidence on the water dissemination of influenza- and coronaviruses is inconclusive.

In the case of plant viruses, there is accumulating evidence of viruses remaining infectious for sufficiently long periods where they become a threat and can infect important plant crops. Although values vary according to species, virions found in water can remain viable for days and weeks, possibly facilitated by aggregation and/or adsorption to solid materials (e.g., clays and silicates) or to organic debris, bacteria or algae. *Pepino mosaic virus* (PepMV; *Potexvirus*) remains infectious for up to 3 weeks, *Potato virus Y* (PVY; *Potyvirus*) for 1 week, and *Potato spindle tuber viroid* (*Pospiviroid*) for up to 7 weeks in water at $20 \pm 4°C$ under controlled conditions. In the case of PepMV and PVY, virions were released from the roots of infected plants. Further, it has been demonstrated that *Pepper mild mottle virus* (*Tobamovirus*), one of the most abundant RNA viruses in human feces, can be found in sea and ground waters at concentrations that range from 1.7×10^1 to 1×10^4 genome copies per liter. This indicates not only the persistence of viruses in the environment, but the complexity of interactions that can be established between a nonhost mammal and a plant virus that survived the conditions in an animal's gut as well as water. The findings also demonstrate that the virus is a promising indicator of fecal pollution of water bodies. Other plant viruses detected in environmental waters (lakes, rivers, sea, tap, and irrigation waters) belong to the genera *Alphacarmovirus*, *Cucumovirus*, and other genera of the family *Tombusviridae*. When associated with the outer covering of the aquatic zoospores of the fungus *Olpidium bornovanus*, virions of *Melon necrotic spot virus* (*Tombusviridae*) remain viable for several years.

Soilborne Viruses

Some viruses are extremely stable under soil conditions. In general, virus survival in soils depends mostly on temperature and virion adsorption to soil (sand and clay colloids). Other influential factors include soil moisture content, presence of aerobic microorganisms, levels of resin-extractable phosphorous, exchangeable aluminum, organic matter, and soil pH. TMV (*Virgaviridae*), for instance, remains infectious in soil for several years in living or dead plant debris. Infections have been reported with plants brought into physical contact with soilborne viruses during transplanting. Although the majority of plant viruses are transmitted by arthropod vectors and invade the host plants through the aerial parts, there is a considerable number of plant viruses that infect roots via soil-inhabiting vectors such as plasmodiophorids, chytrid fungi, and nematodes (see later).

Regarding animal viruses, polioviruses (*Picornaviridae*) are stable in soils provided temperatures are not high (e.g., 3 months at 4°C). At low temperatures and a pH of 7.5, some enteroviruses in soil may survive from 110 to 170 days. *Influenza A virus* (*Orthomyxoviridae*) H5N1 can be found in soil-based composts. Similarly, other animal viruses can be found in soil contaminated with feces, urine, other body fluids, and the carcasses of terrestrial organisms. Bacteriophages have been recovered from saline soils. Virus titers are low in hyper arid desert soils and bacterial lysogens levels are high, in cold desert soils. Most viruses found in the former, however, have yet to be identified and represent uncharacterized virus phylogenetic lineages.

Transplantation, Anastomosis, and Grafting

Of equal significance is the transmission of viruses via allotransplantation. A variety viruses may be transmitted by this route and include *Human immunodeficiency virus 1*, *West Nile virus*, *Human betaherpesvirus 5*, *Rabies lyssavirus* as well as hepatitis B, C, and E viruses. With xenotransplantation (cross-species transfer from pigs to humans), there is a risk of transmitting **p**orcine **e**ndogenous **r**etro**v**iruses (PERVs), **p**orcine **c**yto**m**egalo**v**irus (PCMV), HEV genotype 3, **p**orcine **l**ymphotropic **h**erpes**v**iruses (PLHVs), and **p**orcine **c**irco**v**iruses (PCVs). Only HEV, however, is known to infect humans *in vivo*, while HEV, PCV2, and PERV have been reported to infect human cells *in vitro*.

Similarly, plant viruses are often transmitted by grafting and budding. These centuries-old techniques are used in the vegetative propagation of fruit trees and, in recent decades, vegetable crops. In grafting, the upper shoot (scion) of one plant grows on the root system (rootstock) of another plant. In the second method, a bud is taken from one plant and grown on another. Vascular continuity is eventually established, in both instances, resulting in a genetical composite that functions as a single plant. On one hand, grafting or budding onto resistant rootstocks serves as a principal tool in disease management, but on the other hand, contact between the stock to bud or scion enables the spread of viruses, even without a complete graft union.

Certain parasitic plant species form connections to their hosts, similar to graft junctions, and are able to facilitate the dissemination of virus pathogens. No less than 67 plant viruses and viroids including, but not limited to CMV (*Bromoviridae*), TMV, *Tomato mosaic virus* and *Tobacco rattle virus* (*Virgaviridae*), PVY (*Potyviridae*), *Tomato yellow leaf curl virus*, and *Beet curly top virus* (*Geminiviridae*), along with *Potato spindle tuber viroid* (*Pospiviroidae*) can be transmitted between plants by at least 20 different species of the parasitic plant *Cuscuta* (*Convolvulaceae*) by means of a bridge created between infected and noninfected plant hosts. Parasitic plants produce root-like structures called haustoria which penetrate the host, connect to its vasculature and facilitate the exchange of materials such as water, nutrients, and pathogens between the host and the parasite, and between any plants simultaneously parasitized, even unrelated plant species. *Cuscuta* seems to act in some cases as a passive pipeline between parasitized plants as there is no replication in the parasite. Other viruses, like CMV and *Grapevine leafroll-associated virus 7* (*Closteroviridae*), however, can replicate in *Cuscuta*. Additionally, *Cuscuta* can transmit *Cuscuta*-hosted viruses to the parasitized plant. Finally, parasitism by *Cuscuta* species presumably increases the susceptibility of plants to virus infection.

Zoonosis in the Spread of Viruses

A zoonosis refers to any disease that is naturally transmitted from animals (mostly vertebrates) to humans, or the other way around, and hence represents cross-species transmission of viruses. A vector may be involved, or the virus is transmitted by contact with the infected host or with its direct consumption, or a derived animal product, its fluids or even a vaccine aimed at deterring infection in the intended host. As with all other pathogens, zoonotic spillovers require that the virus overcomes a hierarchical series of barriers in order to establish an infection. The probability of spillover is determined by: (1) amount of available virus (pathogen pressure), (2) dose of exposure, and (3) characteristics (genetic, physiological, and immunological) of the recipient host, that together with (2), determine the severity of the infection. Following cross-species exposure of a recipient host, the within-host barriers determine the likelihood that an infection will establish. Physical barriers include the skin, mucous membranes, mucous, stomach acidity, and absence of virus receptors. Other barriers that may block infection in both infected and neighboring cells include the innate immune response of the host (see

Chapter 10: Host–Virus Interactions: Battles Between Viruses and Their Hosts) and the molecular compatibility between the host and the virus (e.g., lack of host cytoplasmic products required for virus replication and within-host transmission of the virus). If the virus is able to overcome these barriers to replicate and spread in the new host, the outcome of the infection may be the death of the new host and a dead-end spillover infection or sustained human-to-human transmission. Among the zoonoses of viral etiology that affect humans we find rabies, Ebola, Influenza (H1N1), SARS, and Yellow fever, and more recently, Chikungunya and Zika.

Virus Transmission by Vectors

Besides these modes of transmission, some viruses also use a shuttle mechanism involving vectors. A vector is broadly defined as any organism, invertebrate or vertebrate, that functions as a carrier of an infectious agent between organisms. In most cases, acquisition and inoculation of the infectious agent occurs during vector feeding. Common vectors of viruses are found among the arthropods. Those with a piercing-sucking feeding behavior such as mosquitoes (or other blood-feeding dipterans) and ticks are especially significant for vertebrate viruses, and the aphids, whiteflies, thrips, and hoppers for plant viruses. Other animal vectors of viruses include nematodes, bats, rodents, flying foxes, and horses. Species of plasmodiophorids and fungi also vector viruses.

Different modes of virus–vector interactions have been identified by Animal and Plant Virologists. Animal virologists recognize two major categories of virus–vector relationships; viruses are said to be either mechanically transmitted or biologically transmitted by their vectors. Mechanical transmission refers to the nonspecific transmission of viruses by the vector. That is, viruses acquired externally by the vector during normal feeding behavior on an infected organism, are inoculated during the next feed on another organism. On the other hand, biological transmission is characterized by a specific association of a virus with a particular arthropod species or genus and, more importantly, the virus is ingested by the vector and is able to propagate within the vector before transmission to another host can occur. These classifications are still in use today. Plant virologists, however, have developed a more elaborate framework to represent the types of plant virus–vector interactions during transmission. Two major categories of virus–vector relationships have been defined over the years that relate to the acquisition and inoculation periods, retention periods, and latent periods (the time between ingestion of the virus and the ability of the insect to inoculate a host). Namely, circulative and noncirculative. In circulative vector transmission, the virus acquired during vector feeding on an infected host, is ingested, crosses the intestinal barrier and invades the salivary glands. From there, the virus is inoculated into a new host during feeding. If the virus replicates in one or several tissues and vector organs, the interaction is referred to as circulative propagative; if not, it is termed as circulative non-propagative. In noncirculative transmission, the virus binds only to the mouthparts of the vectors (or legs). Neither internalization nor replication in the vector occurs. Noncirculative transmission is further classified as nonpersistent (virus on mouth-parts of vector leading to short term transmission) or semipersistent (movement of the virus to the foregut).

Apart from choosing an appropriate vector, another strategy used by some viruses to guarantee the success of their transmission involves the deliberate manipulation of both the vector and the host. For example, plant viruses may modify the composition of volatiles emitted by infected hosts (CMV and its host *Cucurbita pepo*). This alters the host's perception by the vectors, attracting or repelling vectors. It is reported as well that plant viruses—especially those that circulate through the vector body or infect the vector—modify vector behavior to enhance transmission (*Tomato spotted wilt orthotospovirus* and its thrips vector, *Frankliniella occidentalis*). The modification of vector behavior by vertebrate-infecting arboviruses has also been investigated and in some instances has been found to be related to changes in saliva protein composition, caused by viral infection of the vector (*Dengue virus* type 2 and its mosquito vector, *Aedes aegypti*).

The sections that follow provide further details on these host–virus–vector interactions. For the sake of brevity, and hopefully clarity, the terminology commonly used by plant virologists is used, including where relevant, the characteristics associated with animal virus-vector transmission. The reader is encouraged to review the articles referred to at the end of the chapter for fruitful and thought provoking discussions on the definition of a vector, the categorization of plant/animal virus–vector interactions and mechanics of vector transmission.

Noncirculative Transmission of Viruses

In noncirculative, nonpersistent stylet-borne mode of transmission, plant virions are retained on the animal's stylet (acrostyle in aphids) during feeding and are released upon feeding (and the secretion of saliva) on another host. Transmission of these viruses results not from mere contamination of virions on aphid stylets, but from specific interactions. In the case of potyviruses, a viral encoded proteinase (HC-Pro) facilitates retention of the virion by acting as a bridge between the **c**oat **p**rotein (CP) and aphid protein(s) associated with the aphid stylet. Aphid transmission of *Cauliflower mosaic virus* (CaMV) is more complex and requires three CaMV-encoded proteins. While noncirculative, nonpersistent transmission of plant viruses is only so far found among viruses transmitted by aphid vectors, several aphid, whitefly, and leafhopper-transmitted viruses show a noncirculative, semipersistent transmission relationship. Here, viruses are also not internalized within the insect vector, but they are retained on chitin-lined areas for longer time periods (Fig. 5.1). *Bemisia tabaci*-transmitted *Lettuce infectious yellows virus* (LIYV; *Closteroviridae*) shows this type of transmission relationship. LIYV requires the CP for its whitefly vector transmission. Virus release is achieved by regurgitation instead of salivation due to the fact that, unlike aphids, the foregut in whiteflies is physically separated from their stylet and salivary ducts.

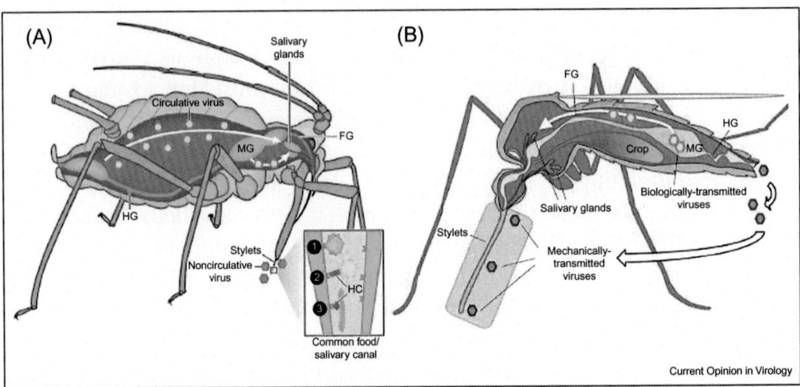

FIGURE 5.1 Different routes of viruses in their arthropod vectors. (A) Pictures the situation for plant viruses where emblematic vectors are aphids, but where other sap-feeding hemipteran insects have similar relationships with the virus they transmit. The gut is represented in *blue* and the salivary glands and salivary duct in *brown*. The white arrows represent the cycle of propagative and circulative viruses (*green hexagons*) within the aphid body, across the gut epithelium to the hemolymph and/or other organs, and ultimately to the salivary glands. While propagative viruses replicate in these organs, circulative viruses are supposed to pass through cellular barriers in a replication-independent manner. Noncirculative viruses appear at their specific attachment sites at the tip of the stylets as *red hexagons*. FG: foregut; MG: midgut; HG: hindgut. The inset at the bottom right represents the common food/salivary canal located at the tip of the aphid maxillary stylets. Noncirculative viruses interact with putative receptors embedded in the cuticle. (1) In the capsid strategy, viruses directly bind putative receptors via a domain of their capsid protein (case 1, for example the genus *Cucumovirus*). (2) In the helper strategy the virus-receptor binding is mediated by additional viral proteins designated *helper components* (HC, *blue*). Best-known cases are the genera Potyvirus (where the HC is designated HC-Pro, case 3) and *Caulimovirus* (where the HC is designated P2, case 2). (B) Pictures the situation for viruses of vertebrates where emblematic vectors are mosquitoes, but where other blood-feeding dipteran insects, or even ticks, have similar relationships with the virus they transmit. The gut is represented in *blue* and the salivary duct and glands in *brown*. The *white* arrow represents the cycle of biologically transmitted viruses (*green hexagons*) within the vector body, across the gut epithelium to the hemolymph and/or other organs, and ultimately to the salivary glands. Biologically transmitted viruses are all believed to replicate in these organs. Mechanically transmitted viruses are thought to be retained with residual blood meal in the mouthparts at undefined locations (*pink* rectangle region of the proboscis). FG, foregut; MG, midgut; HG, hindgut. The *red virions* at the rear of the insect illustrate cases where infectious virus units can be excreted with the feces at (or close to) the feeding sites and may thus contaminate the wounds induced upon vector feeding. Source*: (A) Reproduced from Blanc, S., Gutiérrez, S., 2015. The specifics of vector transmission of arboviruses of vertebrates and plants. Curr. Opin. Virol. 15, 27–33. Available from: https://doi.org/10.1016/j.coviro.2015.07.003.*

This mode of transmission via arthropod vectors is also used by animal viruses and is known as *mechanical transmission.* As indicated above, neither internalization nor replication in the vector occurs. Mechanical transmission of viruses involves the transfer of virions by a vector to a person or animal through direct contact with mouthparts, legs, and/or the body of an arthropod. Unlike plant viruses, however, details on the animal virus proteins involved in these interactions are less known. However, the mechanical transmission of, e.g., *Lumpy skin disease virus* (LSDV, *Poxviridae*) by *Aedes aegypti*, but not by other known similar vectors (*Anopheles stephensi*, *Culex quinquefasciatus*, *Culicoides nubeculosus*, or *Stomoxys calcitrans*) suggest the potential of a complex interaction between the virus and the vector.

Characteristics of the vector–virus relationship for viruses transmitted by soil-inhabiting vectors such as plasmodiophorids, chytrid fungi, and nematodes are less well defined. For the most part, they exhibit features of noncirculative, nonpersistent transmission but also circulative nonpropagative in other cases (see later). Most viruses transmitted by soil-inhabiting organisms are ssRNA(+) viruses, with notable exceptions of members of the genera *Varicosavirus* (*Rhabdoviridae*) and *Ophiovirus* (*Ophioviridae*) with ssRNA (−) genomes. *Beet necrotic yellow vein virus* (*Benyviridae*), transmitted by plasmodiophorids, causes rhizomania disease, characterized by a massive proliferation of lateral roots and rootlets along with severely stunted taproots. *Potato mop-top virus* (*Virgaviridae*), on the other hand, which is also transmitted by plasmodiophorids, causes brown arcs or rings in potato tuber flesh. In the majority of cases, the effects of soilborne virus infection are generally observed on the aerial parts of plants. Plasmodiophorids and fungi, obligate parasites confined to various types of root cells, carry viruses within or on the surface of their zoospores and resting spores. They are transmitted to the plant host, or acquired from it, while the organism is growing within the plant cell. There is no evidence that the viruses can multiply within the vectors. Similarly, nematodes transmit viruses on their stylets, but it also appears that virus can accumulate within the animal. In the latter instances, there is the gradual release of virus during feeding over prolonged periods. *Grapevine fanleaf virus* (GFLV) and *Arabis mosaic virus* (ArMV) are examples of viruses transmitted by two different species of *Xiphinema* nematodes, *X. index* and *X. diversicaudatum.*

Circulative, Nonpropagative Transmission of Viruses

Circulative viruses, by definition, enter the insect body and disseminate to various tissue systems prior to their transmission to plant hosts. This mode of transmission would be referred to as biological transmission in animal virology. In this mode of transmission, circulative, nonpropagative, the virus circulates through the food canal after acquisition, spreads through the midgut, the hindgut, and the hemocoel presumably without replicating in the insect vector. The virus then crosses accessory salivary glands, via the saliva canal, and is transmitted to a new host in the saliva upon feeding. This mode of transmission has been reported for the plant virus families *Geminiviridae*, *Nanoviridae*, and *Luteoviridae* (Fig. 5.1). Aphid transmission of the latter group, luteoviruses, is highly specific. Virus CP is actively involved in the acquisition and transcytosis of virions through the gut to the hemocoel. Another virus protein (the CP read through domain translated from a suppression of termination of the CP gene) allows for the interaction with and passing through membranes of accessory salivary glands of the aphid vector. Phloem host proteins apparently are required for virus uptake and transmission by aphid vectors. Limited examples of this route of transmission have been noted with animal viruses. One example is *Thogoto virus* (*Orthomyxoviridae*) that is vectored by ticks (*Amblyomma variegatum*). Another interesting example is the LSDV (*Poxviridae*), an economically important disease of cattle that occurs across Africa and in the Middle East. The primary mode of transmission of LSDV is via *Rhipicephalus* tick species. Although the virus invades various tick organs, no evidence has been obtained for replication.

Circulative, Propagative Transmission of Viruses

In other cases, a persistent virus completes a similar cycle within the vector's body, but replicates within the gut, the salivary glands and sometimes other tissues of the insect prior to transmission to new hosts (Fig. 5.1). This is designated as the biological mode of transmission of animal viruses par excellence. In plants viruses, it is the exception more than the rule.

Both phytoarboviruses and animal arboviruses more or less follow the same general pathway of transmission. The virus, present in the plant sap or the animal blood meal after feeding, accumulates at high titers in the midgut cells from which they are subsequently released into the hemocoel. Secondary infections involve other tissues, including reproductive tissues (which allow for the vertical transmission of the virus to the progeny, see later). After infection of salivary glands and release of infectious virions into salivary secretions, the virus is transmitted horizontally to new hosts. Competence for transmission depends on both the virus and the vector(s), but the molecular determinants are not fully known. Evidence indicates that there must be cellular receptors, along with virus ligands, that allow internalization of the virus for its ensuing replication in the vector. Few plant-infecting viruses are transmitted by this mode and belong to the families *Phenuiviridae*, *Reoviridae*, *Rhabdoviridae*, *Tospoviridae*, and *Tymoviridae*. Some members (for instance, *Reoviridae* and *Rhabdoviridae*) are able to infect either animal or plant hosts.

In some virus–vector associations, the virus is believed to directly impact on several components of the vector fitness such as longevity, growth rate, and reproduction, as well as feeding behavior. The response elicited by the virus in host not only influences the insects' fitness, but also virus transmissibility. Additionally, insect competence in transmitting a virus is influenced by its own microbiome.

Vertical Transmission: Parent to Offspring Transmission of Viruses

Vertical transmission refers to generational transmission of viruses from parents to their offspring. HIV-1, e.g., can be acquired *in utero* (via breaks in the placental barrier or transcytosis of cell-associated virus), during delivery (intrapartum), or via breastfeeding. Approximately 20% of viral plant pathogens are known to be seed transmitted. Seed transmission is commonplace in the *Potyviridae*. However, the mechanism by which the virus enters the seed is unknown. There is some evidence in *Pea seed-borne mosaic virus* (*Potyviridae*) that the virus may directly invade the embryo via the suspensor. On the other hand, evidence for the indirect invasion of the embryo via invasion of reproductive meristematic tissue early in plant development has been demonstrated in *Barley stripe mosaic virus* (*Virgaviridae*). At the other extreme are "*vertically transmitted*" viruses that live in symbiotic or commensal associations with their hosts. When temperate bacteriophages achieve a lysogenic state, they are propagated by vertical transmission to the next host generation. That is, transmission to both daughter bacteria is by cell division.

Some viruses utilize both horizontal and vertical routes to transmit and maintain levels in a host population. Bee viruses are one example. Virus transmission in honey bees appears to involve foodborne transmission, venereal transmission, vector-borne transmission, and mother-to-offspring transmission. *Zucchini yellow mosaic virus* (*Potyviridae*) can also be transmitted both horizontally by aphids and vertically by seeds, but the predominant method is by horizontal transmission via aphids in a noncirculative manner. *Tomato yellow leaf curl virus* (*Geminiviridae*) is vectored between susceptible plant hosts by whiteflies (horizontal transmission); the virus is also passed from female vectors to males during copulation (also horizontal transmission) and from females via her eggs to the next generation (vertical/transovarial transmission). Although transmission of mycoviruses is typically achieved by the spread of contaminated mycelia or by hyphal anastomosis, vertical transmission also occurs through mitotic, and sometimes meiotic, spores. Very recently, however, it has been demonstrated that *Sclerotinia gemycircularvirus 1* (*Genomoviridae*) can extracellularly infect its fungal host (*Sclerotinia sclerotiorum*) vectored by the mycophagous fly *Lycoriella ingenua* (Diptera: Sciaridae), whose progeny are also viruliferous. Most probably this is not an isolated case of a mycovirus transmitted by a vector and confirmation of more cases is pending. Understanding transmission dynamics of viruses is particularly important, not only for the identification of key hosts and modes against which interventions could or should be targeted, but also to anticipate potential unintended consequences (positive and negative) that may occur in response to the selective pressures that elimination efforts exert on these systems.

FURTHER READING

Andika, I.B., Kondo, H., Sun, L., 2016. Interplays between soil-borne plant viruses and RNA silencing-mediated antiviral defense in roots. Front. Microbiol. 7, 1458. Available from: https://doi.org/10.3389/fmicb.2016.01458.

Blanc, S., Gutiérrez, S., 2015. The specifics of vector transmission of arboviruses of vertebrates and plants. Curr. Opin. Virol. 15, 27–33. Available from: https://doi.org/10.1016/j.coviro.2015.07.003.

Fenner, F., McAuslan, B.R., Mims, C.A., Sambrook, J., White, D.O., 1974. The Biology of Animal Viruses, second ed. Academic Press, New York, p. 834.

Flint, S.J., Enquist, L.W., Racaniello, V.R., Skalka, A.M., 2003. Principles of Virology: Molecular Biology, Pathogenesis, and Control of Animal Viruses, second ed. ASM Press, Washington, DC, p. 850.

Gibson, K.E., 2014. Viral pathogens in water: occurrence, public health impact, and available control strategies. Curr. Opin. Virol. 4, 50–57. Available from: https://doi.org/10.1016/j.coviro.2013.12.005.

Hull, R., 2001. Matthew's Plant Virology, fourth ed. Academic Press, London, UK, p. 1056.

Lequime, S., Paul, R.E., Lambrechts, L., 2016. Determinants of arboviral vertical transmission in mosquitoes. PLoS Pathog. 12, e1005548. Available from: https://doi.org/10.1371/journal.ppat.1005548.

Mehle, N., Ravnikar, M., 2012. Plant viruses in aqueous environment – survival, water mediated transmission and detection. Water Res. 46, 4902–4917. Available from: https://doi.org/10.1016/j.watres.2012.07.027.

Plowright, R.K., Parrish, C.R., McCallum, H., Hudson, P.J., Ko, A.I., Graham, A.L., et al., 2017. Pathways to zoonotic spillover. Nat. Rev. Microbiol. 15, 502–510. Available from: https://doi.org/10.1038/nrmicro.2017.45.

Wilson, A.J., Morgan, E.R., Booth, M., Norman, R., Perkins, S.E., Hauffe, H.-C., et al., 2017. What is a vector? Philos. Trans. R. Soc. Lond. B Biol. Sci. 372. Available from: https://doi.org/10.1098/rstb.2016.008520160085.

Yeh, Y.-H.-, Gunasekharan, V., Manuelidis, L., 2017. A prokaryotic viral sequence is expressed and conserved in mammalian brain. Proc. Natl. Acad. Sci. 114, 7118–7123. Available from: https://doi.org/10.1073/pnas.1706110114.

Chapter 6

Viruses as Pathogens: Plant Viruses

Paula Tennant[1], Augustine Gubba[2], Marcia Roye[1] and Gustavo Fermin[3]

[1]*The University of the West Indies, Mona, Jamaica,* [2]*University of KwaZulu-Natal, Pietermaritzburg, South Africa,* [3]*Universidad de Los Andes, Mérida, Venezuela*

[Regarding plant viruses] It is likely that the interaction between the host defense system and the viral defense suppression system is the major determinant of both host range and symptom expression.

Roger Hull

In just over one hundred years since the description of the first plant virus, the number of plant virus species has seemingly increased exponentially. Indeed, more than 1000 virus species are presently known to infect cultivated field, vegetable, and fruit crops worldwide. The exponential increase is attributed not only to the discovery of new viruses, but also to the improved diagnostic testing and the correction of cases where virus infections were mistaken for other conditions. Increased international movement and globalization of trade have also influenced the introduction and proliferation of viruses across countries.

Nonetheless, there is mounting interest in the increasing number of viruses that cause disease epidemics in agricultural crops. Virus infections, after fungal infections, can substantially decrease the yield and quality of crops (less than 10% to more than 80%–90%), even in instances where latent virus infections exist and plants appear healthy. Plant virus infections do not normally cause plant death. Plant viruses are biotrophic pathogens that require living tissue for their multiplication and survival. Moreover, they are able to induce productive infection only in those plants that have not developed specific defensive responses to their virulence factors.

The main and first response of plants to virus attack is the **h**ypersensitive **r**esponse (HR). HR is mediated by the action of the plant resistance

Viruses. DOI: https://doi.org/10.1016/B978-0-12-811257-1.00006-1

genes (*R*) and is characterized by the development of an area of cell death, at the point of attempted ingress, that prevents further spread of the virus throughout the plant. A more specific defense response involving Dicer-type RNases belonging to the Argonaute-type protein family, among others is also mounted against virus infections. This antiviral response, which is one of the manifestations of a complex set of cellular processes known as RNA silencing, is covered in Chapter 10, Host–Virus Interactions: Battles Between Viruses and Their Hosts. Should the virus be successful in maneuvering these layers in the plant's defense system by the acquisition of virulence factors that counteract or suppress the defense, it is in a position to trigger infection and to initiate and maintain replication in the plant host. However, another set of hurdles must then be tackled. Viruses must face multiple tissues and cell types that differ in physiological and biochemical properties. They must subvert host protein synthesis functions and regulate mRNA translation. Virus genomic nucleic acids, after they have been uncoated, translated, and copied within replication complexes (Chapter 3, Replication and Expression Strategies of Viruses), subsequently journey to neighboring cells and establish systemic infection via the plant's vascular tissue. Interactions of viral and cellular factors facilitate these steps and help to establish optimum infection conditions, but they may also indirectly affect various physiological processes of the host. Four patterns of infection, namely acute, chronic, persistent, and endogenous, have been identified.

Acute viruses cause disease in crop and ornamental plants. These viruses are transmitted horizontally, often by insect vectors, and are occasionally transmitted vertically through seed. They are either specialists, like *Barley stripe mosaic virus* that only infects a few closely related plants, or generalists, like *Cucumber mosaic virus* (CMV) that has a host range of about 1200 species. Their infections are resolved in one of three ways: death of the host, recovery of the host, or conversion to chronic infections. The difference between acute and chronic plant viruses is subtle and relates to the duration of infection. Chronic plant viruses remain with the host plant for extended periods, and may or may not cause observable disease symptoms. Much less is known about persistent viruses which were previously referred to as cryptic viruses. Persistent viruses are mainly transmitted vertically and remain with the host without eliciting symptom development. Similarly, endogenous viruses may not be pathogenic. These viruses integrate into the genome of the plant host. Sequences derived from both groups of plant DNA viruses—the ssDNA geminiviruses and dsDNA viruses of *Caulimoviridae*—have been detected within plant genomes. Integration of viral sequences is most likely due to illegitimate recombination because these viruses do not encode an integrase function and do not require integration into the host genome as part of their replication cycle. Endogenous viruses can exist as benign components of plant genomes, but in some cases they are pathogenic or, conversely, provide viral immunity.

Finally the contributions of various agricultural practices to the provision of ideal environments for the perpetuation of plant viruses, and their vectors, cannot be underscored. Continuous, large quantities of genetically similar hosts and high plant density further aggravate the problem of plant virus diseases. Since most plants are inherently resistant to numerous pathogens, it seems reasonable to assume that advancing knowledge of the molecular and cellular basis of host–pathogen interactions will identify the means not only to engineer or to select durable resistance, but also to produce effective and environmentally acceptable methods of managing diseases. Indeed, the use of resistant varieties, conventional or transgenic, has been the most economical and practical measure; however, effective management of virus diseases necessitates an integrated approach, including the integration of several tactics centered on the avoidance of the sources of infection, avoidance of the vector, modification of cultural practices along with the use of resistant varieties. Development of such integrated management practices requires correct identification of the virus(es) associated with the disease and an understanding of the ecology of the virus(es) and vectors involved.

The sections that follow examine selected viruses of important crops—their distribution, biological and molecular characteristics, and the approaches that manage the diseases they elicit. The viruses were chosen based on their potential impact on food security. They differ considerably in their biological properties and dissemination, but all can cause major epidemics and crop losses, which are particularly devastating in developing countries that depend on a few staple crops for food security and income.

GEMINIVIRUSES: CASSAVA MOSAIC GEMINIVIRUSES, *MAIZE STREAK VIRUS, WHEAT DWARF VIRUS*

Geminiviruses infecting tuber and cereal crops belong to the largest family of plant viruses, the *Geminiviridae*, and are either assigned to the genus *Begomovirus* (*African cassava mosaic virus*, ACMV) or *Mastrevirus* (*Maize streak virus* (MSV) and *Wheat dwarf virus* (WDV)). Approximately a century ago the etiological agents of two diseases, cassava mosaic and maize streak, were recognized as threats to crop production, and they continue to cause serious epidemics today. WDV is regarded as an important pathogen of wheat and barley.

Cassava mosaic geminiviruses

Cassava **m**osaic **d**isease (CMD), first reported in East Africa in 1894 and in Madagascar, Uganda, and Tanzania in the 1930s–1940s, now affects every cassava growing area in Africa. CMD is caused by a complex of seven begomoviruses, collectively referred to as **c**assava **m**osaic **g**eminiviruses (CMGs) that include ACMV, *East African cassava mosaic virus*, and a complex of recombinants. CMGs are frequently present in mixed infections which have contributed to genetic exchange via recombination and the emergence of different strains and novel virus species with increased virulence. Adding to the complexity is the acquisition or exchange of satellite DNAs between CMGs. Satellite DNAs are about 1.3 kb and are of two types: the nanovirus-like DNA 1 or alphasatellites, and the DNA B-like DNA β or betasatellites. Alphasatellites are capable of independent replication, but depend on the CMGs (i.e., a helper virus) for their movement and encapsidation, whereas the betasatellites depend on their helper virus for replication, movement, and encapsidation. Further, betasatellites affect symptom induction and enhance helper virus pathogenicity by increasing viral DNA levels in host plants and suppressing plant antiviral defense systems. Alphasatellites are regarded as nonessential for disease development and, in some cases, have been shown to attenuate disease symptoms. CMD-infected plants typically exhibit misshapen and twisted leaflets and present a pale green or yellow mosaic (Fig. 6.1). The size and number of tubers are reduced. The viruses are transmitted by the whitefly, *Bemisia tabaci*, and when infected cuttings are used for propagation. *Betasatellite* and *Deltasatellite* are the new genera comprising the novel family of plant viruses, *Tolecusatellitidae*.

CMGs are characterized by quasi-isometric geminate particles measuring 30 × 20 nm. Virions contain circular, bipartite, ssDNA genomes encapsidated by many copies of a single **c**oat **p**rotein (CP) subunit of 30 kDa. The two genomic components of CMGs, referred to as DNA A and DNA B, are each about 2.7 kb. Both are required for infection. Both are distinct in the number

FIGURE 6.1 Symptoms typical of **c**assava **m**osaic **d**isease (CMD) in cassava. CMD-infected plants typically exhibit misshapen and twisted leaflets and present a *pale green* or *yellow mosaic*. Source*: Photograph courtesy of Claude Fauquet.*

and function of the genes encoded, except for an **i**ntergenic **r**egion (IR) comprised of about 200 nucleotides. DNA A encodes six gene products: the CP which also is the determinant of whitefly transmission, the AV2 protein that has been implicated in viral movement, the Rep protein that is essential for replication, the **tr**anscriptional **a**ctivator **p**rotein (TrAP) that functions in viral ssDNA encapsidation and the suppression of gene silencing, the **r**eplication **en**hancer (REn), and the AC4 protein which serves as an important symptom determinant. DNA B genes encode the **n**uclear **s**huttle **p**rotein (NSP) and the **m**ovement **p**rotein (MP), both of which are involved in the cell-to-cell movement of virus particles. The **o**pen **r**eading **f**rames (ORFs) are organized bidirectionally in both genome components, separated by the IR which contains key elements for replication and transcription of the viral genome (Fig. 6.2).

CMGs are usually introduced into host plant cells during colonization and phloem feeding by whiteflies. Upon release from virions the genomic ssDNA is copied to dsDNA forms, called the **r**eplicative **f**orms (RFs) which are organized as minichromosomes in the nucleus. Transcription by host RNA polymerase II allows for the production of the Rep protein. Rep initiates rolling-circle replication which is similar to that used in ssDNA bacteriophages (refer to Chapter 3, Replication and Expression Strategies of

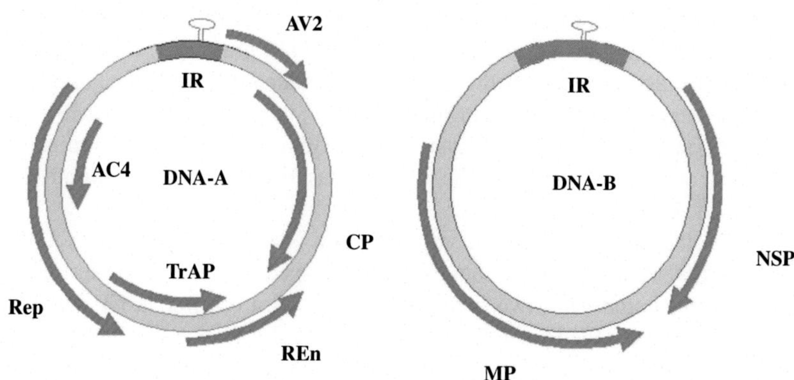

FIGURE 6.2 Typical genomic organization of bipartite member of the *Begomovirus* genus (*Geminiviridae*) such as *African cassava mosaic virus* (ACMV). The genome of ACMV consists of two circular, single-stranded DNA components of about 2700 nucleotides each. The two components that designated DNA A and DNA B are distinct except for the 200 nucleotides **i**ntergenic **r**egion (IR), which is almost identical between the two components. Both DNA components are required for infectivity in the host plant. *IR*: intergenic region with the stem loop. The **o**pen **r**eading **f**rames (ORFs) on DNA A virion sense: *CP*: coat protein; *AV2*: movement protein; complementary sense: *Rep*: *rep*lication-associated protein; *TrAP*: *tr*anscription *a*ctivator *p*rotein; *REn*: replication *en*hancer protein; *AC4*: *ac4* gene. DNA B virion sense: *NSP*: nuclear shuttle protein; complementary sense: *MP*: movement protein.

Viruses). In the later stages of infection, Rep represses its own transcription, causing the activation of TrAP expression, which in turn activates the expression of CP and NSP. Circular ssDNA is subsequently encapsidated by CP into virions, which are available for whitefly acquisition or ssDNA journey to neighboring cells, initially with the assistance of NSP across the nuclear membrane, and then across plasmodesmata with MP.

CMGs are best managed by phytosanitary measures that include rogueing of infected plant materials, propagating cassava cuttings that are disease free, and the planting of resistant varieties. The primary source of resistance used in conventional breeding programs is from the wild relative of cultivated cassava, *Manihot glaziovii*. Later efforts have focused on using cassava landraces with single CMD resistance genes. Unfortunately, most of the CMD resistant materials are susceptible to *Cassava brown streak virus* (*Potyviridae*), another economically important virus affecting cassava production. Transgenic materials using viral *rep* genes or RNA interference technology have had some success in producing materials with different levels of disease resistance, and are currently being tested under field trials.

Maize streak virus

Maize **s**treak **d**isease (MSD) is a major threat to cereal crops in sub-Saharan Africa. Of the 11 known strains of the etiological agent, MSV, only the MSV-A strain currently causes economically significant streak disease in maize. The other strains, designated as MSV B to MSV K, infect barley, wheat, oat, rye, sugarcane, millet, and various wild grass species. Epidemics are typically associated with drought and irregular rains at the beginning of the rainy season in West Africa and Kenya. Yield losses due to MSV are in the region of 30%–100% in East Africa. Although MSD is predominantly a disease of maize in Africa, the disease has been reported in South and South East Asia.

Unlike the previous group of viruses, MSV is a single, covalently closed, circular, ssDNA molecule of about 2.7 kb encapsidated in 22 × 38 nm geminate particles comprising two incomplete icosahedra of 32-kDa CP subunits. Four proteins are coded for CP, MP, and two **rep**lication-associated proteins (Rep and RepA). Two common regions, intergenic short and long, are present. These regions contain sequence elements necessary for viral replication and transcription. MSV is transmitted by at least six leafhopper species in the genus *Cicadulina*, but mainly by *C. mbila* and *C. storeyi*. Within days of infection, host plants display circular, pale spots on young leaves. As the disease progresses, mild to severe chlorotic streaks develop along leaf veins and stunting occurs. Maize plants infected at an early stage produce undersized, misshaped cobs or give no yield at all.

For now the best means of limiting the impact of MSD on maize yields involve the use of conventionally bred resistant varieties, coupled with sound crop management practices. Cultural practices such as the timing of planting and the removal of remnants of the previous crops which can act as reservoirs may reduce the incidence of MSV. Disease avoidance practices such as barriers of bare ground between early and late-planted maize fields to reduce leafhopper movement and subsequent spread of MSV, avoidance of maize plantings downwind from older cereal crops and the use of crop rotations that minimize invasion by viruliferous leafhoppers are also suggested. Chemical control has been effective in decreasing leafhopper migration when seedlings are most vulnerable. Currently available MSD varieties have limited resistance to MSV. Resistance in these varieties is typically manifested as reduced symptom severity combined with low virus titers. Some success has been reported with the development of transgenic resistance using mutated and truncated versions of the *rep* gene. However, the transgenic plants exhibited stunting and infertility.

Wheat dwarf virus

WDV is another mastrevirus cited as a limiting factor for cereal producing areas in many European countries. WDV disease outbreaks have caused yield losses in Europe (UK, Italy, Germany, and Poland), Africa (Tunisia and Zambia), and Asia. Disease incidences of up to 90% are observed in bread wheat fields of Syria, Iran, and Sweden. WDV is regarded as a grass generalist since it infects wheat, other cereals such as barley, oat and rye, and several wild grasses. Five strains, WDV A to WDV E, have been described. Most WDV isolates from wheat belong to the strain WDV E. Although it is widely accepted that wheat-infecting isolates of WDV are unable to infect barley, and barley-infecting WDV are unable to infect wheat, host range studies indicate that, at least occasionally, both strains can infect the other's host.

WDV is transmitted by the leafhopper, *Psammotettix alienus*, in a persistent manner. Symptoms appear as stunted and tufted growth, including the development of streaks of yellowing on leaves, and reduced number of spikes that are often sterile. Suppressed heading and root growth in infected plants lead to yield reductions. Very few options are currently available for the management of WDV. These include cultural practices that reduce the population of leafhoppers coupled with the removal of grassy weed reservoirs. Late sowing and/or the use of cultivars with rapid development are suggested. Applications of chemical insecticides, which target the leafhoppers when seedlings emerge and later when secondary infection may occur, have helped. Assessments of wild wheat relatives, in particular *Aegilops tauschii*, are revealing interesting results of a recovery phenotype and may be useful as a genetic resource for the improvement of resistance to WDV in bread wheat. RNA interference technologies are being investigated.

POTYVIRUSES AND CRINIVIRUSES: *SWEET POTATO FEATHERY MOTTLE VIRUS, SWEET POTATO CHLOROTIC STUNT VIRUS,* AND *SUGARCANE MOSAIC VIRUS*

Virus infection is the main limiting factor in sweet potato cultivation, which is ranked as the seventh most important food crop worldwide. The crop is vegetatively propagated and viruses can persist over successive crop cycles. Often infection in the field develops because of multiple viruses interacting in a complex. One such complex results in the **s**weet **p**otato **v**irus **d**isease (SPVD). The disease is caused by the synergistic interaction between *Sweet potato feathery mottle virus* (SPFMV) and *Sweet potato chlorotic stunt virus* (SPCSV). Synergism refers to an interaction between viruses that causes more severe symptoms than infection with either alone, and in which multiplication of at least one of the viruses is enhanced by the other. SPCSV infected plants display vein clearing, chlorosis, small deformed leaves, severe stunting (Fig. 6.3), and an almost 99% reduction in tuber yield. SPVD was first reported in Nigeria in 1976 and is now reported in almost every sweet potato growing area worldwide.

FIGURE 6.3 Symptoms of **s**weet **p**otato **v**irus **d**isease (SPVD) include (A) vein clearing, (B) mild mosaic and leaf purpling, and (C) plant stunting and severe leaf deformation. Source: *Tennant, P., Fermin, G., 2015. Virus Diseases of Tropical and Subtropical Crops. CABI, Wallingford, 264 p.*

SPFMV, a member of the genus *Potyvirus* in the family *Potyviridae*, has flexuous particles (830–850 nm in length) composed of a single CP (38 kDa) and contains an ssRNA(+) genome of about 10.6 kb. Initially, four strains of the virus were recognized: **E**ast **A**frican (EA), **r**usset **c**rack (RC), ordinary (O), and common (C). However, strain C, redesignated as *Sweet potato virus C* (SPVC), is now considered a different potyvirus. Strains RC, O, and SPVC are distributed worldwide; the EA strain is geographically more widespread than previously thought. Recombination occurs frequently between the strains and presumably is a driving force in their evolution. One isolate of strain O, for example, is a triple recombinant between O, EA, and RC strains. The genomic organization of SPFMV is typical of the *Potyvirus* genus, consisting of a large open reading frame coding for a polyprotein (Fig. 6.4). The polyprotein is subsequently cleaved to yield 10 mature proteins (P1, HC-Pro, P3, 6K1, CI, 6K2, VPg/NIa-Pro, NIb, and CP). The first

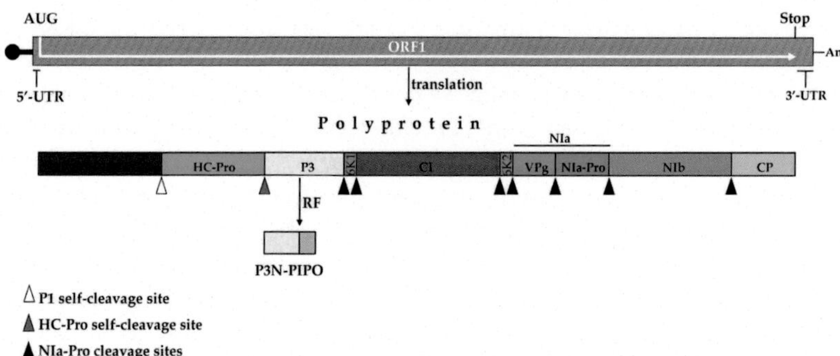

FIGURE 6.4 Schematic representation of a typical *Potyvirus* genome. Potyviruses are polyadenylated at their 3′ termini [A(n)] and possess a covalently linked VPg protein (*black circle* at the 5′ end). The genome has one **o**pen **r**eading **f**rame (ORF) which is translated into a large polyprotein (*internal arrow within the box* starting from the ATG start codon to the stop codon). The polyprotein product is proteolytically processed into individual proteins by the P1 (*white triangle*), HC-Pro (*gray triangle*), and NIa (*black triangles*) virus coded proteases. An additional protein, P3N-PIPO, has been identified and is obtained as the N terminal end of P3 fused to a translation frameshift product derived from a "*hidden*" *pipo* cistron. *Sweet potato feathery mottle virus* (genus *Potyvirus*, family *Potyviridae*) contains another embedded ORF, **p**retty **i**nteresting **s**weet **p**otato **p**otyvirus **ORF** (PISPO). PISPO is of 230 codons and is specific to some sweet potato-infecting potyviruses, while PIPO, of 66 codons, is present in all potyviruses. (P1 proteinase, HC-Pro Helper component-proteinase, required for aphid transmission, P3 component of viral replication complexes, 6K1 involved in virus replication, CI cytoplasmic inclusion protein, 6K2 involved in virus replication, VPg a virus genome-linked protein, NIa nuclear inclusion protein a; proteinase, NIb nuclear inclusion protein b; virus replicase, and CP coat protein; virus genome encapsidation, aphid transmissibility, virus translation and replication, cell-to-cell and systemic movement, host-specific pathogenicity determinant).

three cleavage products are P1, HC-Pro, and P3. An additional short ORF, called PIPO, overlaps the P3 region of the polyprotein ORF. Another ORF, **P**retty **I**nteresting **S**weet **P**otato **P**otyvirus **ORF** (PISPO), has been detected in P1. Interestingly, PISPO, rather than HC-Pro in other potyviruses, is a suppressor of RNA silencing in SPFMV.

SPFMV is transmitted by several species of aphids (*Aphis gossypii*, *Aphis craccivora*, *Myzus persicae*, and *Lipaphis erysimi*) in a nonpersistent manner. It is not seed-transmitted in sweet potato. The host range is narrower than most potyviruses and is mostly limited to the family Convolvulaceae, and especially to the genus *Ipomoea*. Leaf symptoms of SPFMV infected plants are generally mild and may consist of the classic irregular chlorotic patterns (feathery mottle), chlorosis of older leaves and vein clearing. Some strains of SPFMV cause necrotic lesions on the root exterior (RC disease), while other strains produce symptoms on the root interior (internal cork disease). The main economic loss due to SPFMV occurs during its synergistic association with SPCSV in the SPVD complex.

SPCSV, the other virus in the SPVD complex, belongs to genus *Crinivirus* of the family *Closteroviridae*. In single infections, SPCSV can reduce yields by 50%, causing mild stunting, combined with slight yellowing or purpling of older leaves. However, SPCSV has the ability to break down the natural resistance of sweet potato to other viruses and mediate severe synergistic viral diseases with other sweet potato viruses. The most common and severe of these diseases is SPVD. Other viruses belonging to the genera *Potyvirus*, *Carlavirus*, *Cucumovirus*, *Ipomovirus*, and *Cavemovirus*, from diverse virus families, can result in synergistic diseases, and severely affect sweet potato yield upon coinfection with SPCSV. In many cases, infection of sweet potato by two or more different viruses causes greater damage than does infection by each of the viruses separately. In SPVD-affected plants the titer of SPFMV is raised and SPFMV is readily acquired by aphid vectors only from such plants.

Based on molecular and serological analyses, SPCSV can be differentiated into **E**ast **A**frican (EA) and **W**est **A**frican (WA) serotypes. Outside West Africa, strain WA seems to have a wide geographical distribution whilst isolates of the EA strain have been largely restricted to countries in East Africa. The virus is phloem-limited and is transmitted by grafting or by whiteflies (*Bemisia tabaci* biotype B, *Trialeurodes abutilonea*, and *Bemisia afer*) in a semipersistent manner (refer to Chapter 5, Host Range, Host–Virus Interactions, and Virus Transmission). The host range of SPCSV is limited mainly to the family Convolvulaceae, and the genus *Ipomoea*. SPCSV has flexuous particles of 850–950 nm in length and 12 nm in diameter. The bipartite genome of SPCSV, consisting of RNA1 (9407 nucleotides) and RNA2 (8223 nucleotides), is encapsidated by a 33 kDa major CP. After *Citrus tristeza virus*, SPCSV appears to have the second largest genome of all plant viruses. Its RNA1 contains two overlapping ORFs that encode the complete repertoire of proteins responsible for virus replication. These include a papain-like cysteine proteinase, methyltransferase, helicase, and polymerase domains. In addition, SPCSV encodes an RNase III enzyme and a protein, P22, from the 3′ terminal ORF on RNA1. Evidence suggests that RNase III has dsRNA-specific endonuclease activity that enhances the RNA-silencing suppression activity of p22. Presumably the synergistic effect of SPCSV on other viruses is due to the activity of these two proteins to ablate the host's RNA-silencing defense. RNA2 contains the *Closteroviridae* hallmark gene array needed for viral movement, encapsidation and vector transmission: a heat shock protein homologue (Hsp70h) and the CP. Several mechanisms, including the production of coterminal sgRNAs, frameshifting, and polyprotein processing, have been shown or proposed to be used to facilitate virus genome expression (refer to Chapter 3, Replication and Expression Strategies of Viruses).

Effective and durable SPVD disease management is based on prevention. No single management tool provides adequate control against the natural viral complexes that infect sweet potato. A series of cultural methods such as weed control, intercropping, and rogueing of infected plants have proven effective in minimizing losses. Weeds play an important role in the incidence and spread of sweet potato viruses as they possibly serve as alternate hosts for insect vectors and viruses. Removal of reservoir weed hosts, especially wild *Ipomoea* species, in a wide area around a crop may relieve the inoculum pressure. At present the best way to manage viral diseases in sweet potato is to supply growers with virus-indexed propagation material. Such material can be produced by the meristem tip or shoot tip culture techniques, which are based on the propagation of the youngest tissues of the shoot apex that have uneven, low virus titers. Recently the combination of meristem tip culture with cryo- or thermotherapy was shown to drastically enhance the efficiency of virus elimination in sweet potato. The development of resistant sweet potato varieties is the most promising means of controlling viral disease in the long term. Wild relatives of sweet potato (e.g., *Ipomoea trifida*) have served as sources of resistant genes. Breeding programs in Uganda work at combining SPVD resistance with desirable agronomical traits such as yield, earliness, and acceptable culinary quality in sweet potato cultivars. Although progress has been made, it remains to be seen if these cultivars will retain their resistance when challenged by differing strains of the SPVD, which may occur in different geographical locations. Attempts have been made to obtain resistance using the pathogen's own genetic material. Some protection was obtained with transgenic sweet potato transformed with the CP gene of SPFMV and/or SPCSV. However, given the multiplicity of viruses occurring under field conditions, this approach has had limited success. Other attempts at multiple virus resistance involve transforming plants with DNA segments derived from CP segments of SPCSV and SPFMV linked to a viral DNA "*silencer*." Transgenic sweet potato plants showing varying levels of resistance to SPFMV and *Sweet potato virus G* were obtained.

Sugarcane mosaic virus

Another potyvirus, *Sugarcane mosaic virus* (SCMV) either alone or in combination with other viruses, infects many species of Poaceae including sugarcane and maize. The virus consists of an ssRNA(+) genome of 9.6 kb covalently linked to a virus genome-linked protein at its 5′ terminus and a poly(A) tail at its 3′ terminus, and is encapsidated by a single type of CP in flexuous rod particles (~750 nm × 13 nm). The genome encodes a large polyprotein that is cleaved by three viral proteases (P1, HC-Pro, and NIa) to produce 10 functional proteins (P1, HC-Pro, P3, 6K1, CI, 6K2, NIa-VPg, NIa-Pro, NIb, CP) and a small frameshift-derived peptide (PIPO).

SCMV is an important virus pathogen, especially in European and Chinese maize production, causing serious losses in grain and forage yields in susceptible cultivars. The virus has been reported in at least 25 countries around the world. It was first reported as a pathogen of sugarcane in Java in 1882 with its characterization as a viral entity infecting sugarcane and other grasses confirmed in 1919 in Puerto Rico. The virus has also been reported as a pathogen of maize in sub-Saharan Africa since the 1930s. Although the natural host range is restricted to members of the Poaceae family of vascular plants, cultivated cereals such as wheat, barley, rye, and rice are rarely infected naturally. Common symptoms of SCMV in different hosts like sorghum, maize, sugarcane, and other grasses include mosaic patterns, streaks, stunting. Coinfections with other viruses are very common. SCMV coinfection with *Maize chlorotic mottle virus* (MCMV), which results in the economically important maize lethal necrosis disease, is the most common of these coinfections. Serious yield loses due to SCMV and MCMV infections on maize have been reported. Losses of between 20% and 80% have been recorded in China. The extent of the losses varies depending on type of germplasm cultivated as well as the prevailing climate and soil conditions. SCMV, like most potyviruses, is transmitted by several aphid species in a nonpersistent manner. These species include *Myzus persicae*, *Myzus euphorbiae*, *Schizaphis graminum*, *Aphis fabae*, *Acyrthosiphon pisum*, and *Rhopalosiphum padi*.

An integrated approach involving controlling weeds, avoiding contaminated host planting near and within the crop, reducing vector populations, and using SCMV-free plants derived from tissue culture and thermotherapy approaches and/or tolerant hybrids are regarded as the best ways of managing SCMV. Attempts have been made in the transfer of resistance from wild relatives as well as mutation breeding. However, conventional breeding is difficult because of not only the complexity of the genetic background of the crop, but also the extensive screening against the major diseases of sugarcane that is required by the industry prior to the release of an improved variety. Sugarcane is host to some 120 pathogens of viral, bacterial, fungal, and phytoplasmal origin; any new variety is required to demonstrate a high measure of general resistance against this range of pathogens. Suffice to say, RNA silencing–based approaches are under investigation. High levels of resistance are observed with transgenic plants transformed with the SCMV CP gene constructs or **h**air**p**in (hp) RNA-induced silencing simultaneously targeting CP and HC-Pro.

BABUVIRUSES: *BANANA BUNCHY TOP VIRUS*

Banana bunchy top disease is one of the most economically important diseases in many banana- and plantain-producing regions in tropical and subtropical Asia, Australia, Europe, and Africa. The disease was first reported in Fiji in 1879, followed by Taiwan (1890), Egypt (1901), and Australia in 1913. It has not been confirmed in the Americas or the Caribbean. Listed among the world's top 100 worst invasive species, infections with the etiological agent, *Banana bunchy top virus* (BBTV), can cause high losses anywhere from 90% yield losses to almost complete destruction of the crop.

Banana bunchy top disease was so named after the distinctive symptoms exhibited during the advanced stages of infection. At this stage the apical leaves are narrow and upright, giving the top of the plant a rosette or a bunched appearance (Fig. 6.5A). The diagnostic symptom is, however, the "*Morse-code*" or "*green dot-dash streak*" pattern on the underside of the leaves or the appearance of dark green streaks of varying lengths on petioles, midribs, and leaf veins during the initial stages of infection (Fig. 6.5B). Plants infected at an early stage remain stunted and do not produce fruits. Bunches may partially develop on infected plants, prematurely bursting through the sides of the pseudostem. Lateral shoots (referred to as suckers) that develop from the rhizome of an infected mother plant are generally severely stunted, show the typical bunched appearance of leaves at the top of the pseudostem, and do not produce fruits. Some banana varieties, such as the Cavendish types, are more readily infected with the virus.

Spread of BBTV is associated with the movement and distribution of infected vegetative planting materials (suckers, rhizome, tissue culture plantlets) and is exacerbated by transmission via the banana aphid (*Pentalonia nigronervosa*, Hemiptera: Aphididae) in a persistent-circulative, nonpropagative manner. Hosts other than *Musa* spp. that are also able to sustain feeding by *P. nigronervosa*, have been described as possible reservoirs of the virus and include *Canna indica* (Zingiberales: Cannaceae) and *Hedychium coronarium* (white ginger, Zingiberaceae). BBTV is not mechanically transmissible.

BBTV is the type member of the genus *Babuvirus* of the family *Nanoviridae*. The genus also contains two other species, *Cardamom bushy dwarf virus* and *Abaca bunchy top virus*, which infect large cardamom and *Musa* spp., respectively. Virus particles are isometric, measuring 18–20 nm

FIGURE 6.5 Banana exhibiting bunchy top disease symptoms. (A) Narrow, upright apical leaves giving the top of the plant a rosette or a bunched appearance. (B) Diagnostic "*Morsecode*" or "*green dot-dash streak*" pattern on the underside of the leaves. Source: *Photograph courtesy of John Hu.*

in diameter. The genome is complicated and distinctive among viruses. The genome of BBTV is multipartite, consisting of six circular, ssDNAs, each about 1 kb in size that are separated by distinct capsids. The multipartite genome organization of viruses in the family *Nanoviridae* suggests a complicated evolution. It has been suggested that nanoviruses at some point switched from plant hosts to a vertebrate host and then, on recombination with a vertebrate-infecting virus, gave rise to circoviruses.

BBTV's DNA components are monocistronic and encode a single protein, with the exception of one component which encodes two proteins. The six components, named according to their functions, include DNA-R (replication), DNA-S (encapsidation), DNA-C (cell cycle modulation), DNA-M (cell-to-cell viral spread), DNA-N (transporting DNA to and from nucleus), and DNA-U3 (unknown). All components share the same basic organization of a single ORF in the virion sense adjacent to a "*common region stemloop*" or CR-SL sequence which is very similar to that found in geminiviruses, and a **m**ajor **c**ommon **r**egion (CR-M). The ability of the DNA-M protein to induce severe symptoms in plants was recently demonstrated, and the protein was implicated as a pathogenicity determinant and a silencing suppressor with a similar function to the DNA B-MP of geminiviruses. Curiously the protein also displays fungicidal properties against *Fusarium oxysporum* f. sp. *cubense*; BBTV-infected bananas are more resistant to the fungus. The gene product of DNA-C, Clinck, also has the ability to counter RNA silencing. Babuviruses replicate their genomes in the nuclei of infected cells via the rolling-circle mechanism initiated by the virus-encoded replication initiation protein (Rep). The mechanism is similar to that used by geminiviruses and ssDNA bacteriophages. Additionally the virus encodes (DNA-C) a replication enhancer protein, designated as Clink. Clink is able subvert the cell cycle control of the host and force cells into DNA synthesis or S phase, making them more permissive for viral replication. The mechanism is similar to that found in mammalian tumor viruses; both mammalian tumor viruses and plant geminiviruses (Rep) have common cellular targets and employ similar strategies to modulate the host's cell cycle control. Nucleotide sequence analysis of DNA-R, DNA-N, and DNA-S reveals two distinct groups of BBTV isolates that are designated as the Asian and South Pacific groups. The Asian group includes isolates from the Philippines, Taiwan, and Vietnam while the South Pacific group contains isolates from Australia, Burundi, Egypt, Fiji, India, Tonga, and Western Samoa.

Management of banana bunchy top disease is primarily through the use of virus-free planting materials and strict rogueing; that is the regular removal and destruction of infected plants that can serve as pathogen inoculum and contribute to the spread of the disease. Controlling aphid-vector populations is recommended as a strategy against BBTV in some regions; however, the aphid vector is widespread and difficult to control via systemic insecticides. Regular field sanitation, involving deleafing, desuckering, and weeding, is recommended for maintaining aphid colonies under the threshold at which more winged aphids are produced. Engineered resistance is attractive for the management of BBTV as natural sources of resistance are not known. Recent studies have demonstrated the possibility of generating transgenic bananas resistant to BBTV through an RNAi-mediated strategy. In two studies, bananas were transformed with different constructs of the viral *Rep* gene, given importance of the BBTV DNA-R in the infection and replication processes. Transformation with constructs such as a mutated *Rep* sequence, antisense strand of *Rep*, inverted repeat of *Rep* sequence, and **hairpin** (hp) RNAs have generated BBTV resistant lines. Whether transgenic plants could be an alternative strategy for managing BBTV in the field remains to be tested.

CUCUMOVIRUSES: *CUCUMBER MOSAIC VIRUS*

CMV is the type species of the genus *Cucumovirus* within the family *Bromoviridae*. CMV has a single-stranded, tripartite RNA genome composed of RNA1, RNA2, and RNA3 (Fig. 6.6). Each molecule carries a 5′ end cap and contains a tRNA-like structure at the 3′ end. The molecules are packaged in separate, nonenveloped, icosahedral particles ca. 29 nm diameter. A subgenomic RNA (RNA4) is also packaged with RNA3. Thus mature CMV consists of three spherical particles. All three are required for virus infectivity.

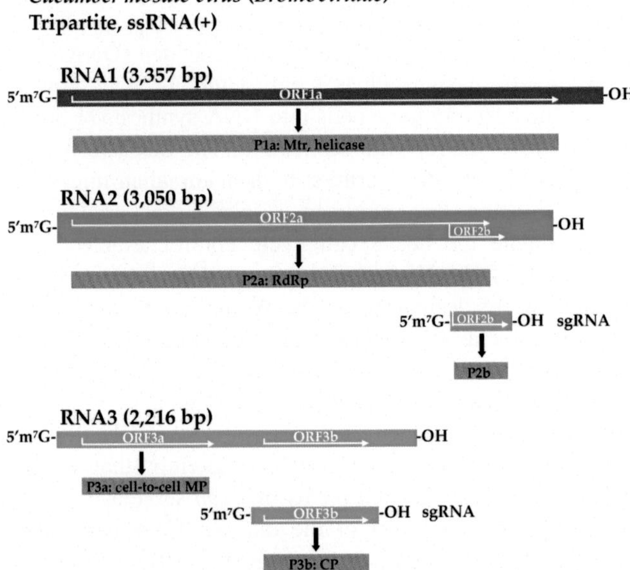

FIGURE 6.6 Genome organization of *Cucumber mosaic virus* (CMV). The genome of CMV is composed of three ssRNA molecules of positive polarity. Each molecule carries a 5′ end cap and contains a tRNA-like structure at the 3′ end. Genomic RNA1 encodes a replicase protein (P1a) harboring **m**ethyl**tr**ansferase (Mtr) and helicase domains. Genomic RNA2 encodes a second replicase protein (P2a) with a polymerase domain (RdRp). Both P1a and P2a, along with host factors, are required to initiate replication. A third nonstructural protein P2b is expressed, as **sub**genomic **RNA** (sgRNA), from RNA2 and functions in cell-to-cell movement and posttranscriptional gene silencing. Genomic RNA3 encodes a fourth nonstructural protein, the **m**ovement **p**rotein (P3a, cell-to-cell MP), and the structural **c**apsid **p**rotein (P3b, CP) that is expressed via subgenomic RNA (RNA4).

CMV expresses five proteins from three positive-sense, genomic RNAs and two subgenomic RNAs. RNA1 is monocistronic and codes for protein 1a, which together with protein 2a, make up the virus' replicase complex. RNA2 also encodes the 2b protein which is derived from a subgenomic RNA designated as RNA4A. Protein 2b is a multifunctional protein involved in long-distance movement through the host, symptom induction, and it acts as a virulence determinant by suppressing gene silencing. RNA3 encodes the 3a or MP and the CP. The latter protein is translated from a subgenomic RNA4. CP is also involved in cell-to-cell movement as well as virion assembly and aphid-mediated transmission. As with all RNA(+) viruses, CMV coordinates the use of the infecting viral genomic RNA as a template for translation and replication; once genomic strands are translated, minus (−) strand RNA intermediates are synthesized which then act as template for the synthesis of genomic and subgenomic plus (+) strands.

Based on nucleotide sequence data, CMV is divided into subgroups, I and II. A notable feature that distinguishes CMV strains of subgroup II (e.g., Q-CMV) from those strains of subgroup I (e.g., Fny-CMV) is the presence of an additional RNA species, referred to as RNA5. RNA5 is a mixture of the 3′ terminal 307 and 304 nucleotides regions of RNAs 2 and 3, respectively.

CMV is distributed worldwide, in both tropical and temperate regions. Isolates of subgroup II are frequently reported in temperate regions, whereas those of subgroup I are found in East Africa, the Mediterranean, California, Brazil, and Australia. CMV's host range is reputed to be very wide; in fact the virus is regarded as possessing the widest host range of all plant viruses. Although originally detected in 1916 in cucurbits (Cucurbitaceae), CMV has been found infecting more than 1200 plant species belonging to at least 100 different families. Among these are peppers, beans, tomatoes, potatoes, onion, spinach, banana, and ornamentals, as well as semiwoody and woody plants and weeds. Transmission between hosts occurs by various routes; mechanically, by aphids that feed on infected plants, by seed, by methods of vegetative propagation, and also by the parasitic plant, *Cuscuta* sp. Nonetheless, the most effective vectors of CMV are aphids, represented by more than 80 different species in 33 different genera. The most important and studied is the green peach aphid, *Myzus persicae*. Aphids transmit the virus in a nonpersistent manner, and their efficacy is linked to virus titer and plant defense. During the early stages of CMV infection, plants appear to promote the growth and development of aphids. Concomitant to the increase in virus titers however is the induction of the salicylic acid signaling pathway which triggers an effective chemical defense against the aphids. As a result, the numbers of aphids decrease. Winged forms that are subsequently produced migrate from the infected plant, leading to virus transmission.

Although some CMV-infected plants are symptomless (e.g., alfalfa), the vast majority of the virus' hosts display an array of symptoms that can include stunting, mosaic (a pattern of alternating yellow, or light green, and dark green areas on leaves surface), leaf vein yellowing, leaf deformation, and flower color breaking. Symptomatology varies according to growing conditions, host, and specific strains of the virus and the presence of **satellite RNAs** (satRNAs). Some isolates of CMV contain small linear, single-stranded RNA molecules, ranging in size from 332 to 405 nucleotides, known as satRNA. These small RNAs are dependent on CMV for replication, encapsidation, and vector transmission. While most CMV satRNAs attenuate or do not modify CMV symptoms, in tomato two main types can be distinguished: those that attenuate or do not modify CMV symptoms (nonnecrogenic satRNAs), and those that aggravate systemic necrosis (necrogenic satRNAs).

Coping with CMV is not an easy task. Common management strategies include the use of virus-free plants, strict aphid control, removal of reservoir hosts, rogueing, and use of resistant plants or tolerant varieties. Resistant varieties of tomato, pepper, legumes, and a few cucurbits are available. Natural resistance genes of either cultivated crops or related wild species were utilized in their development. The efficacy of transgenic resistance for the management of the disease has also been demonstrated with virus-derived genes (CP, satRNA, antisense RNA, replicase) and nonviral genes coding for RNases, ribozymes, and pathogenesis-related proteins.

TUNGROVIRUSES AND WAIKAVIRUSES: *RICE TUNGRO BACILLIFORM VIRUS* AND *RICE TUNGRO SPHERICAL VIRUS*

Rice tungro was reported for the first time in the 1940s, but it was years later that the disease was attributed to a viral infection and not to a nutritional deficiency. Rice tungro, which means "*degenerated growth*" in a Philippine dialect, is a devastating disease caused by coinfection with two morphologically and genetically different viruses, *Rice tungro spherical virus* (RTSV) and *Rice tungro bacilliform virus* (RTBV), depicted in Fig. 6.7. The viruses are common in mixed infections in South and South East Asia where they cause economic losses, amounting very often to 100%. The disease affects cultivated rice, but it is also prevalent in wild rice and other plants belonging to the Poaceae family. Infected plants show symptoms of discoloration, stunting, and a reduced number of tillers as well as grains that are either sterile or partly filled. Individually, these viruses elicit mild symptoms. If infected only with RTBV, plants develop milder symptoms of yellowing and reddening of the leaves and stunted growth. If infected only with RTSV, very mild stunting, with no yellowing of the leaves, occurs. The interaction between the two viruses is described as synergistic, involving partial dependence. That is, RTBV on its own cannot be transmitted from plant to plant, but it can do so with the help of RTSV that encodes an insect transmission factor. Both rice tungroviruses are transmitted by leafhoppers (e.g., *Nephotettix virescens*) in a

(A) ***Rice tungro bacilliform virus* (*Caulimoviridae*)**
Monopartite, dsDNA-RT

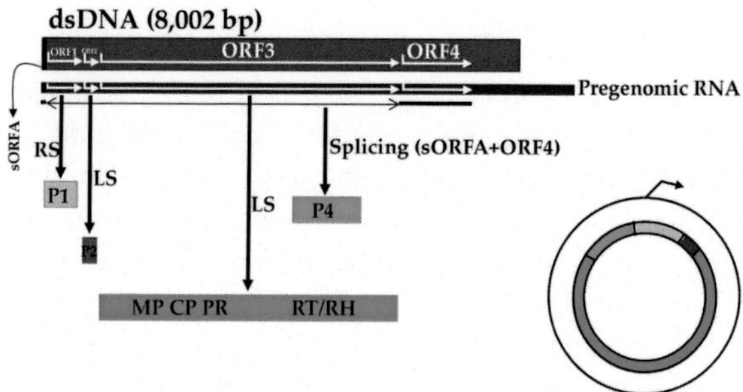

(B) ***Rice tungro spherical virus* (*Secoviridae*)**
Monopartite, ssRNA(+)

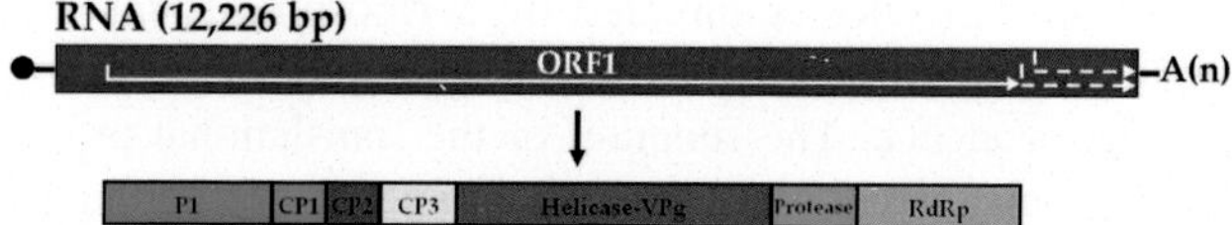

FIGURE 6.7 Genome organization of rice tungro disease viruses. (A) The genome of *Rice tungro bacilliform virus* (RTBV, *Caulimoviridae*) consists of a noncovalently closed circular, single dsDNA molecule; each strand contains a single discontinuity at specific locations. Only one strand (plus-strand) contains the coding sequence. There are two intergenic regions: one large region between ORF4 and ORF1, and a smaller region between ORF3 and ORF4. The pregenomic RNA promoter, the polyadenylation signal and the minus-strand primer-binding site are all located in larger intergenic region. Upon transcription, a single pregenomic RNA is generated, and hence, is terminally redundant (circular genome at the bottom right). This pregenomic RNA serves as template for the translation of ORFs 1−3: ORF1 codes for a protein (P1) of unknown function that is translated by a **r**ibosome **s**hunting (RS) mechanism, while ORF2 and ORF3 are translated by a **l**eaking **s**canning (LS) mechanism. The product of ORF2 (P2) is a virion-associated protein, and the product of ORF3 (P3) is a polyprotein with domains for the **m**ovement **p**rotein (MP), **c**apsid **p**rotein (CP), aspartic **pr**otease (PR), and reverse transcriptase with a ribonuclease H activity (RT/RNase H1). Functional products are generated by the action of the viral protease. Splicing of pgRNA leads to fusion between the first short ORF in the leader sequence at the 5′ end (sORFA) and ORF4 and the generation of a monocistronic mRNA for P4. P4 functions as a suppressor of RNA silencing. (B) The genome of *Rice tungro spherical virus* (RSTV, *Secoviridae*) consists of a single ssRNA(+) molecule with a virus-encoded VPg protein covalently bound to the 5′ terminus and a track of As (poly(A) tail) at the 3′ terminus [A(n)]. The RTSV genome codes for a single large ORF that is translated into a long polypeptide which is sequentially cleaved by a 3C-like protease to yield all final functional products. Waikaviruses code for three coat protein subunits (CP1, CP2, and CP3) of similar sizes, after the leader protein (P1). The other proteins contained in this ORF include the helicase, VPg protein, protease, and replicase (RdRp). Two potential ORFs (*dotted arrows*) at the 3′ end, that probably produce sgRNAs, have been reported.

semipersistent manner. RTSV essentially acts as a helper in the acquisition and transmission of RTBV.

Rice tungro bacilliform virus

RTBV is the type virus of the genus *Tungrovirus* (*Caulimoviridae*). The family *Caulimoviridae* is one of the few virus families that include isometric or bacilliform virions that infect plants or animals (insects). In the particular case of RTBV the virion is a nonenveloped bacilliform particle of 130×30 nm that harbors a circular dsDNA molecule ca. 8 kb that codes for four open reading frames (P1, P2, P3, and P4). Upon delivery into host cells the virus dsDNA is found in the nucleus and is transcribed by the host's RNA PolII—the same enzyme that transcribes protein-coding genes of the host. The resulting transcript, referred to as **pregenomic RNA** (pgRNA), serves as a polycistronic mRNA and the corresponding proteins are synthesized in the cytoplasm by specialized translation mechanisms (Fig. 6.7A). Translation of the first ORF is initiated by a **r**ibosome **s**hunt (RS) mechanism whereby ribosomes access a start codon downstream a long, highly structured leader sequence. The function of the translational product (P1) is not known. ORF2 and ORF3 are translated by a **l**eaking **s**canning (LS) mechanism. ORF3 has similarities with the *gag-pol* core of retroviruses and contains a putative MP, the CP, the aspartate **pr**otease (PR), and the **r**everse **t**ranscriptase (RT) with a ribonuclease H activity. RTBV P2 can interact with the CP domain of the P3 polyprotein and likely participates in RTBV capsid assembly. ORF IV, which is most distal from the pgRNA promoter, is expressed by yet another mechanism, that of splicing. Splicing of pgRNA leads to fusion between the first short ORF in the leader sequence at the 5′ end (sORFA) and ORF IV and the generation of a monocistronic mRNA. Protein P4 is a suppressor of virus-directed RNA silencing. pgRNA, which serves as polycistronic mRNA, is also used as the template for reverse transcription-mediated replication of the viral DNA. In contrast to retroviruses, replication does not involve an integration phase.

Rice tungro spherical virus

RTSV, the other partner of the synergistic interaction, is the type species of the genus *Waikavirus* (*Secoviridae*). Its genome consists of a linear ssRNA(+) molecule of slightly more than 12 kb, capped at the 5′ terminus with a VPg protein and a ca. 50 nucleotides long poly(A) tract at the 3′ end (Fig. 6.7B). The genome is translated into a single polyprotein that is subsequently processed by a virus-encoded protease to at least eight functional products. These include a putative leader protein (P1), the capsid proteins (CP1, CP2, and

CP3), a helicase, VPg protein, 3C-like protease and replicase (RdRp). The function of the two smaller proteins of the 3′ end ORFs is not known. Virus replication occurs in the cytoplasm within viral factories made up of ER-derived membrane vesicles. Virus RNAs are packaged in nonenveloped icosahedrons with diameters of 30 nm. RTSV can infect members of the families Poaceae and Cyperaceae. As indicated earlier, apart from the development of mild stunting in some plant species, infected plants do not develop symptoms of infection.

Several sources of genetic resistance have been reported in rice against RTSV, but there are only a few against RTBV. Additionally, none of the host resistance sources is genetically well characterized. **I**ntegrated **p**est **m**anagement (IPM) programs that include vector avoidance, synchronous growth, crop rotation and the use of nonisogenic lines with varying degrees of resistance to the vector, along with the use of pesticides aimed at reducing vector density in fields affected by the disease, have allowed for some level of protection. Transgenic resistance has yielded mixed results against the disease complex. The levels of resistance vary from complete resistance, symptom attenuation to a delay in symptom development. Both immunity and symptom attenuation have also been observed with transgenic plants carrying full length or truncated versions of the RTSV *rep* gene or an inverted repeat of the gene encoding the RTBV's ORF IV. RTBV has apparently evolved a dual counter-defense strategy to evade siRNA-derived resistance. One involves the production of decoy dsRNA that protects other regions of the viral genome from the action of RISC. These RNAs are generated from the noncoding region just downstream of the PolII transcription start site. In the other strategy, RTBV's protein P4 interferes with the cell-to-cell spread of siRNAs, thus rendering RNA silencing against the virus ineffective. At the same time the maintenance of pararetroviral sequences in the plant genome may have potential benefits for the host plants. It was recently posited that endogenous pararetroviruses may play a role in conferring resistance against pararetrovirus infections. Pararetrovirus-like sequences that are similar to RTBV have been found in the genome of some rice accessions. These **e**ndogenous **RTBV**-like sequences (ERTBV) are highly rearranged and dispersed throughout the rice genome, suggesting that illegitimate recombination occurred when putative virus progenitors integrated. So far, no association between virus disease and endogenous virus sequences has been reported, but it seems that there is an association between ERTBV profiles and the degree of RTBV susceptibility. That is, species with low ERTBV copy numbers tend to be vulnerable to RTBV infection. *Oryza* AA-genome species with low copy number of ERTBVs show extreme susceptibility to RTBV.

FURTHER READING

Adams, M.J., Zerbini, F.M., French, R., Ranenstein, F., Stenger, D.C., Valkonen, J.P.T., 2012. Family *Potyviridae*. In: King, A.M.O., Adams, M.J., Carstens, E.B., Lefkowitz, E.J. (Eds.), Virus Taxonomy: Classification and Nomenclature of Viruses: Ninth Report of the International Committee on Taxonomy of Viruses. Elsevier, San Diego, pp. 1069–1089.

Blomme, G., Ploetz, R., Jones, D., De Langhe, E., Price, N., Gold, C., et al., 2013. A historical overview of the appearance and spread of *Musa* pests and pathogens on the African continent: highlighting the importance of clean *Musa* planting materials and quarantine measures. Ann. Appl. Biol. 162, 4–26.

De Bruyn, A., Harimalala, M., Zinga, I., Mabvakure, B.M., Hoareau, M., Ravigné, V., et al., 2016. Divergent evolutionary and epidemiological dynamics of cassava mosaic geminiviruses in Madagascar. BMC Evol. Biol. 16 (1), 182.

Gibbs, A., Ohshima, K., 2010. Potyviruses and the digital revolution. Ann. Rev. Phytopathol. 48, 205–223.

Hasiów-Jaroszewska, B., Fares, M., Elena, S., 2014. Molecular evolution of viral multifunctional proteins: the case of potyvirus HC-Pro. J. Mol. Evol. 78, 75–86.

Hull, R., 2002. Matthews' Plant Virology, fourth ed. Academic Press, Bath, Great Britain, p. 1001.

Kamitani, M., Nagano, A.J., Honjo, M.N., Kudoh, H., 2016. RNA-Seq reveals virus–virus and virus–plant interactions in nature. FEMS Microbiol. Ecol. 92, fiw176. Available from: https://doi.org/10.1093/femsec/fiw176.

Pooggin, M.M., Rajeswaran, R., Schepetilnikov, M.V., Ryabova, L.A., 2012. Short ORF-dependent ribosome shunting operates in an RNA picorna-like virus and a DNA pararetrovirus that cause rice tungro disease. PLoS Pathogen. 8, e1002568. Available from: https://doi.org/10.1371/journal.ppat.1002568.

Scholthof, K.-B.G., Adkins, S., Czosnek, H., Palukaitis, P., Jacquot, E., Hohn, T., et al., 2011. Top 10 plant viruses in molecular plant pathology. Mol. Plant Pathol. 12, 938–954. Available from: https://doi.org/10.1111/J.1364-3703.2011.00752.X.

Tennant, P., Fermin, G. (Eds.), 2015. Virus Diseases of Tropical and Suptropical Crops. CABI Plant Protection Series, CAB International, Wallingford, Oxfordshire, United Kingdom.

Varma, A., Malathi, V.G., 2003. Emerging geminivirus problems: a serious threat to crop production. Ann. Appl. Biol. 142 (2), 145–164.

Chapter 7

Viruses as Pathogens: Animal Viruses, With Emphasis on Human Viruses

Jerome E. Foster[1], José A. Mendoza[2] and Janine Seetahal[3]
[1]The University of the West Indies, St. Augustine, Trinidad, [2]Universidad de Los Andes, Mérida, Venezuela, [3]Eric Williams Medical Sciences Complex, Champs Fleurs, Trinidad

When you're trying to close in or 'kill' a virus, you're really trying to kill host cell machinery

Paul Roepe

Apart from bacterial pathogens, viruses are among the most familiar pathogens that cause disease in humans. They cause a wide range of effects. Some make us sick for a day or two before going away, while others are lifelong, and like Ebola, can cause life-threatening complications. Because of their impact on our health and quality of life, many human viruses (and related animal viruses) have been studied in detail. In this chapter we look at pathogenic viruses of six families that have significant impact on human health.

To cause disease in humans, viruses must go through several steps that are common to each in some form, though the details differ from virus to virus. First, the virus must find a site of entry to invade. After the virus has breached the outer defenses of the body by invading a susceptible cell it then has to begin replication (the copying of its genome). In order to do so efficiently the virus has to overcome the host immune defenses, transcribe its genes and then translate proteins to make new viruses (see Chapter 10, Host–Virus Interactions: Battles Between Viruses and Their Hosts). Once it has done so sufficiently in the susceptible cell it then spreads to other target cells, usually through the blood and/or lymphatic system. On reaching additional target cells it replicates in these, and eventually exits the current host in sufficient numbers to infect other individuals; i.e., be transmitted. Some viruses share common routes of transmission or target cells, but differ in

Viruses. DOI: https://doi.org/10.1016/B978-0-12-811257-1.00007-3

mechanisms of overcoming the immune response or entry into the body; some may be cytopathic while others may not only not damage the cell but integrate themselves into the cell's genome. The end game for all species is the propagation of the viral genome, and in the case of pathogenic viruses, this usually results in humans contracting some form of illness.

RETROVIRIDAE: HUMAN IMMUNODEFICIENCY VIRUS 1

The family *Retroviridae* is a large diverse group of enveloped RNA viruses which include the pathogens, **h**uman **i**mmunodeficiency **v**iruses (HIV), and **h**uman **T**-**l**ymphotropic **v**iruses (HTLV). They bear the name retroviruses because their RNA genome is (retro) transcribed into linear double-stranded DNA during their replication cycle. This reverse transcription is performed by an enzyme characteristic to the family called **r**everse **t**ranscriptase (RT) which is also known as an RNA-dependent DNA polymerase. During the replication cycle, viral dsDNA is usually integrated into the host genome as a DNA provirus which can remain silent (i.e., latent) or become transcriptionally active to produce virions. The family has seven genera: *Alpharetrovirus*, *Betaretrovirus*, *Deltaretrovirus*, *Epsilonretrovirus*, *Gammaretrovirus*, *Lentivirus*, and *Spumavirus*, of which the lentiviruses are the most relevant in terms of human health. For that reason, we will focus on the genus *Lentivirus*.

The HIV belong to the genus *Lentivirus* within the *Retroviridae* family and are the best characterized members of the family. There are two types of HIV: HIV Type 1 (HIV-1) and HIV Type 2 (HIV-2) both of which can cause the disease **A**cquired **I**mmune **D**eficiency **S**yndrome (AIDS) in humans. In AIDS, the virus infects certain target immune cells leading to a progressive deterioration of the body's defenses. This leads to opportunistic infections such as tuberculosis, HIV-related tumors, and complications with other existing illnesses such as hepatitis B and C. Infection usually begins with an influenza-like illness which resolves after a short period. The first signs of HIV-related illness can take anywhere between 5 and 10 years on average to show, and progression to AIDS can occur 10–15 years after initial HIV infection, if left untreated.

HIV/AIDS is spread through certain bodily fluids which include blood, semen, preseminal fluids, vaginal and rectal fluids, and breast milk. The virus spreads when the fluids come into contact with damaged tissue or mucous membranes such as those found in the vagina, rectum, penis, or mouth. As a consequence, the emergence and spread of HIV/AIDS has occurred largely through sexual engagement, intravenous drug use, and blood transfusion in countries which do not screen donors. The disease has become not just a medical issue, but also a socioeconomic one.

Evidence shows that HIV/AIDS has been around since the early 20th century, but it was only recognized when it became a global epidemic in the 1980s. HIV was first isolated in 1983. Because of the high mortality and morbidity of the disease, and the common routes of transmission, a certain degree of social stigma has become attached to infected persons which negatively impacts on their well-being. In 2016 more than 1 million people globally died from AIDS, making it one of the leading causes of mortality. The significant morbidity and mortality caused by the disease has impacted significantly on the labor supply in endemic regions, especially in Africa. Progression to AIDS is highly likely with an HIV-1 infection, but less so with HIV-2. HIV-1 has a wide global distribution, while HIV-2 is localized to West Africa where it is endemic.

Structure and Genomic Organization

The HIV-1 virion is spherical in shape and has a diameter of ~120–130 nm. The virion has an outer envelope layer, an inner core and between them a matrix. Within the core reside two copies of the ssRNA(+) genome along with enzymes needed for replication. The envelope itself is made up of two glycoproteins embedded in two lipid layers. The first, glycoprotein 120 (gp120), protrudes outward from the lipid layer, and is anchored to the virion by the transmembrane glycoprotein 41 (gp41). Attachment to host cells is facilitated by gp120, and gp41 is vital for cell fusion and entry. The matrix, which lies beneath the envelope, is made of protein 17 (p17). The capsid is situated beneath the matrix.

The HIV-1 genome comprises a total of nine **o**pen **r**eading **f**rames (ORFs) which encode 15 proteins. Of these the three major ORFs are: (1) *gag*, which codes for internal structural proteins of **ma**trix (MA, p17), **ca**psid (CA, p24), **n**ucleo**c**apsid (NC, p7), and p6; (2) *pol*, which codes for RT (p51/p66), **in**tegrase (IN/p32), and **pr**otease (PR, p11); and (3) *env*, which codes for the external structural proteins that make up the envelope, i.e., the **su**rface glycoprotein (SU, gp120) and **t**rans**m**embrane glycoprotein (TM, gp41). There are several other regulatory and accessory genes encoded in the viral genome: *tat* and *rev* encode proteins essential for virus replication; *vif* and *nef* are involved in the infectivity of virus particles; *vif*, *vpu*, and *vpr* influence the assembly and replication capacity of newly synthesized HIV particles.

Genetic Diversity

The ability of HIV-1 to undergo recombination paints a complicated picture with respect to its genetic diversity. HIV-1 genetic subtypes are grouped into four categories each with distinct geographic distribution but with common clinical outcomes: M (main/major), O (outlier), N (non-M and non-O), and P. The M group is responsible for the HIV-1 pandemic and has at least 10 subtypes (A, B, C, D, E, F, G, H, and J), with at least 58 **c**irculating **r**ecombinant **f**orms (CRFs). The African continent historically has had all M group subtypes in circulation, with the B subtype being the main causative agent of HIV-1 infection in Europe and the Americas.

Replication and Pathogenesis

HIV-1 attaches to its target cell through the interaction of the envelope protein gp120 with the CD4 receptor found on the surface of immune cells such as T helper cells, monocytes, macrophages, and dendritic cells. This first binding allows subsequent binding interactions between gp120 and other coreceptors such as CCR5 and CXCR4 which facilitate membrane fusion. Fusion between viral and host cell membranes is then triggered by the actions of gp41 to allow release of the HIV-1 genome into the cell. Once in the cytoplasm the viral core disassembles and the viral RNA genome is reverse transcribed into linear double-stranded DNA by the viral RT. The viral DNA and viral integrase then form a **pre**integration **c**omplex (PIC) that translocates to the nucleus through nuclear pores. There the integrase catalyzes the insertion of the viral DNA into the host genome. The initial step of cleaving the host DNA is performed by the integrase, while ligation of the viral DNA is thought to be performed by host cellular enzymes. Once integrated, the provirus is transcribed by the host cell's machinery (i.e., RNA polymerase II) and viral transcripts, with different levels of splicing, are exported to the cytoplasm for translation. To create virions, two copies of unspliced viral RNA and viral proteins assemble, and the virus exits the cell through budding from the cell membrane. Once outside the cell, the action of the viral protease converts immature virions to mature and infectious: the enzyme breaks down select protein chains in noninfectious virions that allow the virions to rearrange themselves into mature particles.

Without treatment HIV infection progresses rapidly through three stages: acute infection, chronic infection, and finally AIDS. The virus can be spread at any stage, but is most infectious during the acute stage. In this acute stage, which lasts 2–4 weeks the virus spreads rapidly, entering and destroying CD4-bearing T lymphocytes as it replicates. In the chronic stage of infection, the virus replicates at a lower rate and the individual may not exhibit HIV-related symptoms (referred to as clinical latency or asymptomatic HIV infection). This may last up to 10 years, but can be as short as five depending on the individual. In the third stage (AIDS), individuals begin to contract opportunistic infections and/or infection-related cancers because of the long-term damage to the immune system.

Surveillance and Detection

The first tests designed to detect HIV were **e**nzyme **i**mmuno**a**ssays (EIAs) that detect antibodies to the virus produced by the immune system, and indirect methods of virus detection. These antibodies would be those released in response to surface viral proteins such as the capsid (p24) or the gp120 receptor. First-generation and second-generation EIAs were designed to detect IgG antibodies against HIV-1. Third-generation EIAs use "*antigen sandwich*" techniques and can also detect IgM antibodies against HIV-1, which develop earlier after infection. Rapid HIV tests are single-tube EIAs designed with all the necessary reagents for nonclinical and point-of-care settings, provide results in less than 30 minutes and are currently the most widely used method of detection. However, because there is the possibility of false-positive results with these rapid tests, further confirmation is usually required by additional assays like Western blots to isolate the actual protein antibodies, while others may use **r**adio**i**mmuno**a**ssays (RIAs). Other confirmatory tests include detecting the presence of actual viral RNA such as with the HIV NAAT (**n**ucleic **a**cid **a**mplification **t**est), or the viral antigen that antibodies react with, such as the p24/capsid antigen test. Recent developments have led to the use of more sensitive Fourth-generation or combination detection kits that detect both antigen (usually p24) and antibody simultaneously.

Treatment

There is no cure for HIV, but the advent of **a**nti**r**e**t**roviral therapy (ART) in the 1990s has led to treatment regimens that successfully prevent the progression of HIV infection to AIDS, and also reduce transmission of the virus. This has led to successful long-term management of the disease in infected individuals, with HIV/AIDS now considered a serious but chronic illness rather than the acute condition first observed in the 1980s, especially in developed countries. However, ART is far from being universally accessible in the poorer developing nations, and in these situations HIV/AIDS can still be considered an acute illness.

ART involves the use of several medications designed to interfere with specific points of the HIV life cycle (Fig. 7.1). There are six classes of HIV drugs: nucleotide and **n**ucleoside **r**everse **t**ranscriptase **i**nhibitors (NRTIs), **n**on**n**ucleoside **r**everse **t**ranscriptase **i**nhibitors (NNRTIs), **p**rotease **i**nhibitors (PIs), **in**tegrase strand **t**ransfer **i**nhibitors (INSTIs), fusion inhibitors, and entry inhibitors. NNRTIs and NRTIs prevent reverse transcription of viral RNA to DNA by inhibiting the viral RT. INSTIs block the integration of the synthesized viral DNA into that of the cell. PIs block the action of the viral protease, thus preventing production of mature virions, and are usually used in combination with pharmacokinetic enhancers. Fusion and entry inhibitors

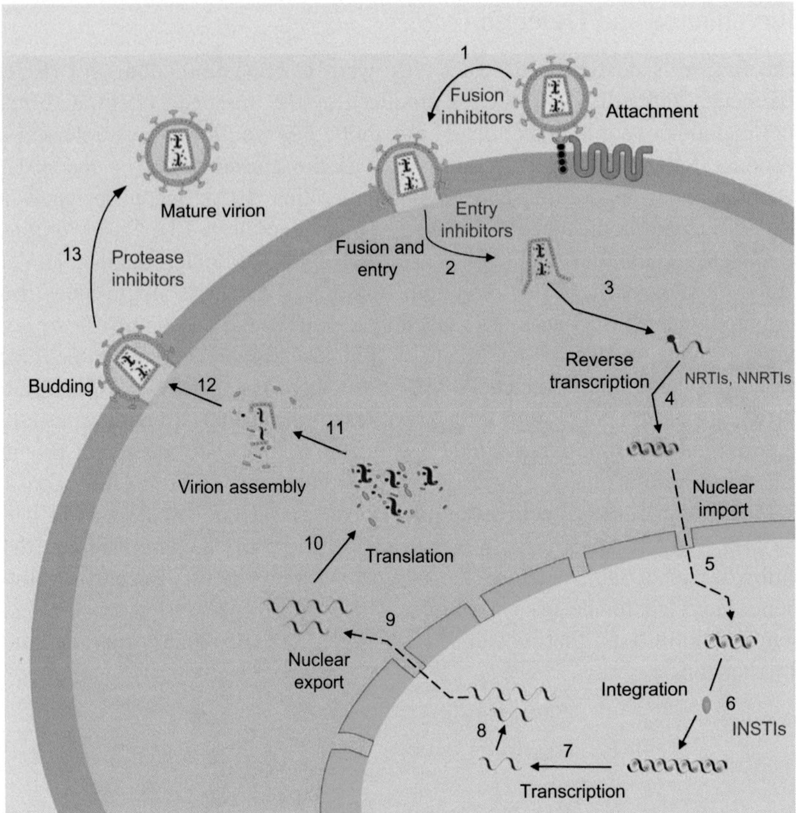

FIGURE 7.1 HIV replication cycle and points of actions of antiretroviral drugs. (1) Virion attaches to CD4+ cell receptor through gp120, aided by CCR5 or CXCR4 coreceptors (2) virion fuses with cell membrane and nucleocapsid is released into cytoplasm (3–4) viral RNA is released and undergoes reverse transcription (5) viral dsDNA is imported into nucleus and (6) viral integrase creates provirus (7–8) transcription of viral genes produces RNA transcript and (8) viral RNA genome copies (9) transcripts are exported from nucleus and (10) viral proteins are translated (11) viral proteins and ssRNA genomes assembly and (12) virion buds from the host cell; (13) finally the action of viral protease creates mature virions.

prevent the gp120 and gp41 viral proteins from binding to their target receptors, thus preventing the virus from entering the cell. Recommended treatment regimens consist of a combination of at least three of these drugs from two different classes, as the antiviral impact on the suppression of viral replication seems to be synergistic or additive. In some instances, the drugs are used as a prophylactic measure. Despite advances in HIV-1 treatment in recent years, the development of drug resistance is still an issue that leads to cases of treatment failure—a direct effect of the high evolutionary rate of HIV-1 including its genetic variation due to recombination.

FLAVIVIRIDAE: DENGUE VIRUS AND OTHER MEMBERS

The *Flaviviridae* family consists of four genera: *Flavivirus*, *Hepacivirus* (*Hepacivirus C*), *Pegivirus*, and *Pestivirus* (*Classical swine fever virus* and bovine diarrhea viruses). These four genera exhibit diverse biological properties and show no serum cross-reactivity, but are similar in terms of morphology, RNA organization and replication strategies. The genus *Flavivirus* consists of 53 viruses, most of which are considered arboviruses (arthropod-borne viruses), with some transmitted by mosquitoes (such as *Yellow fever virus*, the prototype virus of this family, along with *Zika virus*, *Dengue virus*, and *West Nile virus*); others by ticks (e.g., *Tick-borne encephalitis virus*), and still others are considered zoonotic (transmitted between rodents or bats in which the vector arthropod is unknown).

Structure and Genomic Organization

Flaviviruses are spherical, ssRNA(+) viruses of ~40–60 nm in diameter. The flaviviral genome is covered by a capsid which in turn is surrounded by a lipid bilayer that has envelope glycoproteins on its surface. The genome has a 10 Kb ORF located between two noncoding regions (the 5′-UTR and the 3′-UTR), both of which are of great importance in RNA replication. The RNA translation product is cleaved posttranslationally at specific sites by both cellular and viral proteases (Fig. 7.2). This RNA encodes three

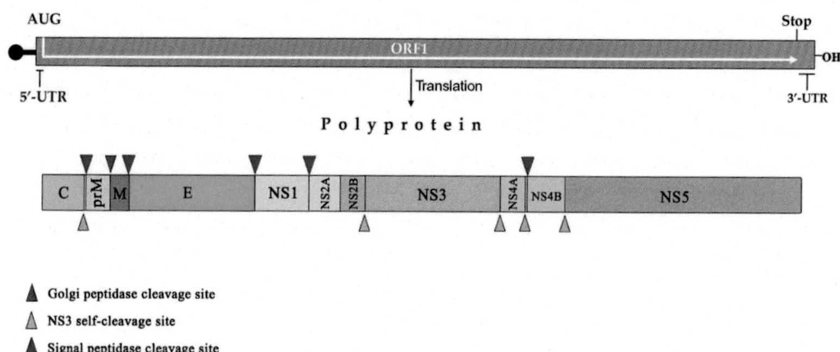

FIGURE 7.2 Genome organization of *Dengue virus* (DENV). DENV ssRNA(+) acts as both the viral mRNA (for translation) and the viral genome. The ssRNA codes for three structural proteins (C capsid, prM/M premembrane/membrane associated protein, E envelope protein) and seven nonstructural proteins responsible for viral replication and cellular processes (NS1, NS2a, NS2b, NS3, NS4a, NS4b, NS5) and has untranslated regions at either end (5′ and 3′ UTRs). The genome consists of a single ORF which is translated into a polyprotein. This is then modified posttranslationally by various cellular peptidases and the virus protease NS3 (self-cleavage) resulting in the various viral proteins necessary for replication.

structural proteins (capsid, premembrane/membrane, and envelope) and seven nonstructural proteins (NS1, NS2a, NS2b, NS3, NS4a, NS4b, and NS5).

Dengue virus

Dengue is an acute febrile disease caused by the **d**engue **v**irus (DENV) which is transmitted by mosquitoes (*Aedes aegypti* or *Aedes albopictus*). The disease is caused by four serotypes of the virus: DEN-1, DEN-2, DEN-3, or DEN-4. DENV infection can range from asymptomatic to bleeding and shock (termed as severe dengue) which may lead to death. In up to 80% of cases, the disease can be mild or even asymptomatic. The clinical picture is characterized by fever of up to 7 days, of nonapparent origin, associated with the presence of headache, retroocular pain, myalgia, arthralgia, exanthema, and prostration, with or without hemorrhage. Severe dengue is due to plasma leakage, leading to shock or fluid accumulation (edema) with difficulty breathing, with or without severe bleeding and severe involvement of the liver, heart or **c**entral **n**ervous **s**ystem (CNS). It has been estimated that 50–100 million people are infected annually in tropical and subtropical regions, where more than 2.5 billion people are at risk, accounting for one-third of the world's population. Of these annual infections, up to 2 million develop severe manifestations, with an estimated mortality rate of ~25,000 per year for the disease.

Genetic Diversity

There is significant genetic diversity within each of the 4 DENV serotypes with several intraserotypic subtypes, some of which have shown associations with disease severity. The gold standard for identifying subtypes is DNA sequencing (partial or full genome), whereby isolate sequences that diverge by more than 6% belong to different subtypes. As with other viruses, phylogenetic analyses are regularly used to not only ascertain subtypes, but genetic lineages within those subtypes. Such DENV genotypes and lineages have been shown to undergo different types of selective pressures in both endemic and epidemic cycles.

Pathogenesis

Dengue is a systemic inflammatory disease in which the vascular endothelium is affected. The formation of antiviral antibodies devoid of a protective role induced by the previous invasion of a heterologous serotype of the DENV leads to enhanced infection. Neutralizing antibodies in adequate numbers serve to neutralize DENV infection, but there is evidence that when the concentration of these antibodies falls below this threshold they can instead promote entry of DENV into cells which express the Fcγ receptor through the process called antibody-dependent enhancement. It is thought that in the case of secondary infections the presence of heterologous but nonneutralizing antibodies derived from primary infections enhance secondary infections, in this manner leading to the possibility of increased disease severity.

Cell damage by direct action of the virus induces apoptosis and potential necrosis that can affect cells such as hepatocytes, endothelial cells and neurons. Antibodies react against DENV nonstructural proteins resulting in a cross-reaction against coagulation proteins such as fibrinogen, as well as complement-activated release of anaphylactoxins which increase vascular permeability. Endothelial cell lysis by cytotoxic T cells and the action of cytokines released by monocytes and T lymphocytes also cause alteration in vascular permeability leading to fluid leakage, generating hypovolemia, shock state, and noncardiogenic pulmonary edema in severe cases. Polyclonal proliferation of B cells induced by the secondary infection also leads to IgM production. These antibodies cause cell lysis through complement activation. There is also the inhibition of platelet aggregation caused by cross-reaction between viral antigens and platelet molecules, which contributes to hemorrhaging. In summary, there is likely to be an abnormal and exaggerated reaction of immunity, resulting in an altered antibody response to elevated levels of viremia and circulating antigens and thereby exacerbation of cytokine production, T lymphocyte activation and alteration in the elimination of apoptotic bodies.

Diagnosis

Diagnosis in endemic areas is usually established from clinical and epidemiological data. Confirmation of the diagnosis requires demonstration of the presence of the virus indirectly or directly. The definitive method of diagnosis is performed by viral isolation, which is carried out with a blood sample during the first 72 hours after the onset of symptoms. The virus can be isolated by inoculating it into mosquitoes and/or cultures of mosquito or vertebrate cells, followed by identification by immunofluorescence or PCR. Standard or real time PCR may be employed for direct detection from blood of virus RNA or by detection of viral NS1 by immunoassays. There are also indirect methods such as detection of dengue IgM in blood after the fifth to ninth day of disease or dengue IgG (demonstration of seroconversion by paired samples separated by 15–20 days).

Treatment and Prevention

There is no specific antiviral treatment for dengue. Treatment is symptomatic and preventive of possible complications, and patients should be classified into groups according to their risk of complications. At the moment, a certified dengue vaccine is not available, due to the complications associated with the enhancing nature of DENV antibodies. Tetravalent vaccines designed to confer an immune response and subsequent protection against all four serotypes simultaneously have been trialed, but with less than satisfactory results due to the nature of the pathogen and antibody-dependent enhancement events. Preventive measures include avoiding mosquito bites and eliminating larval breeding sites, as well as communicating risks through mass media and promoting preventive behavior in the population.

Yellow fever virus

Yellow fever is a severe illness caused by the homonymous *Flavivirus* (*Yellow fever virus*, YFV). The disease only occurs in Africa, South America, Central America, and the Caribbean; i.e., in areas with a tropical climate. It is transmitted by the bite of the mosquito *Aedes aegypti* and other mosquitoes of the genera *Aedes*, *Haemagogus*, and *Sabethes*, which abound in humid and tropical areas, around stagnant waters. Transmission of this virus occurs primarily in forest areas (enzootic cycle) between wild animals (reservoirs). The virus can pass accidentally to man (epizootic cycle) and cause epidemics by transmission between humans with the participation of *A. aegypti* as the main transmitting vector (urban or epidemic cycle).

Genetic Diversity

The genetic diversity of YFV has been under-investigated mostly due to the availability of a safe and effective vaccine for over seven decades. Despite this, the virus continues to emerge periodically in endemic regions. Using analyses of select genomic regions, including the prM/M, E, and 3′-noncoding regions, up to seven genotypes have been proposed: West African genotypes I and II, East African genotype, East/central African genotype, Angolan genotype, and South American genotypes I and II.

Clinical Features

Following the bite by an infected mosquito, symptoms usually develop after an incubation period of 3–6 days. Yellow fever can vary in degrees of severity. The mild form is not very characteristic and is only suspected in endemic areas and especially during epidemics. It starts abruptly with high fever, chills, and headache. There may also be muscle and joint pain, hot flashes, loss of appetite, vomiting, and jaundice. It usually lasts 1–3 days and resolves without complications. Whereas the severe or classic form of the disease is characterized by an initial period similar to that previously described, and there may also be nosebleeds and bleeding from the gums. Most people recover at this stage, but others can progress to the third stage within 24 hours which is the most dangerous (intoxication stage). Fever appears again, jaundice (100% of cases) along with multi-organ dysfunction.

This may include hepatic, renal and cardiac insufficiency, bleeding disorders, and brain dysfunction including delirium, seizures, coma, shock, and death.

Diagnosis

Diagnosis in endemic areas is usually established from clinical data. Confirmation of diagnosis requires demonstration of the presence of the virus indirectly or directly. The presence of IgM and IgG antibodies may be detected by EIAs or by immunofluorescence (most commonly used tests), while viral isolation is useful in cases of suspected Yellow Fever in viraemic patients (early consultation) or PCR using blood samples.

Treatment and Prevention

There is no specific treatment for yellow fever. In severe cases, symptomatic and supportive treatment is indicated, including the use of blood products for severe bleeding, dialysis for renal insufficiency and administration of intravenous fluids. The best method of control is vaccination in endemic areas (and travelers to these regions). The vaccine is an attenuated virus (Yellow Fever 17D) that confers lifelong immunity in 99% of those inoculated. Control measures based on the isolation of ill individuals are also effective, and as far as possible, those individuals who have been exposed to the vectors. Insect control, along with the use of protective clothing, repellents, and nets to reduce contact with the vectors are also recommended.

Zika virus

The *Zika virus* (ZIKV) is another flavivirus that is transmitted by the bite of the mosquito *A. aegypti*, the same vector that transmits dengue and YFVs, and is responsible for an acute febrile disease associated with a skin rash. The virus circulates predominantly in wild primates and arboreal mosquitoes such as *Aedes africanus*. Nevertheless, the main vectors identified nowadays are *A. aegypti*, *A. hensilli*, *A. albopictus*, and *A. polynesiensis*. ZIKV is transmitted mainly by the bite of the female mosquito. The global expansion of *A. aegypti* and the increase in international trade and travel have favored the spread of pandemic ZIKV.

This virus was accidentally discovered in 1947 in a sentinel primate placed in the Zika Forest in Uganda; the first human cases were described 6 years later. Seroprevalence studies have shown virus circulation in Africa and Asia for decades; however, only 13 cases were recorded until their appearance in Micronesia with the Yap epidemic in 2007. Since then, ZIKV has continued its eastward expansion in the Pacific in 2013, and in the Americas in 2015. The cumulative number of people infected was about 2 million by March 2016. Its rapid global expansion and association of ZIKV infection with neurological complications such as microcephaly and

Guillain-Barré syndrome (GBS) forced the World Health Organization to declare a "*status of international public health emergency*" in 2016.

Genetic Diversity

Study of ZIKV sequences highlights two lineages of African and Asian origin, the latter being responsible for the current pandemic observed in Micronesia, French Polynesia, and the Americas. Recently, the Asian lineage has also been imported to China and with the ZIKV showing signs of significant genetic evolution over the past half century. A notable feature is the genetic stability between strains with only 12% of nucleotide variations since the onset of the pandemic and less than 1% since its introduction into the Americas.

Clinical Features and Complications

The infection is asymptomatic in the majority of cases. When symptoms occur, they usually occur moderately or acutely, after an incubation period of 3–12 days, and include fever, conjunctivitis, headache, malaise, arthralgia (mainly hands and feet), rash, and inflammation of lower limbs. These symptoms last between 4 and 7 days. Due to its similarity to dengue and chikungunya fever, it can be easily confused with these diseases.

During the French Polynesia epidemic in 2013, the first neurological complications were described, in particular GBS. GBS is a motor and speech polyneuropathy, with ascending sensory neurological deficits. Cases of meningoencephalitis have also been reported. The risk of ZIKV infection with GBS appears moderate, considering that less than 50 cases with GBS were reported in 28,000 cases of zika. In 2015, however, Brazil announced an alert on the abnormal increase in cases of microcephaly with the concomitant zika epidemic. Microcephaly is the consequence of brain damage and is associated with irreversible neurological abnormalities. The risk is higher in the first trimester of pregnancy, although it persists to the end. A recent study carried out in Brazil using data from the outbreak in Bahia in 2015, considered the risk of zika and microcephaly between 0.88% and 13.2% in infected pregnant mothers. Maternal-fetal transmission has been demonstrated formally several times and in different tissues in the fetus, especially in the CNS.

Treatment and Prevention

There is no vaccine or specific treatment for zika. Therefore the treatment is essentially symptomatic. The ingestion of plenty of fluids is recommended, and in cases with disabling pain, the use of analgesics. Treatment of GBS involves intravenous immunoglobulin therapy and the use of respiratory intensive care in cases of respiratory muscle paralysis. Like dengue and chikungunya, preventive measures focus on the elimination and control of

A. aegypti mosquito breeding sites. Although the primary transmission route of the virus is via the *Aedes* mosquito, sexual transmission is possible. The use of barriers (i.e., male and female condoms) is recommended for persons returning from areas where transmission of ZIKV is known to occur and for those living in an area with active ZIKV transmission.

Hepacivirus C

Hepacivirus C (HCV) is the major causative agent of non-A, non-B, parenterally transmitted hepatitis. It is the leading cause of chronic liver disease worldwide affecting 130–150 million people worldwide, with 3–4 million new infections annually and more than 700,000 people estimated to die from hepatitis C-related liver diseases in 2013 alone.

HCV is transmitted mainly through parenteral exposure to blood of chronic carriers and evidence has shown certain groups to be at high risk including persons transfused with blood and blood products prior to 1992 in developed countries, hemodialysis patients, hemophiliacs, healthcare workers, infants born to infected mothers, **i**njection **d**rug **u**sers (IDUs), persons involved in acupuncture and/or tattooing with unsterile tools, and high-risk sexual behavior. HCV/HIV coinfection is also an increasing problem in countries with concentrated HIV epidemics and IDUs, as progression to chronic hepatitis is accelerated in these patients. Groups at risk for HCV infection are also those at risk for tuberculosis as that disease is endemic in countries that do not screen blood products routinely for blood-borne viruses.

Genetic Diversity

The virus genome exhibits significant genetic heterogeneity and at least seven genotypes (1–7) exist, each of which can be further subdivided into more than 80 different subtypes. The "*epidemic*" subtypes are found globally at high prevalence and include subtypes 1a, 1b, 2a, 2b, 2c, and 3a. "*Local epidemic*" subtypes occur at high prevalence but in localized areas; e.g., subtype 4a is found almost exclusively in the Middle East. The "*endemic*" subtypes are less prevalent and are usually found with restricted geographic distributions; e.g., genotype 6 subtypes are found almost exclusively South East Asia.

Clinical Features and Pathogenesis

Acute HCV infection is often asymptomatic and roughly 85% of all cases progress to chronic hepatitis. Untreated chronic hepatitis can lead to the risk of developing serious sequelae such as cirrhosis (15%–30% over 20 years) and hepatocellular carcinoma (2%–4% per year).

In an HCV infection, the first response by the hepatocyte involves various **p**attern **r**ecognition **r**eceptors (PRRs)—proteins of the innate immune system that recognize virus. These then activate pathways leading to interferon (IFN) production. IFN in turn triggers the JAK-STAT pathway which leads to the activation of several hundred genes which serve to block viral replication at different phases of the viral replication cycle. One main PRR is toll-like receptor 3 (TLR3) which recognizes double-stranded viral RNA and induces the production of inflammatory cytokines. Simultaneously another protein, retinoic acid-inducible gene I-like receptor (RIG-I) recognizes the poly-U sequence of the HCV 3′-UTR, inducing production of IFNs. Genetic variations in the locus for IFNλ have been associated with successful IFN-based treatment and spontaneous HCV clearance, but the mechanisms remain poorly understood.

The NS2 protease activity counters the innate immunity through its inhibition of various cellular kinases involved in cytokine expression; NS3/NS4A and NS4B have also been shown to block host cell IFN production through the cleavage of adaptor proteins in IFN pathways. All these contribute to the ability of HCV to mount chronic infections in individuals.

Treatment and Prevention

There is no vaccine for Hepatitis C and therefore prevention is based on reducing the risk of transmission. In some cases the immune system may clear the virus, and in others chronic infection does not result in serious symptoms. When treatment is necessary, the most widely used regimen includes pegylated IFN- alpha (PEG interferon -alpha) with ribavirin (nucleoside inhibitor).

Knowledge of the infecting subtype is crucial because, despite showing no correlation with disease severity, subtype identity seems to impact on predicting patient response to, and length of, medical treatment. The effectiveness of this treatment has also been limited due to expense, and prolonged (24–48 weeks) and noted side effects. Both IFN and ribavirin can cause hematologic toxicity. Ribavirin is also teratogenic, while IFNs can lead to mental and neurological disorders such as depression and anorexia.

More recent treatments have included the use of direct-acting viral agents such as inhibitors of the viral NS3/4a protease in combination with PEG-IFN and ribavirin, but these have also proved limited due to issues such as IFN resistance, drug toxicity, and emergence of PI resistance. Recent WHO guidelines recommend varying dosages and combinations of IFN and ribavirin, depending on the stage and severity of the chronic HCV infection being treated.

TOGAVIRIDAE: CHIKUNGUNYA VIRUS

Chikungunya is a febrile disease that is transmitted to humans by arthropod (mosquito) vectors, mostly *Aedes aegypti* and *Aedes albopictus*. The causal agent of this disease is *Chikungunya virus* (CHIKV), an arbovirus of the genus *Alphavirus* from the *Togaviridae* family, which also includes the monospecific (*Rubella virus*) genus *Rubivirus*. The *Alphavirus* genus contains 31 members that are separated into New World and Old World viruses. The New World viruses are, among others, *Eastern equine encephalitis virus*, *Venezuelan equine encephalitis virus*, and *Western equine encephalitis virus*. Old World alphaviruses evolved independently and members include *Semliki Forest virus* (SFV), *Sindbis virus*, *O'nyong-nyong virus* (ONNV), and CHIKV. Human infections by these viruses are often associated with fever, rash, and arthralgia. CHIKV was originally isolated in 1952 from a febrile patient in southern Tanzania. In fact, "*Chikungunya*" is a word from the makonde language meaning "*to bend,*" alluding to the aspect that patients adopt because of joint pains generated during the symptomatic course of infection.

Characteristics and Diversity

CHIKV belongs to the Semliki Forest Complex of Alphaviruses, which includes ONNV, *Ross River virus* (RRV), and the SFV itself. All Alphaviruses are spherical particles with a diameter of ~70 nm. The CHIKV genome is packaged by the C protein to form the nucleocapsid which is surrounded by a host-cell derived lipid bilayer with two inserted glycoproteins, E1 and E2. The viral genome resembles eukaryotic mRNAs in that it possesses 5′ cap structures and a 3′ poly(A) tail (Fig. 7.3). It also comprises both 5′ and 3′ proximal nontranslatable regions (UTR). Downstream from the 5′-UTR, four different nonstructural genes (nsP1–nsP4) and the noncoding junction region are found in the first ORF, which is initially translated as a nonstructural polyprotein precursor, P1234. This nonstructural precursor is later cleaved by host proteases and the self-contained viral protease nsP2 to form mature nsP1–nsP4 proteins that orchestrate replication. A second ORF further downstream is transcribed to give a subgenomic positive-sense mRNA (26S RNA) via a negative-strand RNA intermediate. The viral structural proteins (C, Capsid; E2, E3, E1, and 6K proteins) are translated from this subgenomic RNA as a structural polyprotein precursor that is posttranslationally cleaved into the individual proteins. Viral RNA replicates within phagocytic mononuclear blood cells as well as endothelial cells, where infection is associated with dysfunction of these cells.

CHIKV exists as three distinct genotypes based on origin: East/Central/South African, West African, and Asian genotypes.

Clinical Aspects

Chikungunya fever is characterized by sudden onset (2–8 days following infection) of fever between 39 and 40°C, rash, headache, and nausea, usually accompanied by joint pains or poly-arthralgia (frequently very painful).

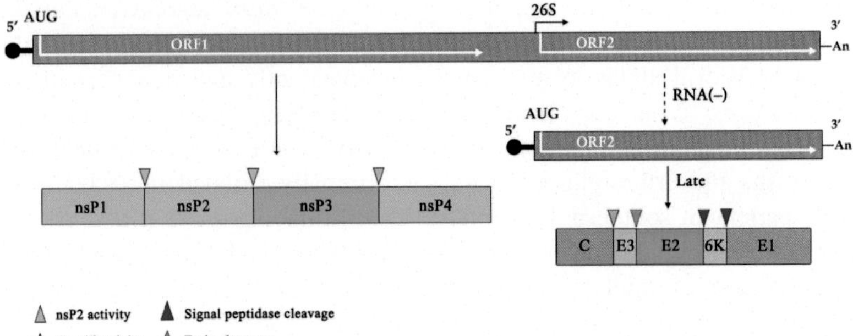

FIGURE 7.3 Genome organization of *Chikungunya virus* (CHIKV). The CHIKV ssRNA(+) resembles that of eukaryotic mRNA with a 5′-cap and 3′-poly(A) tail with proximal 5′- and 3′-UTRs. Downstream of the 5′-end is the first ORF, which encodes the nonstructural proteins, nsP1−nsP4. It is initially translated as a polyprotein precursor, which is later cleaved by activity of the self-contained nsP2 to produce mature nonstructural proteins. The second ORF is transcribed by the virus polymerase complex (nsP2 and nsP3) to a subgenomic mRNA (26S) via a negative-strand RNA intermediate, and encodes the structural proteins (C, E3, E2, 6K, and E1) as a polyprotein precursor. This structural polyprotein is later cleaved, first by the viral capsid (self-cleavage), and then by host furin and signal peptidases, to produce mature structural proteins C (capsid), E1 and E2 proteins (transmembrane glycoproteins), and the accessory protein, 6K.

Arthralgia can become disabling, but usually goes away in a few days. These symptoms are often very similar to those caused by DENV, and concomitant infections may occur in some cases. Most patients recover completely, but in some individuals arthralgia can last for several months or even years. Occasional cases with cardiac and neurological complications have been described. Severe complications of chikungunya fever are very seldom, but in some patients can lead to death.

Transmission and Epidemiology

The factors determining the risk for chikungunya epidemics are poorly understood. Chikungunya fever has been detected in many countries in Africa, America, Asia, and Europe. The arthropod vectors usually sting during the daytime, although their activity reaches its maximum expression at the beginning of the morning and at the end of the afternoon. Both species bite outdoors, although *A. aegypti* can also do so indoors. Epidemic determinants include reservoir hosts, mosquitoes' density, and climatic conditions for transmission. In Africa, human infections have been relatively rare for several years, but there was a major outbreak in the Democratic Republic of the Congo in 1999–2000. There was another outbreak in Gabon in 2007. A major outbreak in the Indian Ocean islands began in 2005, and many cases were imported to Europe, particularly in 2006, when the epidemic was at its peak in La Reunion Island. In 2013, France confirmed two autochthonous cases in the island of St. Martin. Since then, local transmission has been confirmed in other Caribbean islands, becoming the first documented outbreak of chikungunya fever in the Americas. This outbreak soon became an epidemic, spreading across 45 countries in North, Central, and South America, and causing more than 2.9 million suspected and confirmed cases in the region during the next 3 years.

Diagnosis and Treatment

Serological tests, such as EIAs, can confirm the presence of IgM and IgG antibodies against *Chikungunya virus*. The highest concentrations of IgM are 3–5 weeks after the onset of the disease, and may persist up to 3 months. Samples collected during the first week after the onset of symptoms should be analyzed using serological and molecular methods (real time PCR [RT-PCR]). The virus can be isolated from a blood sample in the first days of infection.

There is no specific antiviral treatment for chikungunya fever. Treatment consists mainly in relieving symptoms, including joint pain, with optimal analgesics, antipyretics, and fluid replacement. There is no commercial available vaccine against this virus.

FILOVIRIDAE: EBOLA AND MARBURG VIRUSES

The family *Filoviridae* is a family of enveloped, ssRNA(−) viruses comprising three different genera: monospecific *Cuevavirus*, *Ebolavirus* (with five recognized Ebola viruses, or EBOV), and monospecific *Marburgvirus* (MARV). Filoviruses get their name from their filamentous or thread-like virions which exhibit extreme pleomorphism: virions may appear U-shaped, 6-shaped, or circular. Their lengths vary (with some measuring up to 14,000 nm), but they generally have a uniform diameter of ~80 nm. Both MARV and EBOV cause severe hemorrhagic fever in humans and nonhuman primates, and are thought to be zoonotic in origin, though the natural reservoirs of either virus have not been discovered to date.

Marburg Marburgvirus, the first filovirus to be isolated, was discovered during an outbreak of severe hemorrhagic fever among laboratory workers in Germany and Yugoslavia who had handled tissues from African green monkeys. There was a total of 31 cases with a death toll of 7, and the virus isolated was named after one of the German sites—Marburg. The virus was not observed again until 1975, when a traveler determined to have been exposed in Zimbabwe became sick in Johannesburg, South Africa. Since then, most instances of MARV have been sporadic cases, except for two notably large epidemics in the Democratic Republic of the Congo in 1999 and Angola in 2005.

EBOVs were first discovered in 1976 when two separate unrelated outbreaks of **E**bola **h**emorrhagic **f**ever (EHF) occurred in Zaire (*Zaire ebolavirus*, ZEV) and Sudan (*Sudan ebolavirus*, SUDV). These outbreaks proved to be highly lethal with ~50% of the cases in Sudan and 88% of those in Zaire being fatal. Since its discovery there have been small to medium outbreaks of the EBOV sporadically in Africa in countries such as Gabon, Uganda, and the Democratic Republic of the Congo (DRC) over the past few decades with the largest occurring in the DRC in 1995 and 2012, and in Uganda in 2000 and 2008. However, this changed dramatically with the largest documented outbreak of EHF to date between 2014–2015 in western Africa in which there were 28,000 cases with over 11,000 deaths across four countries starting in Guinea, then spreading to Sierra Leone, Liberia, and Mali. This was found to be caused by a new strain of Zaire ebolavirus—EBOV Makona—which was also imported into Nigeria, Senegal, Spain, the United Kingdom, and the United States.

Structure and Genomic Organization

The typical filoviral particle has a lipid envelope covered with glycoprotein spikes of about 10 nm in length. Under this lipid envelope is the helical nucleocapsid. The genome is ~19 kb in length and consists of seven structural protein genes flanked by 5′ and 3′ untranslated regions arranged in the following manner: 3′-NP-VP35-VP40-GP-VP30-VP24-L-5′. Viral protein NP is the major component of the viral nucleocapsid, and along with VP35, VP30, and L form the helical nucleocapsid structure; L is also the RNA-dependent RNA polymerase (RdRp) that performs virus transcription and replication; GP is the surface viral glycoprotein that mediates entry into host cells via receptor-binding and fusion; VP24 and VP40 are membrane-associated matrix proteins. A unique feature of filoviruses is the presence of a conserved pentamer sequence (3′-UAAUU-5′) located at the 5′ and 3′ ends of all start and stop signals for transcription, respectively. MARV and EBOV differ in the sizes of their genomes (19.1 kb vs 18.9 kb, respectively), their serology and also in the organization of their genes, which tend to have overlaps in ORFs.

Genetic Diversity

The *Margurburgvirus* genus contains only a single species, *Lake Victoria marburgvirus*, which has two strains, **Mar**burg **v**irus (MARV) and **Rav**n **v**irus (RAVV). *Ebolavirus* genus comprises five known species: **Bundib**ugyo ebola**v**irus (BDBV), *Reston ebolavirus* (RESTV), SUDV, *Taï Forest ebolavirus*

(TAFV), and *Zaire ebolavirus* (here Ebola virus, EBOV)—all named after their site of discovery. Of these species, only RESTV lacks pathogenicity in humans, though it causes disease in nonhuman primates and has been shown to be asymptomatic in pigs. EBOV has proven to be the most lethal species in humans. To date there has only been one human infection of TAFV: an individual infected while performing an autopsy on an infected chimpanzee from the Taï Forest, Ivory Coast. The individual recovered fully after treatment.

Replication Cycle

Filoviruses enter cells through the interaction of GP, the sole viral glycoprotein in the envelope, which has two distinct regions—GP1 and GP2. GP1 acts in the manner of a Class I membrane fusion protein and has a **r**eceptor **b**inding **r**egion (RBR). This RBR has a high amino acid identity between MARV and EBOV (~47%) and attaches to a cellular receptor thought to be shared by the two filovirus genera.

Virus enters the cell by endocytosis where cysteine proteases cleave the GP1 region of the viral glycoprotein. This allows GP1 to bind to an internal endosomal protein receptor called **N**iemann-**P**ick**C1** (NPC1). Once this occurs, fusion of the viral membrane with that of the endosome is facilitated by the GP2 region and the viral genome is released into the cytoplasm. Transcription of the viral genome into mRNA is initiated by the binding of viral nucleocapsid protein VP30 and starts at the 3′ end.

The viral protein L acts as the RdRp and, in tandem with its cofactor VP35 and the DNA topoisomerase of the host cell, orchestrates replication of the genome. Translation of the viral genome leads to the accumulation of viral proteins, and specifically VP35 and NP initiate the production of antigenomes. These antigenomes are in turn used as template for producing additional viral genomes. Viral proteins then accumulate at the cell membrane, where the viral glycoprotein G spikes become inserted into the host cell membrane. Expression of the NP protein leads to inclusion bodies in the host cell which in addition to NP also encapsulate proteins VP24, VP30, VP35, and L, and associate with the G proteins embedded in the cell membrane. This then leads to budding of the characteristic filamentous virions from the cell, mainly under the influence of VP40.

Pathogenesis and Clinical Manifestations

MARV and EBOV are pantropic viruses, affecting several tissues and organs. Upon infection, incubation of the virus can take from 3 to 16 days. The onset of the hemorrhagic fever is sudden and is accompanied by fever, chills, headache, myalgia, and anorexia. Other symptoms that may follow include abdominal pain, nausea, vomiting, sore throat, diarrhea, pharyngitis, and conjunctivitis. A transient nonitching maculopapular rash may appear

5–7 days after infection. It is around this time that serious hemorrhagic manifestations such as bleeding into the gastrointestinal tract and in mucous membranes in the nose, lungs, and gums occur. Hemorrhage seen in these diseases is linked to widespread damage to endothelial cells that leads to increased vascular permeability. This bleeding along with hypotensive shock from blood volume loss accounts for the majority of EHF fatalities.

Despite the serious threat filoviruses pose to human and animal health, virulence factors and pathogenesis are still not completely understood, though the picture is becoming clearer. The entry of EBOV (and MAV) into cells can be directly tied to some of the symptoms exhibited in infections. The viral glycoprotein GP allows for entry into monocytes and macrophages and subsequent damage to these cells results in the production of cytokines associated with fever and inflammation. It also allows entry of the virus into endothelial cells, and damage to these directly results in a loss of vascular integrity as seen in hemorrhaging. Other viral proteins act as virulence factors at different points of infection. VP35 sequesters viral RNA, helping the foreign genome evade detection by innate immunity. The protein also competitively inhibits the activation of host IFN regulation factors preventing IFN-β production. VP24, VP 30, and VP40 have also been shown to be suppressors of protective host RNAi pathways, further decreasing the innate immune response. EBOV also evades the cell's immune system, one that discriminates between self-RNA (capped) and nonself-RNA (uncapped), by capping and polyadenylating its mRNAs—another process orchestrated by VP35.

Diagnosis and Detection

Methods used to detect EBOV and MARV antibodies include indirect immunofluorescent assays and ELISA kits which have been designed to detect antiviral antibodies of the filoviral glycoproteins or nucleoproteins such as NP, VP40, and GP. It should be noted that in cases where MARV and EBOV become fatal, patients usually die before a sufficient antibody response occurs, and so such methods are only useful in surviving patients. Detection methods based on RT-PCR, usually designed to conserved regions such as in the NP or L genes, have proven sensitive and accurate. Due to the high pathogenicity of filoviruses detection is limited to laboratories that are designed for biosafety level IV.

Treatment and Prevention

There are no approved virus-specific chemical treatments or vaccines for EBOV or MARV, though experimental EBOV vaccines have been in development since the 2014–15 epidemic. Much of the treatment is supportive with the aim to maintain blood volume and electrolyte balance.

ORTHOMYXOVIRIDAE: INFLUENZA VIRUSES

Orthomyxoviridae is a family of segmented ssRNA(−) viruses with five genera: *Influenzavirus A, Influenzavirus B, Influenzavirus C, Isavirus*, and *Thogotovirus.* The family name refers to the fact that these viruses can attach to mucous proteins on cell surfaces ("*myxo*" is Greek for mucous) and are more "*orthodox*" in comparison to the *Paramyxoviridae*—another group of RNA viruses, which includes species responsible for measles and mumps, that also attach to mucus-producing cells. The most significant human disease caused by orthomyxoviruses is influenza (more commonly called flu). These viruses infect respiratory cells of a wide range of animals (both wild and domesticated) and humans. Influenza is highly infectious and spreads by virus-containing respiratory droplets. Disease severity can range from mild cases to death, and compared to other respiratory illnesses such as the common cold, is more severe. Complications of flu include pneumonia, ear and sinus infections, and worsening of chronic conditions such as asthma, diabetes, and heart disease. People most at risk include young children, pregnant women, adults older than 65 years of age and those with chronic medical conditions.

Virus species in the *Thogotovirus* genus are unique from other orthomyxoviruses in that they are transmitted by arthropods (usually ticks). *Thogoto virus*, the type species for the genus, can infect humans and causes disease in livestock and other mammals. Another species, *Dhori virus*, is also capable of infecting humans. Nonetheless, they are not generally important human pathogens because of the rarity of infections. The sole virus species in the genus *Isavirus*, *Infectious salmon anemia virus*, does not infect humans and is exclusively a pathogen of salmon that is transmitted via water.

Given their relevance in human health, this section focuses on virus species of the Influenza A, B, and C genera. Of these three, the best characterized and most significant in terms of health is *Influenza A virus.* Along with *Influenza B virus*, it is the cause of almost-annual epidemics of respiratory illness. In comparison, *Influenza C virus* infection is usually mild or asymptomatic, and does not cause epidemics, and so is not as significant from a public health perspective.

Structure and Genomic Organization

Virus particles are spherical or pleomorphic in shape, ranging from 80 to 120 nm in diameter. Capsids have a tubular helical symmetry with segmented negative polarity RNA. The typical orthomyxovirus genome is segmented and consists of ribonucleoproteins of varying lengths (50–150 nm) surrounded by a membrane embedded with viral glycoproteins and nonglycosylated proteins. The membrane-associated viral proteins on the surface of the viral particle are hemagglutinin (HA, responsible for host cell binding and virus entry), neuraminidase (NA, responsible for virus budding and release), and matrix protein 2 (M2; an ion channel involved in viral genome replication). Within the virion, the RNA genome segments are bound to **n**ucleo**p**rotein (NP), along with several viral proteins that form the viral RNA polymerase complex (RNA polymerase acidic, RNA polymerase basic 1, and RNA polymerase basic 2). Surrounding the RNP is a layer of matrix protein 1 (M1), which is involved in nuclear transport of the RNP. Nonstructural proteins 1 and 2 (NS1 and NS2) are involved in regulating viral protein expression and replication, respectively, and though originally considered nonstructural, NS2 is actually present in very low amounts in the virion.

Influenza A and B viruses have eight genome segments of ssRNA(-), while *Influenza C virus* has seven. RNA segments 1, 2, and 3 code for PB2, PB1, and PB-A respectively (which comprise the viral polymerase complex, P), while segments 4, 5, and 6 encode the HA, NP, and NA proteins, respectively. Segments 7 and 8 encode two proteins each with overlapping reading frames: M1/M2 and NS1/NS2. *Influenza C virus* differs from influenza A and B viruses by lacking the NA gene. Each segment has 5′- and 3′-UTRs vital for transcription, translation, replication, and efficient packaging of segments into new virus particles.

Genetic Diversity

Influenza virus genomes gradually change over time, accumulating new mutations due to the error-prone nature of the **RNA-d**ependent **RNA p**olymerase (RdRp) that catalyzes the replication cycle. This process of gradual change in genes is termed antigenic drift. Given the segmented nature of orthomyxovirus genomes, they can also undergo genome reassortment. Individual genomic segments from different influenza strains in a single coinfected host cell can thus be transferred horizontally during virus assembly, leading to new genetic strains. This sudden change in genetic makeup is termed antigenic shift, and can confer new fitness phenotypes against which there is no herd immunity potentially leading to deadly flu pandemics. Antigenic shift also allows some influenza strains to adapt to new species such as has been observed in recent times with avian influenza.

Influenza strains are typically defined by the antigenic properties of the HA and NA proteins embedded in the viral particle membrane. Type A viruses are divided into serological subtypes according to the nature of hemagglutinin (H1 to H16) and neuraminidase (N1 to N9). The current system of subtype nomenclature includes the host organism of origin, the geographical location of the first viral isolation, strain number, and year of isolation. In addition, the numbers corresponding to the antigenic description of HA and NA proteins may be added, as shown in the following example corresponding to the nomenclature of the virus responsible for the last pandemic: Virus A/California/04/09 (H1N1). By international convention, human viruses do not indicate the host organism, unlike viruses whose origin is other than man, such as: Virus A/Swine/Iowa/15/30 (H1N1).

Host Range and Distribution

Influenza viruses A, B, and C are able to infect humans; but only the A and B viruses are responsible for annual epidemics, and only the type A is linked to the origin of pandemics. The B and C viruses mainly infect humans, but it has shown its presence in seals and sea lions (type B) and dogs and pigs (type C). Type A viruses are also able to infect several species of land (pigs, horses, and minks) and marine mammals (seals and whales); however, these viruses primarily infect birds (avian viruses). In birds, the virus multiplies mainly in the digestive tract and the infection is usually asymptomatic. Wild aquatic birds, in which all viral subtypes A have been found (16 species of H and 9N), are considered the main reservoirs of the genetic diversity of these type A viruses.

Viral Replication

The stages of viral replication are similar for the three types of viruses; therefore the multiplication of the virus A as a model is explained. The first stage consists of viral adsorption to the cell by the interaction between the receptor *N*-acetylneuraminic acid (NeuAc) or sialic acid and viral HA. Human viruses show a preference for NeuAc in which the glycoside bond to galactose is the α2,6 type, whereas avian viruses prefer an α2,3 bond. The virus enters by receptor-mediated endocytosis, facilitated by the HA1 domain of HA (a process termed *viropexis*). After transport to the cell's ER, decreasing pH inside the endosome leads to a conformational change in the HA2 domain of HA; this mediates the fusion between viral envelope and the endosomal membrane with consequent release of the RNP into the cytoplasm. The RNP migrates to the nucleus of the infected cell, a phenomenon driven by several nuclear localization signals in the viral NP protein.

Once inside the nucleus, replication events begin with the synthesis of **c**omplementary **RNA** (cRNA) in the positive sense which serves as a template for synthesis of new **v**iral **RNA** (vRNA) and simultaneously mRNAs for the synthesis of viral proteins. The latter is done via cap-snatching—a process by which the virus appropriates caps (via its PA subunit) from cellular RNAs for use as primers for viral mRNA synthesis. This requires the ongoing production of pre-mRNA by the cellular DNA-dependent RNA polymerase (Pol II). The virus also utilizes the transcript-splicing ability of the Pol II for some viral products as the virus lacks this activity.

The viral proteins responsible for replication and transcription include the three subunits of the transcription complex (PB1, PB2, and PA) and NP. This three subunit complex interacts with cellular proteins in a sequential fashion to synthetize vRNA and cRNA. Another key viral protein is NS1, which is expressed very early and plays an important role in regulating the expression of viral genes, while it is the main mediator in inhibiting the synthesis of cellular mRNA and hence host proteins in the infected cell.

Eventually vRNA associates with newly synthesized RNPs, and these in turn bind to the matrix protein (M1). These complexes are exported to the cytoplasm by the viral NS2 protein. RNP-M1-NS2 complexes are directed to the cell membrane while viral surface proteins (H, N, and M2) when synthesized are located in the endoplasmic reticulum and Golgi, respectively.

Assembly of new viral particles is carried out at the apical pole of infected epithelial cells just eight hours from the onset of infection. M1 protein, which interacts both with the internal queue of the surface proteins and RNP, plays a key role in the budding of viral particles. The mechanism by which eight RNA segments are included in each viral particle is still unknown. NA contributes to the release of the viral particles from the cell surface, thanks to its sialidase activity. It actually breaks the bond between the H and NeuAc, avoiding thus the formation of aggregates. Typically apoptosis is induced by NA, NS1, and PB1 proteins.

Pathogenesis

Influenza viruses are transmitted between humans through respiratory secretions and droplets produced by coughing or sneezing. The points of exit, and entry, are the nasal mucosa, pharyngeal, laryngeal, and conjunctivas. The virus spreads through the upper and lower airways and induces mucosal inflammation and edema of the larynx, trachea, and bronchi. Desquamation of epithelial and ciliated mucosal cells can occur. Local infiltration by inflammatory cells and the production and secretion of antibodies and cytokines hinder gas exchange, consequently resulting in increased difficulty in breathing.

Infection can range from asymptomatic to more or less severe flu, to severe viral pneumonia leading to death. This broad spectrum of clinical outcomes corresponds to the intrinsic characteristics of the virus coupled with a variety of host-dependent factors such as age, immune status, presence of chronic respiratory or cardiovascular disease and pregnancy.

In adults, a typical infection is characterized by symptoms such as headache, chills, dry cough, and fever manifesting rapidly after an incubation period of 2–3 days; all accompanied by myalgia, arthralgia, nausea, and anorexia. Fever can give on the fourth day, to manifest again on the fifth to sixth days and finally disappear. As the infection progresses, respiratory symptoms follow. Cough can be productive and the thickening of the vascular and bronchial interphase. In children, there may be symptoms similar to those seen in adults, although they may be accompanied by gastrointestinal symptoms, epistaxis, drowsiness, and higher body temperatures. Influenza in children can also be associated with inflammatory diseases of the middle ear or followed by complications such as heart rhythm disturbances and febrile seizures.

Treatment

In general, the treatment of influenza is based on relief of symptoms (resting at home, antipyretics, decongestants, and plenty of fluids). Complications require hospital management with antibiotics and respiratory support measures. Specific treatment in the form of antiviral drugs are available and include inhibitors of M2 transmembrane protein activity (amantadine and rimantadine); and inhibitors of the NA surface glycoprotein (zanamivir and oseltamivir).

Primary prevention includes the timely and specific mass information, workshops, roundtables, broadcast media, and newspapers. It should be remembered that influenza viruses are transmitted from person to person, mainly through droplets of respiratory secretions during coughing and sneezing. Besides the known general measures to prevent and control the transmission of respiratory viruses, such as covering the mouth and nose with a handkerchief or disposable tissue when sneezing or coughing; washing hands frequently with soap and water or using alcohol-based gel; avoiding wiping the eyes, nose, and mouth; and avoiding contact with ill individuals; it should be stressed that the influenza virus can remain contagious on nonporous inanimate surfaces for up to 24–48 hours (e.g., plastic and stainless steel). It is transferable by contact with hands within 24 hours and on clothing, paper, and other tissues for a period of 8–12 hours. WHO recommends that all surfaces should be cleaned with disinfectants such as 1% sodium hypochlorite, or 70% ethyl alcohol on surfaces where chlorine should not be applied.

There is an antiinfluenza vaccine to protect humans against this virus that is recommended in populations at risk every year. Antiviral drugs can be used as chemoprophylaxis in those cases where there is direct contact with infected people. It has demonstrated 70%–90% effectiveness in preventing influenza. Due to antigenic drift, the vaccine has to be redesigned annually.

PAPILLOMAVIRIDAE: HUMAN PAPILLOMAVIRUS

Papillomaviruses belong to a complex and diverse family of viruses (*Papillomaviridae*) found in different animal species such as reptiles, birds, and marsupials, and in more than twenty species of mammals. **H**uman **P**apilloma**v**irus (HPV) is a small nonenveloped dsDNA virus, 55 nm in diameter. Its supercoiled circular DNA consists of about 8000 bp associated with cell histones. The icosahedral capsid is composed of 72 capsomeres, each comprising five molecules of a major protein called L1 of ~56 kDa (main antigenic determinant). The capsid also contains 12 to 36 molecules of a secondary L2 protein of about 76 kDa. These viruses are unique as they almost exclusively infect the skin and/or mucous flat coating stratified epithelia. It is this characteristic which has hindered the development of cell culture methods for papillomaviruses and their study in the laboratory with standard methods.

Papillomavirus infection is related to the generation of skin lesions known as skin and genital warts. Persistent infection by a group of high-risk papillomavirus genotypes is associated with the generation of cancer (squamous cell carcinomas and adenocarcinomas), in various species of animals ranging from small reptiles, birds, and large mammals such as dolphins and whales, including humans.

Genetic Organization, Diversity, and Replication

Papillomaviruses have a common genomic organization. One of the two DNA strands encodes for the production of viral proteins, with three ORFs used for protein synthesis. The viral genome consists of seven to nine ORFs and can be separated into three regions: (1) a noncoding **l**ong **r**egulatory **r**egion (LRR) or **u**pstream **r**egulatory **r**egion (URR); (2) early region or "*E*"; and (3) a late region or "*L*". The LRR is of variable size depending on the papillomavirus genotype and contains the viral origin of replication and it is essential for regulating genome expression (site of promoters, activators, and repressors). The E region represents about half of the genome and contains five to seven ORFs (El to E8), coding for proteins involved in stimulation of cell proliferation (E5, E6, and E7), replication of viral DNA (El and E2) or the regulation of transcription (E2), as well as another protein, E4, which is expressed throughout the late phase of the viral cycle and is involved in the maturation of the viral particle. Only five of these ORFs (El, E2, E4, E6, and E7) are found in all papillomaviruses. The L region comprises two genes present in all papillomaviruses encoding the structural capsid proteins, L1 and L2.

There are up to 200 subtypes of HPV, fifteen of which are cancer-causing and considered high-risk subtypes. HPV is the cause of ~5% of cancers globally, and two high-risk subtypes, HPV16 and HPV18, account for up to 70% of cervical cancers. HPV16 is also responsible for over 90% of anal cancers. Other subtypes are considered low-risk and are the cause of the characteristic warts associated with HPV, such as HPV6 and HPV11.

Replication of papillomaviruses takes place only in terminally differentiated keratinocyte epithelia such as skin and some (oral, genital) mucous membranes. It is well recognized that papillomaviruses enter through a small gap or microtrauma in the skin or mucous lining infecting stem cells or basal cells of the epithelium. In the basal cells, viral DNA persists as low-number free episomes (20–100 copies per cell). Only cells of the upper layers ensure an abundant production of viral particles. The expression of early viral genes in the basal and suprabasal layers of the epithelium is responsible for acanthosis and hyperplasia at the onset of tumor formation. In the case of infection by high-risk HPVs, such as HPV16, HPV18 or HVP31, the formation of malignant lesions and transformation of the stem cells in the cervix may be facilitated.

Internalization of viral particles that are attached to the cell membrane can enter the cell via endocytosis and clathrin-coated vesicle formation or even the caveolar pathway. The decapsidation is carried out in the endosomes by the destruction of the intracapsomere disulfide bridges allowing transport of viral DNA to the nuclear pores of the newly infected cell. Inside the nucleus, the early expression of the E1, E2, E6, and E7 genes occurs. Vegetative viral genome replication takes place in the upper layers of the epidermis where the synthesis of L1 and L2 capsid proteins and the maturation of virions are also carried out just before the release of new viral particles to the outside environment. The viral replication cycle is therefore closely linked to the terminal differentiation of keratinocytes.

Oncogenic Potential of Papillomaviruses

Cervical cancer is the second most common cause of cancer in women in developing countries, and the fourth most common cancer in women worldwide. Some HPV genotypes are associated with more than 95% of cervical cancer and about 20% of cases of head and neck cancer. It has been estimated that the annual global incidence of cervical cancer will reach 500,000 new cases and 250,000 deaths. In fact, HPVs resemble other viruses with oncogenic potential, such as Epstein-Barr virus (*Human gammaherpesvirus 4*, HHV 4) and *Hepatitis B virus* (HBV), for whom infection is common, but the progression to invasive carcinoma is only seen in a minority of cases. This suggests that other factors are involved in the progression to malignant transformation. Studies of immunosuppressed patients and experimental animal models indicate that a defective immune response is one of the key cofactors in the malignant progression of diseases linked to HPV infection.

Women who are not capable of eliminating infections with high-risk oncogenic genital HPVs, usually develop cervical lesions which may progress from low-grade lesions, such as cervical intraepithelial neoplasia grade I or II (CINI or II) to a high-grade lesions (CINIII and carcinoma *in situ*) or even the generation of an invasive cancer. Low-grade lesions allow the production of large amounts of virus, whereas high-grade lesions have a poor ability to produce viral particles. One of the main events in the progression from a productive infection to the high-grade neoplasia is, in most cases, integration of viral DNA sequences into the cellular genome, leading to loss of expression of the viral protein E2 (responsible for regulating the expression of other viral proteins such as E6 and E7). This leads to a proliferation of basal and parabasal cells and the inability to repair mutations that can take place in cellular DNA. It has been shown that viral DNA integration is always accompanied with the deletion of viral E2 gene. In addition, *ex vivo* experiments have shown that viral E2 protein is able to induce cell cycle arrest in G1 phase and apoptosis.

Diagnosis and Treatment

Due to the significant role that high-risk HPV subtypes play in cancer development, most diagnostic procedures have been designed to detect these viruses. The gold standard for detection of high-risk subtypes is the **H**ybrid **C**apture **2** (HC2) test. This is based on the use of stable hybridized RNA probes and chemiluminescence to detect DNA of the thirteen most high-risk subtypes (HPV16, 18, 31, 33, 35, 39, 45, 51, 52, 56, 58, 59, and 68). HPV can be detected by PCR methods, which can either be consensus or type-specific, an example being the cobas HPV test which uses real-time PCR and DNA hybridization. Other assays, such as the Aptima tests, are designed to detect mRNA from the E6/E7 oncogene.

HPV vaccines exist for HPV16 and HPV18. Vaccination is recommended by the WHO for girls 11–12 years of age, with catch-up vaccines for unvaccinated women up to the age of 26 and for men up to age 21. There is no treatment after infection, but the problems associated with infection can be treated. Routine Pap tests (every 3–5 years) are recommended in infected women in order to detect cervical precancer, which is treatable.

RHABDOVIRIDAE: RABIES LYSSAVIRUS

Rabies is a highly fatal but preventable zoonotic disease of major public health importance which causes an acute encephalomyelitis in mammals. Rabies has a wide global distribution and it is found on all continents except Antarctica. The disease is caused by a virus from the Order *Mononegavirales*, a member of the family *Rhabdoviridae* (Greek rhabdos meaning rod) and genus *Lyssavirus* (Greek lyssa meaning rage).

The major global burden of rabies is attributed to dog-mediated transmission. However, in some areas such as in Latin America, where intensive control programs have been implemented, significant decreases in canine-transmitted cases have been observed. Conversely sylvatic-transmission is becoming increasingly important in the global epidemiology of rabies, e.g., in the Americas, with the recognition of different rabies virus variants in numerous species of wildlife, particularly bats.

Rabies is estimated to account for about 60,000 human deaths worldwide per year with roughly one death every 15 minutes. The majority of these deaths occur in children and cases mainly originate from the developing world, primarily Asia, Africa, and Latin America. The burden of the disease in developing counties is significant with an estimated DALY (**d**isability-**a**djusted **l**ife **y**ear) score of ~2 million and annual financial cost estimated at US$6 billion.

Structure and Organization

The *Rabies lyssavirus* (RABV) is an enveloped RNA virus with a compact helical **r**ibo**n**ucleoca**p**sid (RNP) core. The RNA genome is ~12 kb, single-stranded, negative-sense, and unsegmented. It encodes for five structural proteins: nucleoprotein (N), phosphoprotein (P), matrix protein (M), glycoprotein (G), and RNA polymerase (L). The N, P, and L components are included in the RNP core which is surrounded by the viral membrane proteins M and G.

Genetic Diversity

The genome associated N and the transmembrane G genes, have been the most extensively studied of the five, with numerous antigenic and phylogenetic studies focusing on partial or complete sequences of these genes. The virus is independently maintained by several species of mammals within the orders Carnivora and Chiroptera and different variants of RABV occur in different reservoir host species. Rabies exists in two cycles, the urban cycle which maintains the cosmopolitan virus and the sylvatic cycle which maintains several viral variants in several species of wild animals (e.g., bats, mongoose, foxes, raccoons, and skunks). Apart from the cosmopolitan canine lineage, the viral lineages are distinct between the Old World and the New World.

On account of their present geographical distribution, rabies viruses can be grouped into seven major phylogenetically defined clades: (1) American indigenous clade, (2) Cosmopolitan clade, (3) India clade, (4) Asia clade, (5) Africa 2 clade, (6) Africa 3 clade, and (7) Arctic-related clade. Bats are only associated with the American indigenous clade and RABV is the only lyssavirus known to circulate in the Americas. The Indian clade is proposed to be the ancestral progenitor for all the other clades which are associated with nonflying mammals. In general, terrestrial rabies viruses show more similarity by geographic origin whereas bat rabies virus lineages, whose reservoir is less limited by geographical boundaries due to their aerial nature, are more defined by species.

Replication and Pathogenesis

RABV enters the host by the introduction of the virus (in saliva or other potentially infectious material such as neural tissue) through broken skin or mucous membranes. Successful transmission is dependent on several factors including the type of exposure, which can be categorized into bite and nonbite. Bite exposure is the most common and significant method of rabies transmission, with the introduction of virus laden saliva directly into the host. More extensive bites are considered higher risk as there is potentially more viral interaction with a wider network of nervous tissue with greater viral dissemination toward the CNS.

Nonbite exposure is mainly attributed to aerosolized virus, splash exposures with handling and butchering of infected carcasses, tissue and organ transplants and contamination of open wounds or mucous membranes by infected material. After inoculation, the virus replicates locally in affected muscle fibers then accesses the peripheral motor axons by binding to acetylcholine receptors at the neuromuscular junction for travel to the CNS. The virus further replicates in motor neurons of the spinal cord and local dorsal root ganglia before ascending to the brain. Centrifugal spread then occurs from the CNS via slow anterograde axoplasmic flow (passive) along peripheral nerves to the salivary glands, skin, cornea, and other end organs. Viremia does not occur for rabies infection.

Clinical Symptoms

The incubation period is highly variable in both humans and domestic animals, ranging from 3 to 12 weeks postexposure in the majority of cases, but can be up to 6 months in domestic animals and over a year in about 6%–7% of human cases. The exact period is dependent on several factors including the severity of the exposure, the route and site of exposure (including proximity to the CNS), and the initial viral load.

The disease generally develops in three phases: initial, excitatory and paralytic. The clinical syndrome of the disease in animals is variable with clinical presentation usually described as either furious or paralytic. Although the brain stem is involved in both clinical forms of the disease, differences in viral tropism at the inoculation site or the CNS, the route of spread, or the triggering of immune cascades in the brain stem, may account for variation in disease clinical manifestation. The clinical manifestation of bat-transmitted rabies is different from that of canine-transmitted rabies with the latter exhibiting more encephalitic symptoms typical of furious rabies as compared to more peripheral symptoms (tremor, myoclonus, local motor, or sensory deficits) with bat-transmitted rabies.

Initial clinical signs are nonspecific (i.e., anorexia, pyrexia, nausea, headache, sore throat, and lethargy) progressing to behavioral abnormalities such as increased aggression or apparent docility (depending on the form of the disease), and cranial and peripheral nerve manifestations such as facial asymmetry, gagging, prolapse of the nictitating membrane, ataxia, paresis, and paralysis. Death usually occurs 5–11 days after the appearance of clinical symptoms with survival times being longer for patients with paralytic rabies.

Prevention and Treatment

The main strategy for human rabies prevention is the vaccination of canines, as dogs are the primary source of human-contracted rabies virus. Preexposure prophylaxis is also recommended for individuals who are at a high risk of encountering rabid animals such as veterinarians. In this strategy, individuals undertake a series of injections of the rabies vaccine in an effort to prepare the immune system for encounter with the virus. Such individuals are given booster vaccines after several years to maintain the readiness of their immune system. Treatment is based on rapid application of postexposure prophylaxis using vaccine and plasma-derived human or equine immunoglobulins for individuals who have been exposed to suspected rabid animals by bites, scratches or saliva. The immunoglobulins provide passive immunity, allowing the body time to develop its own antibodies before the development of the fatal systemic response to the virus.

FURTHER READING

Barre-Sinoussi, F., Ross, A.L., Delfraissy, J.F., 2013. Past, present and future: 30 years of HIV research. Nat. Rev. Microbiol. 11 (12), 877–883.

Houser, K., Subbarao, K., 2015. Influenza vaccines: challenges and solutions. Cell Host Microbe 17 (3), 295–300.

Lowy, D.R., Schiller, J.T., 2012. Reducing HPV-associated cancer globally. Cancer Prev. Res. 5 (1), 18–23.

Weaver, S.C., Charlier, C., Vasilakis, N., Lecuit, M., 2017. Zika, Chikungunya, and other emerging vector-borne viral diseases. Annu. Rev. Med. Available from: https://doi.org/10.1146/annurev-med-050715-105122, 2017 Aug 28.

Chapter 8

Viruses as Pathogens: Animal Viruses Affecting Wild and Domesticated Species

Jerome E. Foster
The University of the West Indies, St. Augustine, Trinidad

We patronize the animals for their incompleteness, for their tragic fate of having taken form so far below ourselves. And therein we err, and greatly err. For the animal shall not be measured by man. In a world older and more complete than ours, they are more finished and complete, gifted with extensions of the senses we have lost or never attained, living by voices we shall never hear. They are not brethren, they are not underlings; they are other Nations, caught with ourselves in the net of life and time, fellow prisoners of the splendour and travail of the earth

Henry Beston

A significant number of the diseases currently affecting humans have originated in sylvatic cycles, and have crossed over at one point or the other due to the ability of viruses to mutate rapidly and adapt to new species, especially in the case of RNA viruses. The same occurs with zoonosis where close spatial and physical association of feral and domesticated species creates the setting for viruses to cross the species barrier, resulting in emergent diseases. Humans have played a major part in such emerging zoonosis throughout the centuries, with activities such as farming, deforestation, and urbanization. It is now recognized that a One Health strategy is needed at the human-animal-environment interface to effectively combat emerging and reemerging zoonoses. The encroachment of human habitation and industry on these ecosystems which harbor sylvatic viruses encourages contact between suitable vectors with related host species (both domesticated and farm animals) and facilitates such crossover events. The introduction of

Viruses. DOI: https://doi.org/10.1016/B978-0-12-811257-1.00008-5

domesticated, but foreign animal species into areas with established sylvatic host–vector cycles can also lead to contraction of sylvatic viruses by these introduced species with unwanted consequences. Extinction events are rare in these epizootics, but local extirpation can occur where populations are already threatened by other factors. Key viral zoonoses of domesticated animals and their origins are summarized in Fig. 8.1.

At times viruses from foreign or domesticated animals can crossover in the other direction into indigenous wild populations with equal lethal impact. An example of this is *Canine morbillivirus*, an ssRNA(−) virus (family *Paramyxoviridae*) normally found in populations of the domestic dog (*Canis lupus familiaris*). The virus causes a disease of the central nervous system

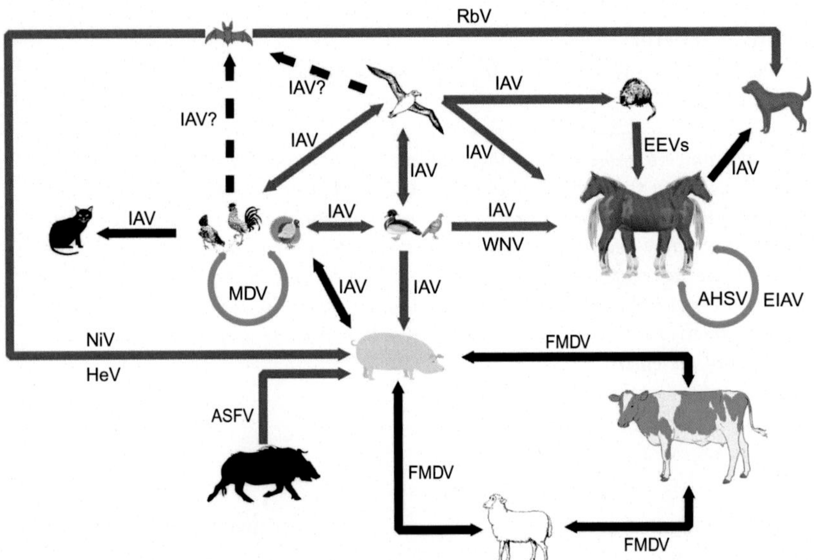

FIGURE 8.1 Key viral zoonoses of domesticated animals and their origins. Wild birds (including shorebirds, seabirds, and wild fowl) are key reservoirs of influenza A viruses (IAVs) that are transmitted when these birds, particularly wild ducks, come in contact with farm animals such as poultry. They can also spread IAVs to other farm animals such as pigs and horses, which have also in the past contracted these IAVs from poultry. These IAVs can spread to other domesticated animals or pets such as cats, dogs, and to other mammals such as rodents. Bats are also found to have IAVs of unknown origin but suspected to come from wild birds also. Horses can also contract equine encephalitides (EEVs) from rodents or birds, and birds can also spread *West Nile virus* to other mammals such as horses. Horses also have several viral diseases such as African horse sickness (AHSV) and Equine infectious anemia (EIAV) in circulation among themselves. Likewise poultry, particularly chickens, have the widespread Marek's disease (MDV) in circulation among populations globally. Bats are reservoirs for many viruses and can spread rabies (RbV) to dogs, and Nipah (NiV) and Hendra (HeV) viruses to pigs. Pigs can also contract the deadly African swine fever (ASFV) from wild boars and are also involved in cycles of *Foot-and-mouth disease virus* (FMDV), which can spread to several other ruminants including sheep, goats, and cattle.

(CNS), gastrointestinal, and respiratory systems and has no known cure, though safe effective vaccines for dogs are available. It is also known to affect several other species such as wild dogs, coyotes, foxes, wolves, and pandas, and more recently lions, with significant mortality rates during infection in these felines.

Other activities outside of agriculture also provide the perfect setting for these epizootics. The global wildlife trade, for instance, has contributed to the emergence and spread of zoonosis. Exotic animals are removed from natural environments and moved to new areas, sometimes carrying equally exotic pathogens with them which may impact native species, yet little has been done to assess the health risks such movement may pose. The risk is even greater with products that are illegally imported as no pathogen surveillance is involved. An example of this is seen in the spread of ranaviruses (family *Iridoviridae*) into populations of wild and farmed amphibians caused by trade. This group of viruses causes massive die-offs in several species including the North American bullfrog (*Rana catesbeiana*).

Another excellent example of newly introduced viruses negatively impacting indigenous wildlife is the introduction of *West Nile virus* (WNV) into North America in 1999. WNV is a flavivirus (family described in Chapter 7: Viruses as Pathogens: Animal Viruses, with Emphasis on Human Viruses) mainly transmitted by mosquitoes of the *Culex* genus, and is indigenous to Africa, Asia, and parts of the Pacific, which since its introduction has become endemic in the North American continent. It causes acute viral encephalitis in many avian species, and humans and other mammalian species are incidental hosts. Wild birds are highly susceptible to infection and experience high morbidity and mortality rates, including several that are endangered such as the Eastern loggerhead shrike (*Lanius ludovicianus migrans*) and the Greater sage-grouse (*Centrocercus urophasianus*). The American crow and other corvids are also particularly susceptible to infection.

Viruses can also have positive effects on an ecosystem. In an ecosystem where a species has become overpopulated, throwing the ecosystem into imbalance, the introduction of viral pathogens of that species can effectively reduce that imbalance. An example of this is the impact of *Rabbit hemorrhagic disease virus* (RHDV), a calicivirus (family *Caliciviridae*, genus *Lagovirus*), on rabbits in Australia. Here the virus has been introduced in an effort to control the invasive European rabbit (*Oryctolagus cuniculus*). Originally imported into Australia as fodder for traditional hunts, the European rabbit population exploded to over an estimated 10 billion, and has brought significant environmental damage including the extinction of some native species, and is officially considered a pest in that country. RHDV infects rabbits causing the deadly rabbit hemorrhagic disease and was accidentally introduced into Australia where it has become endemic, causing a reduction in more than 95% of the rabbit population in some areas of South Australia. In Australia, the virus became a solution to a key economic and ecological issue, though not without controversy. More recently, a cyprinid herpesvirus has been proposed as a potential biological control method for carp, which dominate fish communities throughout many inland waterways in Australia.

The desire to control the spread of such emerging viral pathogens through the culling of vectors is a theoretically sound strategy, particularly when the vector is categorized as a pest, e.g., *Aedes aegypti* mosquito. However, when the transmission cycle involves a vector of a more endangered status, organizations sometimes still fail to consider the impact of such culling on the delicate ecosystems that exist. Mammals such as primates and bats are known to be reservoirs for a number of recently emerged, reemerging, and newly emerging pathogens (causing SARS, Ebola, Nipah encephalitis, and Yellow fever), but these animals also play a crucial role in ecosystems and even agriculture. Bats, e.g., are crucially involved in events such as seed dispersal and consumption (and thus population control) of insect vectors such as mosquitoes. Another facet that has to be considered is the method of vector control: historically, the most cost-effective solutions, particularly in developing economies, involve the use of chemicals in the form of poisons for mammals and pesticides for insects that can negatively impact on other organisms in ecosystems including humans. Policies have to take into consideration all these factors before implementation to ensure that delicate ecosystems are maintained.

In this chapter, we will look more closely at viruses that directly impact animal species of commercial and agricultural importance, and approaches to disease surveillance, prevention, and control.

ANIMAL INFLUENZA VIRUSES

Influenza viruses infect not only humans, but also a wide variety of animals. The focus in this section is on influenza viruses that affect domesticated animals, namely, poultry, swine, and horses, because of their significance to the agricultural and food industry. These animals suffer almost exclusively from infections due to **i**nfluenza **A** **v**iruses (IAVs), with the exception of serological and genetic evidence of the circulation of the influenza C-like viruses in swine. Influenza has been documented in birds since the early 20th century when it was called "*fowl plague*" and is still one of the most important viral diseases of domesticated birds. Wild aquatic birds are reservoirs for these IAVs with all known subtypes isolated from feral species. In the case of these wild bird reservoirs, infections of **a**vian **i**nfluenza (AI) are for the most part asymptomatic or subclinical. It is believed that the virus is maintained in these feral bird populations by constant bird-to-bird transmission on both breeding and wintering grounds with migratory species playing a key role in dispersal of virus into susceptible populations (Fig. 8.1). The virus also shows evidence of long-term persistence in the aquatic environment that may also increase seasonal transmission in the wild birds. Poultry is directly exposed to wild birds in a variety of ways in areas such as ponds, rearing spaces, and live markets where movement of wild species into rearing areas are generally unrestricted. Influenza A infection is the most important cause of respiratory disease in both swine and horses, and the strains that circulate endemically in these species are also thought to have originated in wild birds.

Genetic Diversity and Epidemiology

The structure and genomic organization of IAV is covered in Chapter 7, Viruses as Pathogens: Animal Viruses, with Emphasis on Human Viruses, in detail. Sixteen of the eighteen known HA subtypes (H1–H16) have been isolated in wild birds among the orders Anseriformes (waterfowl and relations, e.g., dabbling ducks), Charadriiformes (shorebirds, e.g., plovers and gulls) and Procellariiformes (seabirds, e.g., petrels and albatrosses) with the most common subtypes being H3, H4, H6, and at times H11. The distribution among the order Galliformes, which includes domestic poultry, is almost as extensive with 13 subtypes detected (H1–H13). The number of IAV HA subtypes observed in mammals (outside of humans) has not been as extensive. All subtypes that are known to infect avian and mammalian species are originated from wild bird reservoirs, with the possible exception of the bat IAVs.

IAVs mutate rapidly through the processes of antigenic drift (due to the error-prone replication of RNA viruses) and antigenic shift (through gene reassortment; see Chapter 7: Viruses as Pathogens: Animal Viruses, with Emphasis on Human Viruses). Despite this potential for genetic variation, studies show that when IAV strains become repeatedly introduced into an animal species, the virus may adapt to the host species and becomes endemic with little reverse transmission in some cases. This is both a major health and commercial concern, as emerging subtypes can potentially jump species barrier and establish themselves in new hosts. The recently emerged human subtype H1N1 strain, H1N1pdm09, is a good example: it caused a regional pandemic of "*swine flu*" in the Americas from 2009 to 2010 and was a reassortant originating in pigs. There are of course exceptions: turkeys, e.g., are very susceptible to classic swine subtype H1N1 (which has circulated in pigs for decades) and the more recent H1N2 and H3N2 swine subtypes in North America.

The emergence of the highly pathogenic H5N1 strain influenza virus in 1996 in China (known as "*bird flu*") is an excellent example of both rapid mutation and species jumping. Before the 1990s, IAV subtype H5N1 was only seen in two outbreaks in 1959 and 1991 in the United Kingdom. The original Asian H5N1 strain was pathogenic only in gallinaceous birds, but evolved in its pathogenicity during a panzootic in avian species (both domesticated and feral birds) across Southeast Asia, Africa, and Europe during 2003–07. This H5N1 virus was highly pathogenic in domestic poultry, but also infected several other animals including cats, dogs, civets, and humans, though the subtype has not become endemic in any other species apart from birds.

Circulating IAV subtypes can also be displaced by emerging ones in populations. A good example of this is seen in horses: subtype H7N7 was the first strain to be isolated and characterized in 1956 in equine populations in Eastern Europe. Several years later in 1963, a second subtype had emerged in horses in North America, characterized as H3N8. After several years of cocirculation of both subtypes, H7N7 has not been isolated in horse populations since 1979, though serological evidence in unvaccinated horses implies it might still be in circulation, affecting horses subclinically. H7N7 was essentially been replaced by the H3N8 subtype which is now largely enzootic in equine populations in Europe and the Americas, despite extensive vaccination programs.

Pathogenesis

The productive and multiple replication cycles of the virus were presented in Chapter 7, Viruses as Pathogens: Animal Viruses, with Emphasis on Human Viruses.

Avian influenza

In birds, **a**vian **i**nfluenza **v**irus (AIV) is transmitted by the fecal–oral route via contaminated water. Infection produces several syndromes ranging from asymptomatic, subclinical infections, to respiratory disease (usually associated with decreased egg production), to systemic diseases with 100% mortality. AIV subtypes are generally classified into two groups based on their pathogenicity in avian species: high and low pathogenicity (HPAIV and LPAIV). LPAIVs can be asymptomatic, but more often cause mild to severe respiratory disease resulting in high morbidity (>50%), but low mortality (<5%). In some cases of LPAIV infection mortality can increase, especially in instances of concurrent or secondary infection with other diseases. The principle site of initial replication is the nasal cavity, and virions are released into the upper respiratory and intestinal tracts. In rare cases, the infection may become systemic causing damage in the kidneys, pancreas, and other tissues that contain cells with trypsin-like enzymes that can cleave the HA viral protein.

HPAIV, in contrast, can cause high lethality in not only gallinaceous poultry but also other avian species, though time to death is usually longer in these other birds. Clinical signs of HPAIV infection usually involve significant damage to major organs including those of the cardiovascular and nervous systems. Gross microlesions can appear a few days after infection in the heart, and occasionally liver, kidneys, and the lungs can be congested and hemorrhagic. The major site of initial HPAIV replication is usually nasal epithelium, where it spreads via the lymphatic or vascular system to replicate in visceral organs, brain, and skin. Death is usually due to a combination of high viral replication in organs (with apoptosis), adverse effects of the immune response (through cytokines and other cell signals), and ischemia from vascular thrombosis.

Swine influenza

The main transmission route of **s**wine **i**nfluenza **v**irus (SIV) is through the respiratory system via nasopharyngeal secretions, and there are year-round cases of acute respiratory syndromes in pigs of all ages. Swine influenza has high morbidity and low mortality, and death is rare in uncomplicated cases. Most mortality is due to coinfections or secondary infections of the respiratory system by other viral or bacterial pathogens of the porcine respiratory disease complex such as *Haemophilus parasuis* and *Porcine reproductive and respiratory syndrome virus*. As in most species, the youngest animals are the most susceptible to serious disease. Symptoms of acute infection include fever, anorexia, coughing, difficulty breathing, muscle stiffness, and "*thumping*" (shaking of the entire body due to efforts to breathe). Fever is the most consistent clinical sign and peaks after 1–2 days of infection. Clinical signs begin from 1-day postinfection (dpi) and cease 3–7 days later, as the pig humoral immune response generates sufficient antibodies to overcome the viral infection. As the disease progresses, cellular immunity clears the virus and protects against reinfection with homologous strains, but there is only limited immunity against heterologous virus.

A characteristic microscopic lesion of SIV infection is necropurulent bronchitis and bronchiolitis: as epithelial cell necrosis occurs due to infection, the bronchi and bronchioles are filled with necrotic and inflammatory cells, which can be observed within the first day of infection as epithelial cell necrosis occurs, and cells accumulate in the airways.

Equine influenza

Equine **i**nfluenza **v**irus (EIV) is extremely contagious and spreads by inhalation of aerosolized virus. Disease is characterized by a rapid acute febrile illness with nasal discharge, a dry hacking cough, anorexia, and lethargy. Morbidity in horses can be as high as 60%–90%, but mortality is very low (1%), and most deaths reported in outbreaks are of neonatal foals or donkeys. In severe infections, complications caused by secondary infections, particularly with bacteria, can lead to bronchopneumonia, allergic bronchitis, and bronchiolitis (7–14 dpi). In severe infections of adult horses, virus-induced apoptosis leads to the loss of cilia in the respiratory airways such as the trachea and bronchi. There is also increased risk of chronic obstructive pulmonary disease. Myositis, myocarditis, and limb edema are also other sequelae associated with EIV infection. Neurological complications in horses have been associated with EIV in one outbreak, but the possible pathogenesis is unknown.

Diagnosis, Treatment, and Management

Diagnostic assays for all three IAVs (AIV, SIV, and EIV) use similar techniques with slight modifications depending on the IAV and host species involved. The clinical signs of SIV infection in particular can be misdiagnosed as hog cholera, enzootic pneumonia, or rhinitis, and so laboratory methods are key to diagnosing SIV in pigs. For equines with possible EIV infection, presumptive diagnosis (based on clinical signs) is confirmed traditionally by the detection of live virus.

Samples are collected using tracheal, oropharyngeal, and/or cloacal swabs in avians, while nasal or nasopharyngeal swabs are sufficient for suids and equines. The preferred method for **v**irus **i**solation (VI) and propagation is inoculation in embryonated chicken eggs. Virus can also be detected in tissue by the use of immunohistochemistry (or immunofluorescence). Serological tests such as the hemagglutination-inhibition tests or ELISA are also used in laboratories. Commercial ELISAs are available for the detection of type-specific IAV antibodies in serum, plasma, and egg yolk. Real-time PCR assays (based on the HA and NA genes) are considered the gold standard for screening due to its high sensitivity and the fact results can be obtained in hours.

Management of IAV is routinely done using vaccination in all three animals. This can be of several types: (1) whole-inactivated/subunit, (2) live-attenuated, and (3) viral-vector-based vaccines. Traditional adjuvanted, whole-inactivated are the most commonly used in all three animals, usually through the route of intramuscular injection. In the case of AI, LPAIVs are usually the base of these vaccines. Live-attenuated AIV vaccines have been shown to elicit sufficient immune response in all three animals, but their use is not recommended due to several factors including the potential for these vaccine viruses to mutate into pathogenic strains through antigenic drift or shift. Virus-vectored vaccination is also available, with genetically modified vectors usually containing the species-specific IAV *HA* gene, but use of these types of vaccines are limited with insufficient field data to justify licencing them in many countries.

Despite the availability of vaccines, potential outbreaks of HPAIV remain a problem in poultry largely due to the difficulty of completely immunizing populations. To control the spread of AIV in infected flocks, poultry are euthanized by various ways including chemical methods (lethal injection with barbiturates, CO or CO_2 poisoning, and inhalant anesthetics) or physical methods such as electrocution, decapitation, or cervical dislocation. The preferred method of mass culling involves the use of CO_2 gas chambers. In equines, there is no specific treatment for those with acute infections and care is supportive. Control measures include quarantine of animals with clinical signs and vaccination.

AFRICAN SWINE FEVER

African swine fever is a contagious and highly lethal hemorrhagic disease of swine and is the greatest obstacle to development of the swine industry in sub-Saharan Africa where it was first characterized in 1921. *African swine fever virus* (ASFV) is the causative agent of the disease and is the sole member of the *Asfivirus* genus within the *Asfarviridae* family. It is a large enveloped, dsDNA virus. ASFV is maintained in a sylvatic cycle involving wild pigs (such as warthogs and bush pigs) and soft tick vectors of the genus *Ornithodoros*. To date, it is the only known DNA arbovirus. Many warthog populations in southern and eastern Africa where ASFV is endemic are infected, and the virus appears to be maintained in transmission cycles between young warthogs and ticks. Older warthogs though persistently infected have low viremia and are generally asymptomatic. Epizootic outbreaks of swine fever arise in two ways. Live warthogs infested with the insect vector may intermingle with domestic swine providing infected ticks with the opportunity to feed and transmit the virus. Domestic swine can also be exposed to the carcasses of infected warthogs, which they feed on, thus contracting the virus directly. The virus spreads effectively by contact via aerosol droplets and blood, feces and other virus-infected tissues. The disease was limited to sub-Saharan Africa until the late 1950–70s when sporadic outbreaks occurred in the Iberian Peninsula, the Caribbean, and South America. The disease also appeared in, and was subsequently eradicated from, Europe in the 1990s via exhaustive culling programs, but from 2007 to 2014 has reemerged in countries in Eastern Europe including Russia, Poland, and Moldova. High mortality of the disease in domestic swine and its rapid spread from its region of endemicity to new virgin territories enzootic outbreaks makes it an important concern to the global swine industry.

Structure and Genomic Organization

The virion is approximately 180–200 nm in diameter and consists of a **n**ucleo**c**apsid (NC) surrounded by lipid and an icosahedral capsid. The genome varies between 170 and 193 kb (depending on the strain of the virus) and is represented by a single molecule of dsDNA, which is covalently closed at the ends in hairpin loops with terminal internal repeats. The genome is predicted to have between 150 and 167 **o**pen **r**eading **f**rames (ORFs) depending on the strain, and encodes for over 50 proteins. Three main regions of importance have been identified from study of completely sequenced ASFV isolates: a highly conserved central region ~125 kb which contains most of the genes necessary for host cell modulation viral replication, and assembly; and two variable regions at either end of the genome called the **l**eft and **r**ight **v**ariable **r**egions (LVR and RVR) which are 40 and 20 kb, respectively. Strain length variation is mainly due to insertions and deletions of multigene families (MGFs) in these variable regions. Two key MGFs are MGF 360 and MGF 503, which reside in the LVR and contain genes such as *1329L* and *A528R*, which are involved in modulating host innate immunity responses. Another important gene is *B646L* that codes for the major capsid protein p72.

Genetic Diversity

There is some controversy as to whether ASFV produces neutralizing antibodies, with evidence for and against, and it has not been possible to categorize strains into discrete serotypes. Virus diversity has been best described as genotypes, with the *B646L/p72* gene being most phylogenetically informative. Other genes, including the central hypervariable region of the *B602L* gene and the *E183L* gene (p54 protein), show similar groupings. Strain comparisons using p72 have identified at least 23 distinct phylogenetic genotypes. Genotype I originates from West Africa and is the strain that originally spread into several European, South American, and Caribbean countries. The other 22 genotypes are all endemic to Eastern Africa and the presence of sylvatic cycles between vector and warthogs is thought to play a role on the greater genetic diversity seen on the continent. The recent reemergence of ASF in Russia and across Eastern Europe has been accredited to the East African strain, genotype II.

Replication

The virus enters host macrophages through several not yet fully known mechanisms, including receptor-mediated and clathrin-mediated dynamin-dependent endocytosis, phagocytosis, and macropinocytosis. Viral proteins p12, p30, and p54 are all involved in the entry process, facilitating virus attachment to several receptors including CD163. Once inside endosomes, the virus decapsulates revealing the virus inner envelope. This fuses with the endosome membrane, allowing release of the naked viral DNA into the cytoplasm where the majority of replication occurs. ASFV virions aggregate in the cytosol near the nucleus, and both strands of viral DNA are used for coding viral proteins. The virus also has its own limited transcription machinery, which enables some independence from the host cell, though it still relies heavily on the machinery of the cell for translation. The nucleus is also involved in viral DNA replication but the specific role is not known. Prior to DNA replication, some early viral genes are expressed via the action of viral enzymes packaged in the infecting virion core, using both copies of viral DNA as template. Once replication is complete, transcription of the remaining genes occurs. Microtubules in the cell are used for both transport of virus to the perinuclear area for replication, thought to be mediated by the viral protein p54. The same microtubule network serves as the viral replication factory, with the replicated DNA and expressed viral proteins aggregating and assembling there. Host cell translation is shut down by apparent recruitment, rather than inhibition, of eukaryotic initiation factor 4f (eIF4F). This along with the pathogen-directed clustering of mitochondria and ribosomes around p72 at the periphery of viral factories enables the virus to take over the translation machinery and stimulates cap-dependent translation of its genome. Viral DNA then becomes encapsidated giving rise to mature virions, and move toward the cell surface where they exit the cell, through exocytosis (budding) or apoptotic bodies.

Pathogenesis

In the vector, ASFV pathogenesis is characterized by a low dose of infection, low mortality (until after the first oviposition), lifelong infection, and efficient transmission between individual vectors and the swine hosts. In warthogs, pathogenesis of the virus is characterized by low viremic titers with unapparent infection, as described earlier. In domestic swine, ASFV infection results in a general reduction of blood cells resulting in lymphopenia, thrombocytopenia, and leukopenia. There is also apoptosis of lymphocytes and mononuclear phagocytes. In infected macrophages, the virus inhibits proinflammatory cytokines such as interferon and **t**umor **n**ecrosis **f**actor (TNF). This widespread inhibition is regulated by a protein encoded by the viral gene *A238LI*, through inhibiting activation of NfKb (a protein complex that is central to innate immunity and the inflammatory response). Strains of ASFV have been shown to differ in their ability to inhibit such expression of proinflammatory cytokines.

Detection, Prevention, and Control

ASFV is routinely detected through PCR amplification of the *p72* gene, and several modifications currently exist including RT-PCR assays. Virus is then isolated and cultured using swine bone marrow or peripheral leukocytes. Positive samples show cytopathic effect several days postinoculation. Other serological techniques, such as ELISA, are also widely used for surveillance. There is no existing vaccine, and development of one is an ongoing issue. Both humoral and cell-mediated antibody responses have been shown to contribute to the immune response to ASFV in swine, and can protect infected pigs from lethal outcomes. However, neutralizing antibodies are insufficient to confer protective immunity and the main method of controlling virus spread is through the isolation and culling of infected animals, and the monitoring of swine imports in disease-free areas.

The control of ASFV transmission in endemic areas becomes complicated due to several factors. These include the ability of the virus to be transmitted, not just by ticks, but, by fresh or cured meat, the persistent infection in some swine that serves as a constant source of infection, and the fact that symptoms of the disease can be confused with other circulating illnesses (such as classic swine fever a.k.a. hog cholera). Nonendemic or ASFV-free regions maintain their status through bans on the importation of live swine or swine products from countries with known infections, and the destruction of waste food material from transport vessels along associated routes.

EQUINE INFECTIOUS ANEMIA

Equine **i**nfectious **a**nemia (EIA) is a disease caused by *Equine infectious anemia virus* (EIAV) that has been reported to infect all equine species including horses, ponies, mules, and donkeys. EIA is endemic in the United States, the Middle East, Asia, parts of Europe, and South Africa and outbreaks have been reported almost worldwide. EIAV is a member of the *Retroviridae* family, genus *Lentivirus*, and is transmitted by bloodsucking insects, such as stable flies (*Stomoxys* spp.) but is most effectively vectored by tabanids such as deer flies (*Chrysops* spp.) and horse flies (*Tabanus* spp. and *Hybomitra* spp.). It is also called swamp fever because transmission is usually high in the summer months in endemic regions, when biting flies are most active. The virus is transmitted mechanically between equines when blood is spread by these biting flies. Transplacental infection of foals has also been documented in mares having clinical signs of disease.

Structure, Genomic Organization, and Diversity

The EIAV virion is structurally similar to that of HIV: it is 100 nm in diameter, and has an outer lipid envelope surrounding an oblong core which contains the nucleocapsid. The genome at 8.2 kb is the smallest and simplest of the known lentiviruses. Genomic organization is similar to that seen in HIV and other lentiviruses (see Chapter 7: Viruses as Pathogens: Animal Viruses, with Emphasis on Human Viruses).

As a lentivirus, EIAV has a rapidly evolving genome due to the low fidelity of the reverse transcriptase and its lack of proofreading ability. The selective pressures of host immunity during persistent infection increases the antigenic variation of the virus, particularly in its Env protein, but significant variation is also observed in other genes such as *gag* and that encoding Rev. Such variation plays a role in disease virulence and persistence and strains vary widely in their virulence and also ability to be grown in cell culture. Distinct strains exist both in the Old and New World and their divergent evolution can be linked to geographic isolation, with New World strains arising in founder equid populations set up by the movement of equid species between regions during eras of colonization.

Pathogenesis and Replication

The main entry point of EIAV is through macrophages rather than other lymphocytes. Rapid and severe anemia is the clearest clinical symptom of infection with the virus. This is caused by a reduction in the red blood cell life span mainly due to hemolysis and enhanced erythrophagocytosis by activated macrophages. A decrease in platelets along with fever is an early clear sign of EIA and directly related to thrombocytopenia. The exact mechanism behind development of anemia and thrombocytopenia is not fully known, but evidence shows complement is involved in red blood cell lysis. Hemorrhaging is thought to be a consequence of the thrombocytopenia. A further complication is depressed erythropoiesis and accompanying decreases in levels of serum iron which exacerbates the anemia. There is also evidence that macrophages infected with EIAV may produce elevated levels of certain cytokines such as TNF which may contribute to febrile episodes during EIAV pathogenesis.

The main neurological symptoms associated with disease include ataxia, meningitis, encephalomyelitis, and encephalitis. These are thought to be caused by EIAV-specific antibody in the CNS, but remain unresolved. Replication-induced mutations in the gp90 segment of the *env* gene are thought to be responsible for the rise of antigenic variants that instigate the disease's characteristic chronic relapses. Horses that survive the acute phase usually become lifelong carriers of the virus; this seems to involve an intact immune response that allows control of the levels of viremia. Horses infected with EIAV rarely exhibit severe suppression of the immune system and thus are not usually susceptible to opportunistic infections. This is in contrast to most other lentiviruses, and the ability for carriers to control replication is still under investigation.

The replication cycle is similar to that seen in HIV and other lentiviruses, which can be reviewed in Chapter 7, Viruses as Pathogens: Animal Viruses, with Emphasis on Human Viruses.

Diagnosis, Prevention, and Control

EIAV is difficult to culture successfully in cell lines to date. An infected animal becomes a carrier for life, and hence serology is the main method of detection. Diagnosis can be performed either by an **e**nzyme **i**mmunoassay (ELISA), or **a**gar **g**el **i**mmuno**d**iffusion (AGID, known as the Coggins test). AGID cannot detect EIAV antibodies during the first 2–3 weeks of infection and can be seronegative up to 2 months. ELISAs are more sensitive than the AGID but also have a greater risk of false-positives. For this reason, positive ELISAs are usually confirmed using the AGID, or subsequent immunoblotting, and a combination of these methods is considered the best diagnostic procedure. Proviral DNA can also be detected by PCR using extract from peripheral blood leukocytes.

Infected horses have to be isolated from susceptible animals, or euthanized. In the same way, foals born to infected mares must be isolated until they can be confirmed as virus-free. During outbreaks, the use of spraying to control vectors has some success in disrupting transmission. As biting insects are strictly mechanical vectors, the virus does not persist outside of the equine host, and the use of disinfectants along with quarantines has been used successfully to prevent spread from outbreaks in countries where the virus is not endemic.

The fact that infected animals successfully control viremia through natural immune response a few months after the acute phase indicates that it should be highly plausible to successfully develop efficient vaccines for EIAV infection. Unfortunately, to date, there has been only moderate success with no broadly protective vaccines available. The earliest vaccines were cell-culture-attenuated vaccines that conferred protection against disease development after virus challenge, but required a period of up to 6 months for protective immunity to develop. The only **l**ive-**a**ttenuated **v**accine (LAV) with a reportedly high protective rate of 90% was developed in China and only trialed there and in Cuba. Lack of access to the study data and the inability to independently verify those results means there is still a need for an efficient LAV globally. The main concern with attenuated virus vaccine use is (1) the inability to differentiate infected from vaccinated animals and (2) the possibility of the virus to revert to virulence given its high mutation rate. Inactivated whole virus preparations provide up to 95% protection against homologous virus challenge (i.e., same EIAV strain) and 100% protection against disease development, but not against heterologous virus strain challenge. Subunit vaccines based on purified viral glycoproteins provide protection against homologous virus infection, but not heterologous virus or development of disease.

EQUINE ENCEPHALITIS

Equine encephalitis encompasses three similar diseases—Eastern, Western, and Venezuelan equine encephalitis—that involve the inflammation of the brain and spinal cord of equid species. These encephalitides are caused by related but genetically distinct alphaviruses (**E**quine **e**ncephalitis **v**irus (EEVs), family *Togaviridae*) named according to the caused illness (see *Structure and Genomic Organization*). The viruses differ in geographic distribution and epidemiology, but all can result in serious and severe neurological syndromes. These alphaviruses are usually maintained in vertebrate hosts such as birds or rodents, and are transmitted to equine species by hematophagous arthropods (Fig. 8.1). They are of great medical import because each can cause potentially fatal zoonosis (i.e., human disease). Initially, infections in equines are subclinical with symptoms of fever, tachycardia, anorexia, and depression. This can become fatal once the virus infects the CNS, and survivors of infection usually experience residual nervous system issues. In humans, there is some similarity with disease characterized by fever, drowsiness, and neck rigidity. This may progress to paralysis, confusion, convulsions, and coma. In addition, survivors of equine encephalitides usually, due to the long-term neurological sequelae experience significant economic burden related to treatment postinfection.

Structure and Genomic Organization

The genomic structure of *Eastern equine encephalitis virus* (EEEV), *Western equine encephalitis virus* (WEEV), and *Venezuelan equine encephalitis virus* (VEEV) are similar to other alphaviruses. They are small, spherical, enveloped, ssRNA(+) viruses ranging from 65 to 70 mm in diameter. The genome is approximately 11.5 kb in length, flanked by 5′ and 3′ **un**translated **r**egions (UTRs) and encodes four **non**structural **p**roteins (NSP1–4) and five structural proteins (E1, E2, E3, 6K, C). The 5′ end is capped, while the 3′ end is polyadenylated. Copies of the capsid C protein, forming the NC, surround the genomic RNA. This nucleocapsid core is further surrounded by a host-derived lipid bilayer embedded with two viral glycoproteins, E1 and E2. Small amounts of the 6K protein are also found in the virion, but E3 is absent.

Pathogenesis and Replication

During infection, the pathogen promotes viral transcription and translation while shutting down that of the host cell. This leads to the attenuation of the host innate immune response, as interferon production by the host cell is markedly decreased, and in the EEVs is effected by the C protein. Infection ultimately leads to cellular apoptosis. Virus enters the host cell through interaction between the E2 protein and cell surface receptors. The E2 protein is thought to have either multiple receptor-binding sites, or the host cell receptor may be one that is ubiquitous, accounting for the broad host range of alphaviruses and the reality may be that both occur. It is also thought the E1 protein may also facilitate the process. Once attachment is established, the virion is endocytosed into the cell via a clathrin-dependent pathway. Acidification within the vesicle causes a rearrangement of the virion E1–E2 dimer, revealing the E1 fusion peptide that fuses with the vesicle membrane releasing the NC into the cytoplasm.

Once in the cytoplasm, the 5′ two-thirds of the viral RNA serves as mRNA for the expression of the **non**structural **p**roteins (NSPs) as a polyprotein. These NSPs (in both cleaved and uncleaved forms) direct the synthesis of a full-length negative strand. Later in the infection, this becomes the template not only for the production of full-length positive-strand genomic RNA, but also transcription of a subgenomic RNA from the 3′ end for structural protein expression in the order C-pE2-6K-E1. NSP1 is thought to perform the capping of viral and subgenomic RNA necessary to facilitate translation in the host cell. NSP2 has helicase activity key to unwinding RNA during replication and transcription, and also has protease activity responsible for cleaving the NS polyprotein. The full function of NSP3 remains unknown though evidence suggests it is involved in minus-strand and subgenomic RNA synthesis. NSP4 displays the RNA-dependent RNA polymerase activity necessary for replication.

Processing of the structural polyprotein begins with the excision of the C protein through an intrinsic serine protease. The remainder of the polypeptide inserts into the **e**ndoplasmic **r**eticulum (ER) where E2 is released in its precursor form pE2 by the action of host enzymes, separating from the 6K protein previously linking it to E1. Subsequently, 6K and E1 are also cleaved and processed in the ER lumen. pE2–E1 dimers form which translocate to the Golgi apparatus and subsequently to the cell surface during which pE2 is cleaved into E2 and E3. Throughout this relocation 6K protein associates with these dimers and forms viriporins (cation-positive channels in the plasma membrane), which aid in virion budding. Conversely, most alphaviruses do not include E3 in the virion, and its main role is thought to be the stabilization of the pE2/E1 dimer. In mammalian host cells, the C protein associates with the genomic RNA copies and each NC arranges itself as distinct entities in the cytoplasm and relocates to the cell membrane to associate with areas containing the E2, E1, and 6K proteins. Virions are then formed by the budding of NCs through these viral protein patches, during which the NC interacts specifically with the cytoplasmic domain of pE2 for assembly.

Eastern Equine Encephalitis

EEEV is found in parts of North America (particularly the eastern coast), Central America, the Caribbean basin, and the northern and eastern coasts of South America. Its main enzootic transmission cycle occurs between passerine birds and mosquitoes, primarily *Culiseta melanura*. Infections in birds are usually asymptomatic, with only a few species such as turkeys, pheasants, and emus showing high mortality rates. EEEV is carried over from this mosquito–bird enzootic cycle to dead-end hosts such as horses and humans by "*bridge vector*" mosquito species such as *Aedes*, *Culex*, and *Coquillettidia*, which feed on both birds and mammals. Due to the breeding habits of the vector, transmission usually occurs around freshwater swamp environments. The virus can also cause infections in other animals such as sheep, cattle dogs, pigs, and goats sometimes resulting in meningoencephalitis and other neurological consequences. In humans, zoonotic infections are deadly with fatality rates between 50% and 80%, and even greater in horses (up to 90%); EEEV is considered the most virulent of the EEVs.

Western Equine Encephalitis

WEEV is found throughout the Americas, from Canada down to Argentina. In the past, the disease caused major epizootic outbreaks in the western regions of North America and hence the name "*western*." Historically, in North America WEEV has caused the death of thousands of humans and hundreds of thousands of equids, with case fatality rate in humans being significantly less than in horses (>3%–7% vs 3%–50%). In recent decades, the incidence of human infections has declined dramatically with no cases reported since the late 1990s. In North America, WEEV is maintained in epizootic cycles involving passerine birds, natural vertebrate hosts and its most common mosquito vector *Culex tarsalis*—a mosquito particularly adapted to irrigated agricultural areas. Transmission to horses and humans is "*bridged*" by several species of mosquitoes, including *Ochlerotatus melanimon*, and both are considered dead-end hosts. In South America, other mammals, such as rodents, bats, turtles, and ungulates, may play a role as significant reservoirs of the virus.

Genetic analyses of WEEV indicate the virus is a natural recombinant of EEEV and a *Sindbis virus*-like alphavirus and is highly conserved across 95% of its genome. Epizootic strains of WEEV are typically more virulent with greater neuroinvasiveness, and North American enzootic strains tend to be more virulent than those found in South America.

Venezuelan Equine Encephalitis

VEEV is a recently reemerging pathogen transmitted between mosquito vectors and vertebrate hosts. Enzootic cycles typically occur in Central and northern South America between small vertebrate hosts (usually rodents), and the *Culex* (*Melanoconion*) species of mosquito. Epizootic cycles occur in horses and the vector involved is usually *Ochlerotatus taeniorhynchus*. During such epizootics in horses, viruses can spillover to humans causing epidemics. VEEV is also capable of infecting via aerosol routes and has been involved in several accidental laboratory infections. For that same reason, it has in the past been developed as a biological weapon in both the United States and the former Soviet Union. Severe encephalitis in humans is less frequent with VEEV than with EEEV and WEEV with a low fatality rate, though neurological sequelae are fairly common. Pregnant women are also at an increased risk of having abortions with VEEV infection.

Diagnosis, Prevention, and Treatment

Detection of all three EEVs can be done through direct assays (virus or nucleic acid isolation), but these take the longest to perform due to the need to culture isolates. More rapid serological assays can be done using cerebrospinal fluid or serum. For serology, IgM is detectable before the second week of infection and so is routinely used in screening for infections with these viruses. IgG is less reliable, especially in horses, as such antibodies are not necessarily predictive of recent infection and can also be due to vaccination especially in the case of EEEV and WEEV. Because cross-reaction is possible between alphaviruses, other tests are also used for confirmation alongside serology. Plaque reduction neutralization tests can distinguish between EEVs and other alphaviruses, but cannot determine subtype. RT-PCR can also be used to rapidly detect viral RNA in the case of all three viruses but comes with the disadvantage of a greater possibility of false-positive results.

Immunity is cross-protective to a degree among EEEV, WEEV, and VEEV, and several single and combination vaccines have been developed that give limited protection in equids, usually having to be administered annually. Most approved EEV vaccines are killed formalin-inactivated whole virus products, though live attenuated and replicon-based vaccines have shown promise and are still being researched. Similar vaccines have a poor rate of seroconversion when used in humans, though there has been limited success in rodents and nonhuman primates. There is no effective antiviral treatment for any of these EEVs and so treatment is generally supportive.

AFRICAN HORSE SICKNESS

African **h**orse **s**ickness (AHS) is a disease that affects all equid species, and is caused by *African horse sickness virus* (AHSV)—an *Orbivirus* of the *Reoviridae* family. The disease is infectious but not contagious and is transmitted by the bite of *Culicoides* midges with the most important vector being *Culicoides imicola*. AHS is devastating in horses, with some AHSV strains causing up to 95% mortality. It is generally characterized by symptoms that develop due to damage to the circulatory and respiratory systems, giving rise to edema in the lungs and intermuscular tissues, and hemorrhaging particularly of serosal surfaces. High fever and inappetence are also common. Zebra is thought to be the natural host and reservoir of the virus, and infected animals rarely exhibit clinical symptoms. The disease is a World Organization for Animal Health (OIE)-listed disease and outbreaks cause economic burden as movement of horses between these and other areas are tightly restricted. Dogs are the only other species that may contract fatal AHS, but almost all known cases have been due to the ingestion of infected horsemeat. They are not thought to be involved in transmission as they exhibit low viremia and the *Culicoides* spp. are not attracted to them. Infections in camels are rare and inapparent and they seem to play no role in AHS epidemiology.

Though AHS is generally limited to sub-Saharan Africa where it is enzootic, with annual outbreaks in southern African countries, there have been past epizootics in more northern African, Asian, and European countries such as Egypt, Morocco, India, Iran, Spain, and Portugal and the vector is present on the European continent. Despite these past outbreaks, the failure of the disease to establish itself outside of tropical and subtropical regions of the African continent to date is likely due to the absence of the natural zebra host in such regions, along with mass vaccination and vector control campaigns. Nevertheless, *Bluetongue virus* (BTV), a similar orbivirus also transmitted by *C. imicola*, has emerged in Europe in recent years due to climate change-induced spread of the vectors. This has raised fears that the highly fatal AHS may potentially cause serious enzootics in the continent.

Structure, Genomic Organization, and Genetic Diversity

AHSV is nonenveloped, icosahedron and about 80 nm in diameter. Its genome consists of 10 linear dsRNA segments of varying size: 3 large (L1–L3), 3 medium (M4–M6) and 4 small (S7–S10) segments. These 10 segments code for 12 proteins including 7 structural viral proteins (VP1–7) and four NSPs (NSP1, NSP2, NSP3/3A and NSP4). The core is surrounded by three capsid layers: an inner capsid shell composed of VP3 dimers, an intermediate capsid composed of VP7 trimers, and finally, a diffuse outer capsid composed of VP2 trimers and VP5. Within the core, VP1 (the viral polymerase), VP4 (the capping enzyme), and VP6 (viral helicase) make up the viral transcription complex and are attached to the inner surface of the VP3 shell. The NS proteins are involved in virus replication (NSP1), morphogenesis (NSP2), and exit (NSP3/3A) from the infected cell. The function of NSP4 has yet to be resolved, but it can bind dsDNA and is thought to possibly play a similar role to the same protein found in BTV of modulating host immunity.

AHSV exists in nine antigenically distinct serotypes, namely, ASHV 1–9. This variation is largely due to the high variability of the VP2 protein of the outer capsid. The RNA segments encoding NSP3/3A and VP5 also show significant variation. All nine AHS serotypes have been documented in eastern and southern Africa, and serotypes 4, 6, and 9 have been detected in outbreaks in the Middle East, Asia, and Europe. AHSV, like other orbiviruses, can not only undergo mutation but gene reassortment, which has serious implications for the efficacy of current vaccines. Viable reassortants introduce unpredictable diversity into the gene pool, which over time can escape and overcome previously acquired immunity, putting animals at risk for future infection.

Pathogenesis and Replication

After entry into the vertebrate host, the AHSV multiplies initially in the lymph nodes (primary viremia) and is disseminated via the circulatory system to the target organs, i.e., lungs, spleen, and other lymph tissue. In these tissues, the virus further multiplies (secondary viremia) and the incubation period from entry to here can take anywhere from 2 to 21 days, though usually it is less than 9. High fever usually coincides with the onset of viremia, which persists until the fever disappears. The disease manifests itself in four forms: pulmonary, cardiac, mild, and mixed. Pulmonary AHS is peracute and characteristic symptoms include pyrexia, depression, and impaired respiratory function such as pulmonary edema, which can lead to respiratory failure and death within 24 hours. Cardiac AHS is subacute with symptoms usually appearing 7–12 dpi. High fever is consistent with progressive dyspnea and edema, and survival rate is usually 50%–70%. Mixed AHS is confirmed after death where horses show signs of both pulmonary and cardiac AHS, while the mild form is subclinical and is the only form seen in zebras and African donkeys.

VP2 is thought to initiate virion attachment to the host cell, aided by membrane fusion-like coils within VP5. Entry into the cell is thought to occur through either macropinocytosis or clathrin-mediated endocytosis, as seen in BTV. Further conformational changes in VP5, due to the low pH in the endosome, lead to membrane fusion and release into the cell. Virus uncoating after entry reveals the inner core, and also appears to induce apoptosis, which may explain some of the more obvious clinical symptoms of the disease such as edema and hemorrhaging. The core contains the machinery which transcribes the 10 genomic segments simultaneously, as well as cap and methylate the resulting mRNAs. The genomic dsRNA segments are never released from the core, enabling the virus to avoid the host cell's defenses. mRNA transcripts are released into the cytoplasm to either be translated into viral proteins (a process promoted by NSP1) or serve as template for dsRNA to be packaged with nascent core particles into virus progeny. Newly translated viral proteins (predominantly NSP2) sequester viral mRNA to form **v**iral **i**nclusion **b**odies (VIBs) in the cytoplasm, which become the site of both dsRNA and further mRNA production. It is believed that these VIBs are the site of genome encapsidation and core assembly, a process directed by NSP2. Maturation of the core particles occurs outside the VIBs by the addition of the outer capsid proteins. Virus particles are then released by budding or direct membrane penetration with apoptosis, with the aid of NSP3.

Diagnosis, Prevention, and Treatment

The classic method for detection of AHSV has been through VI by intracerebral inoculation of day-old suckling mice, though **b**aby **h**amster **k**idney

(BHK21), and African green monkey (Vero) cell cultures are also used as they show cytopathic effects. Subsequent serotyping is then performed by **v**irus **n**eutralization **t**ests (VNTs). These conventional methods can take days to weeks and are mainly used for confirmation. Rapid screening and serotyping methods include indirect sandwich ELISA, probe and RT-PCR assays (based on informative genomic regions such as L2 or L3). Of these rapid methods, RT-PCR is now most widely used due to its ability to detect low levels of antigen in samples without live virus and quick processing time of a few hours.

Immune protection for AHSV is serotype-specific and LAVs are designed which can elicit a protective though incomplete immune response. Such vaccinations are usually recommended annually. More recently, a polyvalent LAV has been developed and put into use in South Africa and adjacent countries. There are however issues with the use of these types of live vaccines such as virus reversion and genome reassortment with other virus strains within vectors which can lead to epizootics. There is no specific treatment for AHS apart from rest. Because the disease is noncontagious, control of disease spread includes restricting the movement of infected animals to prevent new infections foci. The use of proper husbandry to prevent exposure of animals to vectors is also vital, as efficient control of *Culicoides* numbers is difficult due to a lack of complete knowledge regarding their ecology. Movement of horses from countries where HS is endemic to AHS-free countries usually involves strict quarantine and testing to ensure animals are disease-free.

MAREK'S DISEASE

Marek's **d**isease (MD) is a chronic progressive disease of avian species with several variable signs and syndromes. It is caused by *Gallid alphaherpesvirus 2* (GaHV-2), historically called *Marek's disease virus type 1* (MDV-1), a member of the family *Herpesviridae*, subfamily *Alphaherpesvirinae*, genus *Mardivirus*. It is common in domestic chickens, being rarer in turkeys, quail, and geese. MD is estimated to cost the global poultry industry at least US$ 1 billion in losses annually with about half of the world's countries having reported infections.

Structure, Genomic Organization, and Genetic Diversity

GaHV-2 is a dsDNA virus morphologically similar to other herpesviruses. There is an icosahedral capsid core 100 nm in diameter surrounded by an amorphous proteinaceous tegument layer, and finally an outer envelope, which is a lipid bilayer with more than 10 viral glycoproteins embedded in it. The genome is 180 kb in length, with long (UL) and short (US) unique regions, each flanked by inverted repeats of these unique regions aptly named **T**erminal **R**epeat **L**ong (TRL), **I**nverted **R**epeat **L**ong (IRL i.e., inverted repeat of TRL), **T**erminal **R**epeat **S**hort (TRS) and Inverted Repeat Short (IRS, i.e., inverted repeat of TRS).

Translation of GaHV-2 strains has revealed over 100 potential ORFs, the majority of which are homologous to that of other alphaherpesviruses. Other better-studied herpesviruses such as Herpes simplex virus 1 (HSV-1; *Human alphaherpesvirus 1*) have been used to assign putative functions to many GaHV-2 homologues. These include genes encoding for viral replication. The major capsid protein is VP5 (encoded by *UL19*), while the tegument (matrix) has four major virion components: VP11/12, VP13/14, VP16, and VP22, of which the latter (VP22) is conserved among the virus subfamily and thought to be essential for viral replication. There are several envelope glycoproteins including gB, gC, gD, gH, gI, gL, gK, and gp82. Viral protein gB is highly conserved among herpesviruses and is the major envelope component, eliciting neutralizing antibodies in infected birds. Other key proteins include ICP4 which plays a role in maintaining transformed lymphocytes, and origin-binding protein (encoded by *UL9*), the initiator of DNA replication, which have homologues in other herpesviruses. Proteins unique to the genus include the GaHV-2 specific Meq and pp38, which are key modulators of tumor formation and virus latency. Uniquely for a virus, GaHV-2also encodes its own **v**iral **t**elome**r**ase (vTR) located in the TRL and IRL regions, which is thought to maintain transformed cells, playing a central role in both tumorigenesis and metastasis. A number of microRNAs are also coded by the virus genome, which are thought to have roles in both the early cytolytic and latent phases of infection.

The *Mardivirus* genus comprises five species, some historically referred to as serotypes: serotype 1 (*Gallid alphaherpesvirus 2*, GaHV-2 or MDV-1), serotype 2 (*Gallid alphaherpesvirus 3*, GaHV-3 or MDV-2) and serotype 3 (*Meleagrid alphaherpesvirus 1*, MeHV-1); the other two species are *Anatid alphaherpesvirus 1* and *Columbid alphaherpesvirus 1*. MeHV-1 is also historically referred to as Herpesvirus of Turkeys (HVTs). GaHV-2 consists of all known pathogenic strains, which vary in their pathogenic and oncogenic potential, being classified as mild (m), virulent (v), very virulent (vv), and very virulent plus (vv+) GaHV-2. GaHV-3 (e.g., SB-1 MDV) and MeHV-1, are nonpathogenic and nononcogenic.

Pathogenesis

The pathogenesis of GaHV-2 manifests in several, at times overlapping, syndromes. Classic MD commonly involves asymmetric paralysis of one or both legs and wings. It can also involve impairment of the vagus nerve leading to difficulty in breathing and/or dilation of the crop, and there is frequently lymphomatous infiltration in the skin, muscle, and visceral organs. Mortality rarely exceeds 15% in this form. Acute MD has the highest mortality rate (up to 70%) and involves sudden outbreaks in previously unvaccinated and uninfected flocks of mostly young chickens. It usually displays ataxia and paralysis a few days postinfection, and is also characterized by the presence of prominent nerve lesions, though the visceral lymphomas seen in Classic MD are usually absent. Ocular lymphomatosis is rare and involves the graying of the iris of one or both eyes due to the infiltration of transformed lymphocytes in these tissues. This usually results in partial or total blindness. Cutaneous MD is characterized by round nodular lesions of up to 1 cm in diameter, which occur mostly at feather follicles. Nonfeathered areas of the legs may also appear red in color. Additional syndromes include transient brain edema-induced paralysis, atherosclerosis, and degenerative lymphoma.

Virus Replication

Replication involves periods of lytic replication, during which new virions are made and cells die, and latent replication in between, during which the viral DNA is integrated and copied with that of the host cell. Much of the GaHV-2 replication cycle is thought to be similar to that of other herpesviruses. GaHV-2 attaches to the host cell via an array of envelope glycoproteins (gB, gC, gD, and gH) and receptor-mediated endocytosis occurs. Within the endosome, fusion releases the tegument and core of the virus into the cytoplasm. The nucleocapsid is then transported to the nuclear pore via the dynein-microtubule system of the cell where the viral genome is released into the nucleus—processes modulated by the tegument proteins. In latent replication, the viral DNA becomes integrated into host chromosomes of no obvious preference near the telomerase regions, and the viral genome is copied by the cellular replication machinery. The mechanism of integration is not clear, but it is thought to involve the actions of the vTR along with the host telomerase. Currently the stimulus for activation of latent cells remains unclear, and also whether the latter is necessary for the former to occur.

Lytic replication continues with the transcription of genes necessary for the construction of virions. There are three types of genes in the viral genome: intermediate-early, early, and late genes. The intermediate-early genes are the first transcribed—a process transactivated by VP16. Products of these genes promote the transcription of the early genes by the host polymerase II (which are necessary for replication) and protect the virus against the host cell's innate immunity. ICP4 is one such IE protein and is thought to transactivate the *pp38* gene, and thus play a role in maintaining transformed cells. An initial round of genome replication occurs bidirectionally resulting in circular genome amplification. This is then followed by synthesis of linear concatemer copies of the viral DNA by a rolling circle mechanism. Host polymerase II then transcribes late mRNAs, producing viral structural proteins. The glycoproteins become embedded in the nuclear membrane and become the site of the viral factories.

Several UL-transcribed genes (e.g., UL6, UL15, and UL17) are involved in cleaving and packaging the concatemer DNA into monomeric form to make the complete nucleocapsids (termed C-capsids). These C-capsids then aggregate at these nuclear viral factories to exit the nucleus.

Much is still unresolved regarding the mechanism by which GaHV-2 exits the host cell and spreads from cell to cell. The most supported theory is the double-envelopment model, which proposes that the nucleocapsid buds through the inner nuclear membrane forming a first envelope (primary enveloped virion), a process that involves the viral proteins UL31 and UL34. This primary envelope, then moves across the perinuclear space and fuses at the outer nuclear membrane, releasing the nucleocapsid into the cytoplasm. Once in the cytoplasm, the C-capsid binds several tegument proteins (i.e., tegumentation) and is reenveloped through the trans-Golgi network, en route to its final exit via exocytosis. Only these reenveloped mature virions are capable of infection of other cells.

Diagnosis, Prevention, and Treatment

The first effective vaccines for MD were live attenuated GaHV-2 vaccines developed in the 1960s, which reduced the incidence of MD up to 99%. Unfortunately, over time, the evolution of escape mutants (a direct result of the selective pressure of vaccination) leads to increasing outbreaks. In the 1980s, bivalent vaccines combining nonpathogenic GaHV-3 and MEHV-1 strains such as (SB-1 + HVT) were introduced. However, most of these bivalent vaccines have not been very effective against vvMDV and vv+ GaHV-2, and to date only the live attenuated Rispens vaccine (CVI988-Rispens) is effective against these more virulent GaHV-2 strains. However, none of the current vaccines provides sterile immunity but rather limits the severity and mortality of the disease. GaHV-2 transmission is also not inhibited and so the virus not only remains in circulation, but also the risk of increasingly virulent strains emerging that can break through the vaccine-induced immunity remains. For that reason, current strategies include revaccination of birds with a different vaccine from that initially used to garner greater immune protection.

FOOT-AND-MOUTH DISEASE

Foot-and-**m**outh **d**isease (FMD) is a highly contagious, sometimes fatal disease of cloven-footed animals including domestic livestock (such as sheep, cattle, goats, and pigs; Fig. 8.1), and several wildlife species (including buffalo, deer, and bison). The causative agent is the *Foot-and-mouth disease virus* (FMDV), which is a member of the *Aphthovirus* genus, family *Picornaviridae*. FMD is characterized by high fever, which declines rapidly after 23 days, and blisters inside the mouth and on the feet of the infected animal. These blisters extend to the mammary glands of female animals. These blisters may rupture and in the case of foot blisters result in lameness of the animal. Infected animals can also experience dramatic weight loss due to complications in eating due to mouth blisters. As such, mortality from the disease is usually low but morbidity is high in adults, though mortality may be significantly higher in young animals due to issues such as acute myocarditis. The only known natural host for FMDV is the Cape buffalo in which several serotypes (SAT 1–3, see *Structure, Genomic Organization, and Genetic Diversity*) replicate and persist with minimal disease pathology.

Historically, FMDV had a worldwide distribution during the 19th century and was found in regions that farmed livestock except Oceania, but successful eradication and vaccination programs has seen it eliminated from many developed countries, including those in North America and much of western Europe. Currently the disease is still enzootic in much of Africa, Asia, the Middle East, and current disease distribution reflects a lack of economic resources to implement successful prevention and control measures. FMDV persists for several months in infected animals and is moderately stable in the surrounding natural environment, which accounts for the ease of transmission and persistence. Humans, in rare cases, can become infected and human-to-human transmission is rare and insignificant.

Structure, Genomic Organization, and Genetic Diversity

The FMDV virion particle is spherical in shape with a diameter of about 25 nm. The genome is an ssRNA(+) molecule 8400 nucleotides long, and is surrounded by an icosahedral protein capsid. The coding region of the genome is ~7000 bp and is flanked by 5′ and 3′ UTRs involved in regulation of virus replication. The 3′ UTR is polyadenylated, and is a promoter of viral replication. The 5′ UTR has several regulatory structures including an S fragment at its 5′ end, a poly C-tract, several RNA pseudoknots, a *cis*-acting replication element (*cre*) site and an **i**nternal **r**ibosomal **e**ntry **s**ite (IRES). There is also a small viral protein covalently attached to the 5′ UTR at its 5′ end termed the viral protein attached to genome (VPg). The FMDV genome has a single ORF, which reads as follows: 5′ UTR-L-VP4-VP2-VP3-VP1-2A-2B-2C-3A-3B-3C-3D-3′ UTR. The *L* gene encodes two alternate viral proteinases (L_{ab} and L_b), while VP1, VP2, VP3, and VP4 (commonly designated 1D, 1B, 1C, and 1A, respectively) are the four structural proteins that comprise the viral capsomers. Proteins 2A, 2B, 2C, and 3C are also proteinases, while 3A is an inhibitor of cellular protein secretion (along with 2B). Protein 3B encodes VPg found at the 5′ end of the genome, and finally *3D* encodes the core RNA polymerase.

There are seven known serotypes of FMDV that are geographically defined: Asia-1 found only in Asia; SAT-1, SAT-2, and SAT-3 found in

Africa; and types A, O, and C, which are found in Europe and South America.

Pathogenesis and Replication

Routes of transmission of FMDV include mechanical contact (direct and indirect), airborne (droplets and long-range), and oral transmission (i.e., ingestion). Of these, droplet dispersal and inhalation of virus is most common when infected and noninfected animals are in the same space. The incubation of the disease after contact with FMDV is highly variable and depends on the FMDV strain, dose of infection, species infected, and husbandry conditions experienced by animal.

FMDV can enter the host cell through two different pathways. The first involves the binding of the VP1 capsid protein with integrins, followed by clathrin-mediated endocytosis. Once inside the acidified endosome the capsid rapidly dissociates revealing the naked RNA genome. The second pathway occurs through the binding of membrane-embedded heparin sulfate resulting in caveola-mediated endocytosis. The uncoated genome initially undergoes one round of translation of its single ORF resulting in a single polyprotein. In reality, this polyprotein is never seen as various virus proteases cleave it into four primary protein products during translation: L, P1-2A, P2, and P3. L contains the Leader proteinase (L^{pro}) which cleaves the host translation initiation factor eIF4G, preventing translation of host capped mRNAs. This allows FMDV to translate its own viral proteins, using the 5′ UTR IRES and the cleaved fragment of eIF4G. The region contains two different AUG initiation codons resulting in two forms of the L protein, which both are vital to the virus replication cycle. P1−2A is the capsid precursor and is cleaved into VP0 (VP4 + VP2), VP3, VP1, and 2A by protein $3C^{pro}$. Products P2 and P3 are processed to the NSPs, with P2 generating 2B and 2C, and P3 generating 3A, 3B, 3C, and 3D.

The first step in replication is the generation of minus (−) strand RNA, which occurs after the translation of the positive strand. Viral RNA synthesis begins in the cytoplasm once the VPg is uridylylated by the viral polymerase 3D at the *cre* site. Once this happens, VPg can successfully act as a primer for the viral RdRp to initiate transcription of the entire genome. This generates the **r**eplicative **f**orm (RF) of the virus, which serves as the template for new positive strand synthesis. In the final steps of replication, the new plus-strands are encapsidated, with only plus-strands linked to VPg involved to the exclusion of other plus-strands. Viral $3C^{pro}$ cuts polymer P1 into VP0, VP1, and VP3, and these assemble with the viral RNA genome to form provirions. Maturity is attained when the VP0 protein is cleaved to produce mature VP2 and VP4.

Diagnosis, Control, Prevention, and Treatment

Clinically, FMD is indistinguishable from other vesicular diseases of animals, and so diagnosis has to be confirmed by virological assays. The VNT is the "*gold standard*" for FMDV detection, in which antibodies to viral structural proteins are used to detect the virus in cell cultures, and is recommended for conclusive diagnoses. Such cultures are derived from cow or pig cells, which exhibit cytopathic effects when infected. More rapid techniques such as ELISA and RT-PCR are preferred screening methods to VNT in outbreaks, with the latter being a confirmatory tool. RT-PCR (both standard and real-time) has the advantage over ELISA of being more sensitive, reliable, and capable of not only determining serotypes but also virus quantification. Where necessary and cost-effective, laboratories will use a combination of these methods to confirm FMDV. The preferred sample, especially for ELISA, is vesicular fluid or epithelial tissue, though the more sensitive techniques such as VNT and RT-PCR can also use other sample types (e.g., blood, swabs, and feces).

Control of FMD spread requires strict sanitary animal husbandry and/or suitable vaccination strategies. Countries that have experienced FMD outbreaks in the past and are now disease-free, typically relied on control of animal movement, culling, vaccination, and extensive surveillance to reach that status. An important issue with vaccination is that currently there is no validated OIE-approved method that distinguishes vaccinated animals from possible carriers, as they display similarly low viremia. With the high variation of FMDV and the lack of cross-protection between and sometimes within serotypes, vaccination selection is based on FMDV strains in circulation in a given region. Traditional whole-virus inactivated vaccines remain the most common method in cattle, pigs, and sheep. These vaccines have a limited shelf-life and, as explained, usually are serotype- and sometimes strain-specific. These vaccines take several days to elicit an immune response, and require repeat vaccinations to maintain immunity every 4–12 months. Successfully vaccinated animals can also become persistently infected, i.e., carriers. Despite these vaccines not being panacea for different serotypes (because serotypes are largely geographically limited except for type O), proper strategy can overcome that particular issue. Alternative vaccines are in development and have shown promise, including antivirals such as replication-defective human adenovirus delivery system for FMDV proteins and genetically engineered chimeric vaccines, which display a combination of common and serotype-specific structural proteins to elicit an immune response. Recent efforts have also investigated the use of RNA interference to inhibit replication, but these potential advances in vaccination each have issues such as mode of delivery and virus escape that have to be resolved.

FURTHER READING

Carpenter, S., Mellor, P.S., Fall, A.G., Garros, C., Venter, G.J., 2017. *African horse sickness virus*: history, transmission, and current status. Annu. Rev. Entomol. 62, 343–358. Available from: https://doi.org/10.1146/annurev-ento-031616-035010.

Haq, K., Schat, K.A., Sharif, S., 2013. Immunity to Marek's disease: where are we now? Dev. Comp. Immunol. 41, 439–446. Available from: https://doi.org/10.1016/j.dci.2013.04.001.

Jamal, S.M., Belsham, G.J., 2013. Foot-and-mouth disease: past, present and future. Vet. Res. 44, 116. Available from: https://doi.org/10.1186/1297-9716-44-116.

Morens, D.M., Taubenberger, J.K., 2010. Historical thoughts on influenza viral ecosystems, or behold a pale horse, dead dogs, failing fowl, and sick swine. Influenza Other Respir. Viruses 4, 327–337. Available from: https://doi.org/10.1111/j.1750-2659.2010.00148.x.

Weaver, S.C., Powers, A.M., Brault, A.C., Barrett, A.D., 1999. Molecular epidemiological studies of veterinary arboviral encephalitides. Vet. J. 157, 123–138. Available from: https://doi.org/10.1053/tvjl.1998.0289.

Chapter 9

Viruses of Prokaryotes, Protozoa, Fungi, and Chromista

Gustavo Fermin[1], Sudeshna Mazumdar-Leighton[2] and Paula Tennant[3]
[1]*Universidad de Los Andes, Mérida, Venezuela,* [2]*University of Delhi, Delhi, India,*
[3]*The University of the West Indies, Mona, Jamaica*

Perhaps most remarkable is the finding that the genome of pandoravirus salinus contains 2,556 putative protein-coding sequences, of which only 6% have recognizable relationships with genes from known viruses, microorganisms or eukaryotes. Phylogenetic analysis has suggested to some investigators that these viruses form a previously unknown domain, a fourth branch of the tree of life—others have said that it is too early to draw such a conclusion. In any case, the "virus world" is getting more and more interesting.

Frederick A. Murphy

In this chapter, we describe viruses that are associated with life forms other than Animalia and Plantae. These viruses are diverse and successful, like their hosts themselves, and as such the chapter is organized around the classification of the cellular hosts they infect. The hierarchically ranked system known as the **C**atalogue **o**f **L**ife (CoL) is used. This posits a seven-kingdom classification, which includes the prokaryotic kingdoms Archaea and Bacteria, and the eukaryotic kingdoms Protozoa, Chromista, Fungi, Plantae, and Animalia. We are aware that not all scientists agree on this scheme. Nonetheless, adopting this classification makes it easier for readers to relate a group of living cellular beings with the viruses that infect them. We briefly present examples of viruses associated with Archaea, Bacteria, Protozoa, Chromista, and Fungi.

Historically, viruses infecting members of the superkingdom Eukaryota are nearly always referred to as viruses, while those of the kingdom Bacteria, to a substantial extent, are called bacteriophages or phages, and viruses of Archaea described as Archaeal viruses or Archaeal phages. That said, new discoveries of viruses crossing kingdoms blur this distinction.

Viruses. DOI: https://doi.org/10.1016/B978-0-12-811257-1.00009-7

To further add to the complexity is the interaction between bacterial viruses, their bacterial hosts, and the animal host of the bacterial microbe. Although some bacterial viruses can be a part of the microbiome of animals, they are really hosted by a bacterium. The phage does not infect the animal, but the virus coexists (most of the time) under mutually beneficial terms with its animal "*host,*" under the radar of its immune system.

Deep metagenomics studies support an emerging view that species of the kingdoms Archaea, Bacteria, Protozoa, Chromista, and Fungi harbor remarkable viral diversity, and highlight how little of the virosphere has been explored. Viruses belonging to families with only animal or plant hosts represent more than half of all described viruses. However, recent studies have revealed that it is actually prokaryotic viruses that predominate our planet and that they outnumber their hosts by at least an order of magnitude. As more and more studies are available, it is becoming clear that these viruses also exhibit remarkable genomic flexibility facilitated by frequent recombination, lateral gene transfer between the viruses and their hosts, gene gain and loss, and complex genomic rearrangements. In most cases, the infections they induce are cryptic. They range from mutualism, involving an improvement of the adaption abilities of the host toward biotic and abiotic stress, to severe impairment including irregular growth and reduced reproduction in infected hosts.

VIRUSES INFECTING ARCHAEA

Archaea are now recognized as a major part of global ecosystems, contributing up to 20% of the total biomass on Earth. These organisms are major players in global biogeochemical cycles, such as methanogenesis and ammonia oxidation. The niches where archaea are found are very particular and mostly extreme (e.g., hyperthermic, hypersaline, acid, or alkaline). They are adapted to conditions of chronic energy stress and are thus more successful in extreme environments than members of other kingdoms (with the exception of bacterial extremophiles). However, the archaea display many shared characteristics with bacteria and eukaryotes. For example, the cellular ultrastructure of archaea, devoid of a nucleus or organelles, resembles that of bacteria, whereas the replication and transcription machineries present many similarities to those machineries of eukaryotes. In this section, we present on viruses of the *Sulfolobus* species. These viruses, as with their hosts, show adaptation to extreme environments.

Sulfolobus islandicus rod-shaped virus 1

Various morphologically diverse viruses infect *Sulfolobus* species, a hyperthermophilic archaeon from the phylum Crenarchaeota. Currently under debate is whether the diversity of these viruses reflects the earliest forms of life that predate the split of life domains into archaea, bacteria, and eukaryotes. Among the most intriguing dsDNA viruses that infect *Sulfolobus islandicus* from Icelandic hot springs are the rod-shaped rudiviruses, SIRV1 and SIRV2. Taxonomically, these viruses belong to the order *Ligamenvirales* and family *Rudiviridae*. Rudiviruses are purported progenitors of large dsDNA viruses that infect eukaryotes as they share similar features in genome structure, encoded structural proteins, and DNA replication. The rudivirus SIRV1 has a rigid rod shape. The virion length is typically 830 nm. It is not enveloped and lacks lipid membranes. Plugs are found at each end of the rod to which three tail fibers are anchored. The tube-like super-helical coiled DNA is complexed to a multimeric, viral-encoded **ma**jor **c**apsid **p**rotein (MCP). Viral DNA is encapsidated in the A-form instead of the B-form, probably for protecting the DNA in the adverse environment where its host thrives.

The genome of SIRV1 was among the first full-length sequences reported for viruses infecting archaea isolated from extreme environmental samples. The viral genome is linear, dsDNA of 32.3 kb with 45 **o**pen **r**eading **f**rames (ORFs) situated on both strands (Fig. 9.1). The 3′ ends of SIRV1 DNA strands are linked to the 5′ ends by covalent bonds forming a hairpin loop of approximately four nucleotides, a feature reminiscent of eukaryal viruses that infect algae like *Chlorella* species. Genomic structures of SIRV1 and its closely related SIRV2 suggest the occurrence of recombination, duplications,

***Sulfolobus islandicus rod-shaped virus 1* (*Rudiviridae*)**

Monopartite, dsDNA

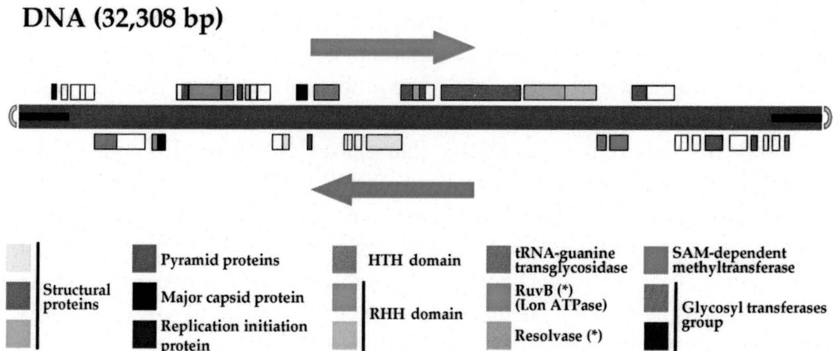

FIGURE 9.1 Genome organization of *S. islandicus rod-shaped virus 1*. Termini of the linear genome (seen as a continuous *red rectangle*) are covalently linked (depicted as *semicircles*). Inverted repeats occur at the 5′ and 3′ ends (seen as solid *horizontal lines* within the *box*), as well as seven direct repeats. The genome encodes at least 45 genes (seen as *boxes* above and below the *genomic rectangle*). Core genes conserved among geographically diverse SIRV including the MCP and various putative transferases are shown as *colored boxes*. Genes to which functions are yet to be assigned are shown as *white boxes*. Gene products associated with Holliday junctions are indicated by *asterisks*. *Green arrows* denote the direction of transcription on each strand. Note that genes on different strands do not seem to overlap.

horizontal **g**ene **t**ransfers (HGTs), and exchange of genes between viruses and hosts. Both SIRV1 and SIRV2 contain **i**nverted **t**erminal **r**epeats (ITR) of about 2 kb length comprising of perfect and imperfect direct repeats interspersed by evenly spaced interrepeat regions. However, in the case of SIRV2, an ITR carries a large insertion, accounting for a genome size of 35.45 kb. The rudiviruses also differ in the presence of a 103 bp ORF present in SIRV2, but not SIRV1. Both SIRV1 and SIRV2 carry multiple copies of 12 bp transposon-like elements within protein-coding regions suggesting transposition as another prominent mechanism for producing genetic variation.

The DNA of SIRV1 replicates via self-priming. Both head-to-head and head-to-tail–linked replicative intermediates have been reported. Short sequences of 21 bp that occur near the termini act as important signals for initiation of DNA replication. DNA replication involves a nick at a specific site (11 nucleotides) from the termini and ensues from a 3′ OH group, another feature reminiscent of dsDNA viruses infecting eukaryotes. However, they do differ in the absence of AT-rich hairpin loops at the termini of the SIRV1 genome. Recombinational events are implicated in SIRV1 based on the occurrence of cruciform structures. These structures are likely resolved by enzymes that recognize conserved sequence motifs common to SIRV1 and dsDNA viruses that infect eukaryotes. Interestingly, the genome of the host, *S. islandicus*, contains at least 126 CRISPRs (clusters of regular interspersed sequence repeats) flanked by 3042 spacer sequences, a unique feature said to underlie host immune responses against invading viruses. At least 10% (or ~302) unique spacer regions within the *S. islandicus* genome resemble SIRV1 sequences suggesting that rudiviruses are major pathogens of these archaea. The success of SIRV1 as a pathogen can be attributed to processes like recombination of CRISPR-associated spacer sequences that enable novel viruses to infect archaeal hosts and sweep viral communities.

SIRV1 is transmitted by passive diffusion through medium. It does not kill its host and there is absence of both lytic and lysogenic mechanisms. The virus is dramatically different from SIRV2 that causes cell lysis. Instead, SIRV1 develops a stable, carrier state with the archaeal host. However, when transformed into nonhost strains of species of *Sulfolobus*, SIRV1 is very unstable and undergoes extensive mutations at the rate of 10^{-9} per replication cycle especially at genomic "*hot spots*" involving virus–host interactions. The host range of SIRV1 includes *S. islandicus* strains like KVEM10H3, REN2H1, KVEM10H1, LAL14/1, and KVE6/3. Its distribution corresponds with the host, that is, sulfuric fields of Iceland. No information is currently available on its economic impact.

Sulfolobus turreted icosahedral virus 1

The tail-less icosahedral, dsDNA virus infecting the archaeon *Sulfolobus solfataricus* was first reported from a hot, acidic geyser (with temperatures above 82°C and pH >2.2) in the Yellowstone National Park, United States. The virus, STIV1, is taxonomically classified under alphaturriviruses of *Turriviridae*, a family that has not been assigned to any taxonomic order. Virus capsids are 74 nm in diameter and contain an inner lipid membrane. STIV is named after its characteristic 12, fivefold vertices with turret-like structures that extend 13 nm above the icosahedral capsid shell. The turret-like structures are used for viral egress from infected hosts and they also aid in viral DNA replication. A 37 kDa **m**ajor **c**apsid **p**rotein (MCP B345) makes up the virus capsid. MCP is linked to a host-derived lipid layer of the inner membrane by electrostatic interactions. MCP is glycosylated. The similarity in the structure of MCP B345 with capsid proteins of *Paramecium bursaria Chlorella virus 1* (*Phycodnaviridae*), the phage *Salmonella virus PRD1* (*Tectiviridae*), and mammalian adenoviruses supports the occurrence of a "*STIV-adeno-PRD1 lineage*." STIV1 was the first lytic virus known to infect members of the *Sulfolobus* genus. Cell lysis is apparently preceded by morphological changes in the cell such as the formation of pyramid-like protrusions on the surface.

The genome of STIV1 is composed of circular dsDNA of 17,663 bp encoding 36 ORFs, many of which have not yet been assigned a function (Fig. 9.2). A viral-encoded C92 protein is currently believed to be responsible for the formation of the pyramid-like structures and is important for replication. Complete information on its replication cycle is not yet known. At least two proteins encoded by the host (a putative DNA-binding protein, Sso7D, and a protein with a vacuolar sorting function, Sso0881) are packaged into virions during maturation. STIV typically packs the dsDNA genome after the virion shell assembles and the process involves a viral-encoded ATPase. Lipids of the viral inner membrane are synthesized *de novo* and co-assemble with the MCP during virion formation. Virus DNA pro-capsids are typically scattered within the host cell. However, after assembly they occur as organized arrays proximal to the pyramidal protrusions. It is believed that this organization of the assembled virions into arrays assists DNA replication and virus egress.

Microarray experiments have revealed that at least 124 host genes are upregulated during STIV1 infection. These upregulated genes are mainly associated with DNA replication, repair, and a number of unassigned functions. Expression of 53 genes is downregulated. All virus-encoded transcripts are detected within 24 hours postinfection including the genes expressed during early infection. Lysis of host cells is usually detected at 32 hours postinfection. Cell lysis is usually preceded by downregulation of host genes involved in cell energetics and metabolism. Viral egress involves modification of the glycosylated host S-layer prior to lysis and dramatic changes in the host morphology with characteristic pyramid-like structures arising from the cell surface. These pyramids lack the S-layer and are composed of distinct lipids and proteins including the viral-encoded C92. The pyramidal

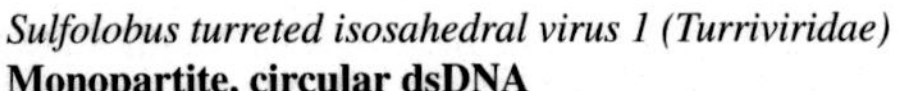

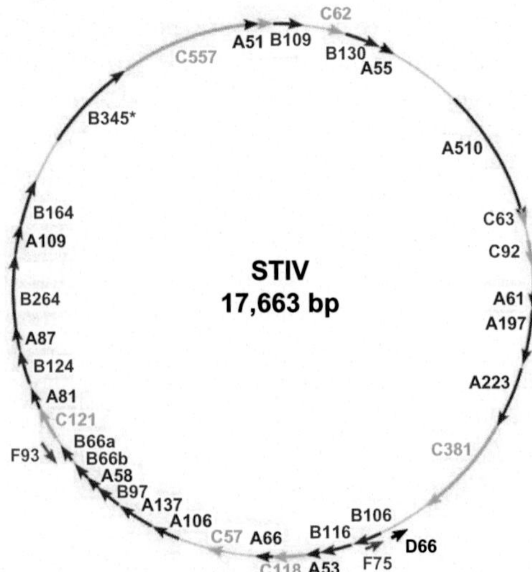

FIGURE 9.2 The circular double-stranded DNA genome of STIV1 showing putative ORFs occurring in multiple reading frames. Proteins encoded on first, second, and third frames of the positive strand are depicted by the colors *red*, *blue*, and *green*, respectively, while those predicted on first and third frames of the negative strand are shown in *black* and *purple* colors, respectively. Viral proteins confirmed by mass spectrometry are B345 (a glycosylated major capsid protein indicated by an *asterisk*), B164, C557, B109, B130, A55, A223, and C381. Not shown is the protein A78 located immediately downstream of A55 but lacking a methionine as the canonical start codon. *Rice, G., Tang, L., Stedman, K., Roberto, F., Spuhler, J., Gillitzer, E., et al., 2004. The structure of a thermophilic archaeal virus shows a double-stranded DNA viral capsid type that spans all domains of life. Proc. Natl. Acad. Sci. U.S.A. 101(20), 7717–7720. Copyright (2004) National Academy of Sciences, U.S.A.*

structures also include cellular proteins like endosomal sorting complexes that participate in virus replication and egress. Egress via pyramidal protrusions is also seen in SIRV2-infected cells.

VIRUSES INFECTING BACTERIA

Escherichia virus P1

Escherichia virus P1 belongs to order *Caudovirales* and family *Myoviridae*. It is commonly referred to as phage P1 and was one of the first bacteriophages to be identified in the coliform *Escherichia coli*. It is the representative species of a small genus *P1 virus* within *Myoviridae*, which also includes *Aeromonas virus 43*. Extensive studies on phage P1 have contributed significantly to pioneering developments in the 1960s in recombinant DNA technologies involving site-specific recombination, restriction modification, and cloning of large fragments of DNA.

The phage P1 has a large icosahedral head of approximately 65–85 nm diameter attached to a characteristic long tail of 220 nm length. A tube-like contractile sheath surrounds the tail that ends in a base-plate and six-kinked tail fibers of 90 nm length. The phage head encompasses a linear dsDNA genome of 93 kb. The complete genome sequence of phage P1 encodes at least 117 genes and contains a large characteristic sequence redundancy at each 5′ and 3′ end. These redundant sequences are variable in length and range from 10 to 15 kb. Upon entry into a host cell, the viral DNA undergoes rapid circularization by recombination between the redundant sequences. This homologous recombination is aided by host-encoded recombinases or by a phage driven, site-specific cre-*lox* recombination system. In the latter case, recombination proceeds between two *loxP* sites located on the terminal redundant regions of the viral genome aided by the phage-encoded "*cre*" protein.

Like other caudovirales, phage P1 is a temperate bacteriophage that can adopt lysogenic or lytic lifestyles. The decision to enter either a lytic or lysogenic phase depends upon the host environment and factors influencing transcription of a monomeric C1 repressor molecule that regulates immunity of phage P1. During lysogeny, the circularized phage DNA (referred to as a prophage) replicates in the host cytoplasm as a low copy number plasmid from an **ori**gin of **r**eplication (*oriR*) using various phage-encoded proteins that also repress the lytic phase. The plasmid prophage is highly stable and inherited by daughter cells following cell division of the bacterial host. Hence replication and partitioning of the phage P1 into daughter cells is tightly regulated. A phage-encoded protein RepA along with iterated repeat sequences at *incA* and *incC* sites participates in replication of the plasmid prophage. Replication is also affected by methylation status of *oriR* sequence on the phage genome and *oriC* sequence on the bacterial genome. Host factors like DnaA, HU, and various chaperones participate in modification of RepA and replication of the circularized prophage plasmid. Like bacteriophages P7 and P22 (that infect *Salmonella* sp.), phage P1 employs two repressor molecules (C1 protein and a C4 RNA) to suppress the lytic phase. Other regulators include phage-encoded transfer RNA molecules, a DNA **m**ethyl**tr**ansferase (MTR), transcriptional antiterminators (Coi and Ant1/2), and a co-repressor protein Lxc. Induction of the prophage is rare but occurs in response to UV damage and nutritional changes in the host's environment. A host transcriptional factor LexA protein is involved in the induction process. The phage P1 uses a different origin of replication (*oriL*) located within the gene encoding RepL protein for the lytic phase. Approximately 37 viral operons are transcribed during lytic phase by the bacterial RNA polymerase. The polymerase holoenzyme is formed in the presence of a phage-encoded Lpa protein whose levels are regulated in turn, by the C1 repressor molecule and a bacterial stringent starvation protein SspA.

An important feature of the virus is "*generalized transduction*," where instead of its own DNA, large fragments of the bacterial host DNA are packaged into the phage head. Unlike the lambda phage, generalized transduction by the phage P1 can mobilize about 100 kb of host DNA between two bacterial host strains. Upon transduction into a recipient bacterial strain, the bacterial genes occasionally integrate into the host genome by site-specific recombination. As a result, the transduced bacterium does not lyse and suffers no toxicity from virus infection.

The host range of phage P1 is *E. coli* that belongs to Gammaproteobacteria and Enterobacteriaceae. Therefore the distribution of this phage reflects that of its ubiquitous host. Transmission of the virus involves uniform penetration of the inner tail tube into the *E. coli* periplasm and shift of the base plate away from the outer membrane during tail contraction. The tail fibers bind to a specific host receptor that is a glucose moiety on the lipopolysaccharide core of the outer bacterial membrane. A lytic trans-glycosylase facilitates penetration of the bacterial cell wall and ejection of phage DNA into the host. A "Sim" protein is rapidly produced that confers immunity to the bacterial host, excluding other invading phages. Upon entry into the host, the introduced DNA remains bound to two phage-encoded proteins called DarA and DarB (a MTR and helicase) that render the viral DNA resistant to digestion by enterobacterial type I restriction endonucleases. Since the *E. coli* glycolipid receptor for phage P1 is common in several Gram-negative bacteria, the virus particle can adsorb onto the outer wall and inject DNA into the cytoplasm of many bacteria. However, it cannot undergo subsequent replication in these bacterial species. Discovery of sequences resembling phage P1 Cre recombinases in metaviromes from complex environments and enterobacteria suggests that further work is necessary to unravel the biology of these phages in different environments and hosts. Comparative genome sequencing of P1-related viruses (e.g., unclassified Punalikevirus viz. phage D6, phage phiW39, phage RCS47, and phage SJ46 identified from various enterobacteria) is likely to reveal insights into novel processes of viral DNA packaging, site-specific recombination, and immunity.

Escherichia virus Lambda

Escherichia virus Lambda is a tailed bacteriophage belonging to the order caudovirales, family *Siphoviridae*, and genus *Lambdavirus*. The virus is also widely known as the lambda phage or bacteriophage λ, and its natural host is *E. coli* K-12. Caudovirales can be easily distinguished from other viruses by the presence of distinct tailed appendages, in addition to an icosahedral head. The tail serves as a conduit for transfer of phage DNA into the infected host and the tail fibers are used to attach to the bacterial host. Caudovirales, like lambda phage, phage Mu, and phage P1, share a conserved three dimensional structure besides a similar genome organization and mode of DNA replication. Genes encoding many essential housekeeping proteins like integrases and recombinases, however, show very little sequence similarity among these phages and resemble diverse viruses. Despite this sequence divergence, all caudovirales exhibit lytic and lysogenic phases in their life cycles. Hence these tailed phages are also known as temperate viruses. Upon entering a lytic cycle, phages multiply and emerge in large numbers after lysis of the infected host. During lysogeny, the phage does not kill the host. Instead, the phage DNA integrates into the host genome as a prophage. The lambda prophage can excise out from the bacterial genome, circularize and replicate as a plasmid until the lytic phase ensues. Integrated prophages show dynamic recombinational events that may confer new phenotypes to their bacterial hosts, including pathogenesis. Lysogeny may be a mechanism that enables temperate phages to persist in complex gut and intestinal environments. The lytic phase is the ancestral state for these viruses, reminiscent of tailed viruses with similar morphology from the early Cambrian era.

The lambda phage linear dsDNA genome contains 48,490 bp with 12 nucleotide single-stranded segments called "*cos*" sites at the 5′ ends; they help in the circularization of phage DNA before replication in the host cytoplasm. The lambda genome encodes 29 essential genes that participate in the lytic phase including the *N*, *Q*, and *cro* genes; regulatory genes like *cI*, *cII*, *cIII*, and *int/xis* that control lysogeny, and at least 38 nonessential or accessory genes (Fig. 9.3). Temporal gene expression patterns drive the early transcriptional promoters P_L and P_R on the phage genome. Bacterial RNA polymerase transcribes the *N* gene from the P_L promoter, while the *cro* gene is transcribed from P_R promoter up to the corresponding terminator regions, t_{LI} and t_{R1}. The *N* gene, the first gene in the P_L operon, is transcribed immediately upon infection. Once levels of N protein start building up in a newly infected bacterial cell, it modifies the host RNA polymerase causing it to read past the terminators t_{L1} and t_{R1}. These transcriptional antitermination events produce proteins that are essential for both lysis and lysogeny. High growth rates and availability of a rich nutrient medium for the bacterial host drive the lytic phase and the release of hundreds of new phages. The N protein further serves as a positive regulator for lysis by forming the N antitermination complex. The *Q* gene, which is situated at the 3′ end of the early P_R operon, produces another transcriptional antiterminator that enables transcription of genes essential for the lytic phase. The *Q* gene is transcribed later than the *N* gene, committing the phage to the lytic phase and ultimately, production of mature infectious virions. The *Q* gene product is a DNA-binding protein that recognizes the −35 to −10 region of the late operon promoter $P_{R'}$. It also binds to a bacterial host-encoded RNA polymerase factor Rho. The lag in transcriptional antitermination events regulated by N and Q proteins is crucial for the switch to the lysogenic phase in lambda phage.

Lysogeny and prophage formation is a characteristic of tailed bacteriophages. Regulatory genes encoding the CI repressor, CII activator, and CIII protein are essential for induction and maintenance of lysogeny. The gene *cI* encodes a repressor molecule (CI) that maintains lysogeny, shuts down lysis by preventing transcription from the early promoters P_R and P_L, and provides immunity to the bacterial host from infections by other related phages. It was

Escherichia virus Lambda (Siphoviridae)
Monopartite, linear dsDNA

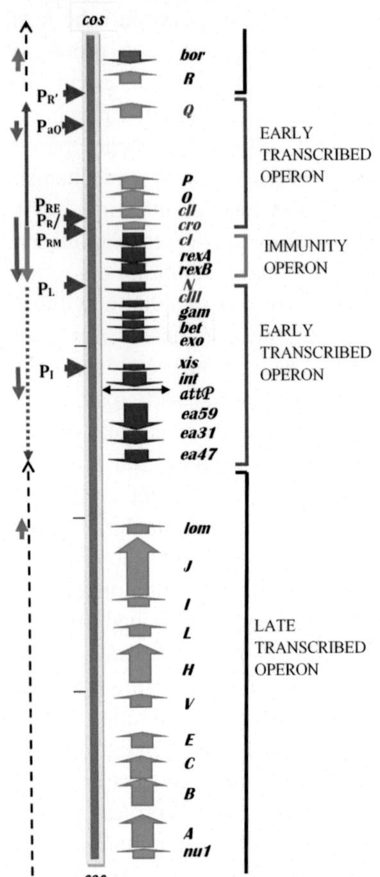

FIGURE 9.3 A map for lambda genome is shown as a linear molecule (*magenta line*) flanked by cohesive sticky ends or "*cos*" sites. ORFs encoding well-characterized viral proteins are shown as blocks of light blue and dark blue arrows pointing in the direction of their transcription. Relative positions of promoters (denoted by the prefix "*P*") are indicated by maroon arrowheads. Transcripts from the late operon are shown as dashed, black arrows. Transcripts from the early operon that originate from P_L and P_R are shown as red, dashed and red, solid lines, respectively. Transcripts associated with virus immunity are shown as a turquoise arrow. Short green arrows denote transcripts produced in prophages, while brown arrows indicate genes expressed at high levels of CII repressor protein. Regulatory genes (*pink letters*) and position of "*attP*" sequence (*double-headed, black arrow*) are indicated. Horizontal bars denote approximate sequence lengths of 10^4 bp.

among the first genes to be identified that auto-regulates its own synthesis. The CI repressor forms concatemers that bind to operator regions controlled by the P_L and P_R promoters causing a loop formation in the DNA between these two promoters and turning off transcription of the *N* and *cIII* genes (from the P_L promoter) as well as the *cro* and *cII* genes (from the P_R promoter). It weakly activates the P_{RM} promoter that continues expression of the *cI* and *rex* genes (that render the host cell immune to phage infections). As a result, the synthesis of CI continues even in the absence of CII and CIII proteins. At high concentrations of CI, transcription of *cI* gene from the P_{RM} promoter is reduced until an optimum level is achieved that represses lysis and infection by other phages. The Cro protein is a weak repressor that can also bind to the P_{RM} promoter, and reduces the transcription of *cI* gene. The ability of CI repressor to shut off transcription of *cro* gene (essential for the lytic phase) and in turn, be regulated by Cro protein (a product of the first gene of the P_R operon) has been referred to as a genetic switch modulating lysogeny or lysis. The CII activator is the major switch for lysogeny that turns on the transcription of *cI* gene from the early P_{RE} promoter and *int* gene (encoding an integrase) from the P_I promoter. CII also activates a third promoter P_{AQ} that produces a small antisense RNA molecule from the *Q* gene that keeps late (transcriptional) operator shut off, preventing onset of the lytic phase. Several host factors influence activity of the CII activator; in addition, CIII protein is important for the stability of the CII activator. It inhibits a host-encoded protease complex from degrading the CII protein. In bacterial cells growing in a limited nutrient environment, protease activity is low, making the cells stable and the invading phage more prone to a lysogenic lifestyle.

The phage genome integrates into the host DNA by site-specific recombination at the *att* sites (present in both the virus and the bacterial host) mediated by a virus-encoded integrase. This viral protein is part of an Integration Complex that also includes host-encoded proteins like IHF (**I**ntegration **H**ost **F**actor) and F_{IS}. The IHF is multifunctional. In addition to facilitating the protein–DNA interaction during recombination at *att* sites, it also influences levels of *CII* translation and *cIII* transcription. Prophages can be induced to excise out of the bacterial genome by UV and chemicals like mitomycin C. During excision, CI repression is abolished and the prophage is released from the site of integration. The phage DNA circularizes at the *cos* sites and replicates within the host. Replication initially proceeds bi-directionally from a single point of origin, making many copies of each strand. Subsequently, replication proceeds by a rolling circle mode before the phage genomes are packed into virions. The phage genes *O* and *P* participate in DNA replication. Host proteins like DNA helicases are co-opted for the process. This stage corresponds to the release from CI repression and resumption of lysis. The protein Xis is a phage-encoded protein that reverses the process of integration. The levels of both Xis and Int proteins influence the continuation of the lysogenic or lytic phase of lambda life cycle. Several nonessential lambda phage genes include lysogenic conversion genes, like *bor* and *lom* that are involved in host interactions, and *exo*, *bet*, and *gam* that aid in recombination events. Other exclusion genes such as *rex* and *sie* prevent further infections by phages.

Resumption of the lytic phase is preceded by assembly of virions. The phage head and tail assemble independently and then combine spontaneously to form each virion. The phage gene *E* encodes the MCP. The genes *Nu1* and *A* encode a lambda terminase holoenzyme that participates in packaging of the virion with the host-encoded IHF protein that bends/folds the phage DNA into the virion. Once large numbers of virus particles accumulate in the host cell, lysis occurs and virions are released. Membrane proteins called holins, phage proteins Rz and Rz1 of the spanin complex and S105/S107 participate in the actual lysis and degradation of the bacterial peptidoglycan cell wall. Infection of new *E. coli* host cells involves recognition of the phage J protein by the LamB protein of *E. coli*. The LamB protein is an outer membrane protein normally involved in maltose uptake. It is the main receptor for lambda phage recognition.

Since its serendipitous discovery in 1950s and until the late 1980s, phage λ has been the pioneering and best studied model for studies on gene expression, lytic/lysogeny switches, and assembly of DNA virus particles. Renewed interest in the study of phage λ and related lambdoid phages is attributed to their use as phage display tools, vaccine/hormone delivery agents, and enzymatic nano-reactors (see Chapter 12, Viruses as Tools of Biotechnology). Another promising application of classical and contemporary perspectives into the λ phage lies in its potential to lyse bacterial pathogens.

VIRUSES INFECTING FUNGI

Cryphonectria hypovirus 1

Cryphonectria hypovirus 1 (CHV1) is a fungus-infecting virus (or mycovirus) belonging to the genus *Hypovirus* of the family *Hypoviridae*. It persistently infects the filamentous fungal pathogen, *Cryphonectria parasitica* (Murrill) Barr, of the phylum Ascomycetes, family Sordariomycetes that causes a devastating blight in chestnut trees (*Castanea dentata* in North America and *C. sativa* in Europe). Infections with CHV1 and several hypoviruses can ameliorate disease symptoms caused by fungal infection, resulting in healing of disease cankers on tree trunks. This phenomenon is known as "*hypovirulence*." Infection by CHV1 does not kill the fungal host but reduces its pathogenicity and its ability to cause disease. Hypovirulence has been studied extensively in CHV1-infected *C. parasitica* due to practical applications in biological control of Chestnut blight. Application of *in vitro* culture techniques to studies on compatibility of fungal strains and virus transmission constitute pioneering work into the genetics of fungus-virus interactions. Availability of genome sequence information, infectious CHV1 clones, and reliable DNA transformation protocols for *C. parasitica* has made this a model system to study mycoviruses.

Hypoviruses like CHV1 lack true capsids. Instead the viral genome is encapsulated within host-derived lipid pleomorphic vesicles that are usually 50–80 nm in diameter. A linear dsRNA genome of the first hypovirus to be sequenced was CHV1-EP713. It is 12,712 bp in length without the poly(A) tail and encodes two ORFs (Fig. 9.4A). ORF A encodes a polyprotein that is autocatalytically cleaved into two proteins called p29 and p40. The p29 is a papain-like cysteine protease that initiates viral protein maturation. ORF B encodes a polymerase, a helicase, and an autocatalytic protease p48 that resembles p29. Isolates of CHV1 typically show sequence divergence in these ORFs. CHV1 identified from natural populations of *C. parasitica* infecting chestnut trees includes the isolates CHV1-EP713 (identified from

(A) ***Cryphonectria hypovirus 1 (Hypoviridae)***
Monopartite, ssRNA(+)

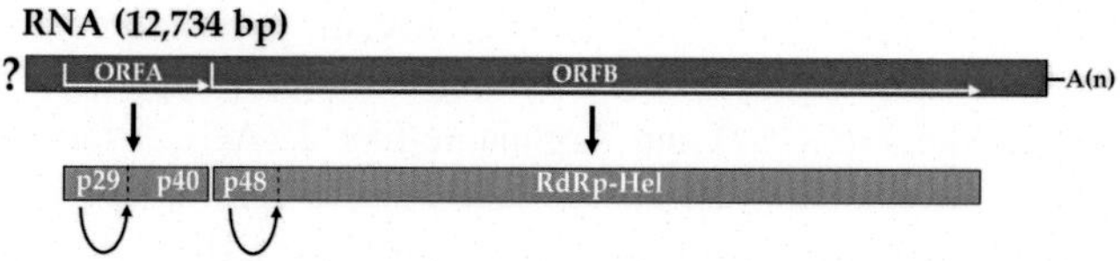

(B) ***Sclerotinia sclerotiorum betaendornavirus 1 (Endornaviridae)***
Monopartite, dsRNA

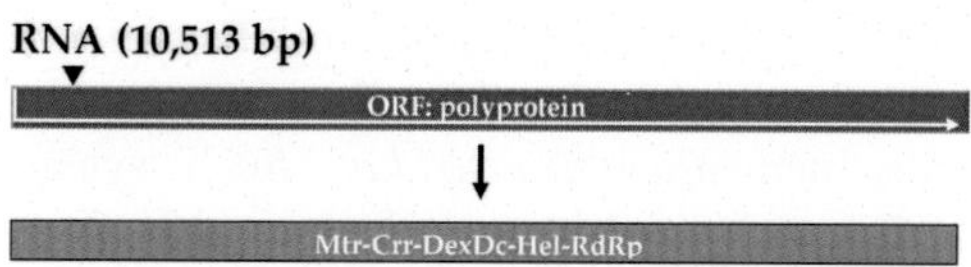

FIGURE 9.4 Genome organization of hypoviruses and endornaviruses. (A) The polyadenylated [A(n)] genome of the hypovirus CHV1 contains two ORFs. ORF A encodes a polyprotein (p69) that is autocatalytically processed into two products, p29 and p40, by the action of a papain-like protease domain located in the p29 coding region. Protein p40 participates in viral RNA amplification and translational control of hypovirus gene expression. Protease p48 is released from the N-terminal portion of the ORF B-encoded polypeptide. The ORF B-encoded polypeptide contains viral polymerase (RdRp) and helicase (Hel) domains. The *curved arrows* depict sites of autoproteolysis carried out by p29 and p48 on their respective precursors. The 5′ terminus of the genome may be capped (?). (B) Members of the *Endornaviridae* family (except SsEV 2) possess a monopartite genome with a single nick (shown as an *inverted triangle*) on the coding strand of the dsRNA genome. The *triangle* illustrates a nick that is common to all other members of the family. The genome contains a single ORF that gives rise to a polyprotein with motifs of a MTR, a CRR implicated in proteolysis, two disparate helicases, DEAD-like helicase (DexDc), and RNA helicase (Hel), as well as viral replicase (RdRp). The putative DNA-binding phytoreo_S7 domain (associated with Horizontal Gene Transfer among phytoreoviruses), upstream to the RdRp motif, is not shown in the polyprotein box.

France), and CHV1-EURO7 (identified from Italy) that differ significantly in their virulence and effects on fitness of their fungal hosts. A large number of unclassified hypoviruses infecting fungi like *Fusarium* sp., *Sclerotinia* sp., *Valsa* sp., *Trichoderma* sp., *Phomoposi*s sp., and *Macrophomin*a sp. have been identified and their sequence information is available. Hierarchical taxonomic relationships among these mycoviruses are currently unknown. The mode of viral replication for CHV1 is cytoplasmic and follows the usual dsRNA virus replication process.

Symptoms of CHV1 infection in the pathogenic fungus include reduced pigmentation, reduced asexual sporulation, reduced conidial germination, loss of female fertility, and reduced virulence. Transmission of CHV1 may be horizontal (by fungal anastomosis or hyphal fusion) or vertical by dispersal of asexual spores called conidia. Horizontal transmission among fungal populations is influenced by the vegetative compatibility/incompatibility of the fungal hosts undergoing fusion. Generally hypovirulence is associated with low diversity of *vic* loci that govern hyphal incompatibility. Fungal hosts typically deploy RNA silencing to combat hypoviruses. Genes encoding **D**icer-**l**ike **2** (DCL2) and **A**r**g**onaute-**l**ike **2** (AGL2) nucleases that participate in RNA silencing are upregulated several fold upon CHV1 infection, an observation currently unique to *C. parasitica*. CHV1 overcomes this fungal defense mechanism by producing a "*suppressor*" from the viral gene p29. Therefore, like its potyviral counterpart HC-Pro, p29 contributes to autocatalytic proteolysis of viral polypeptides as well as the suppression of host-mediated RNA silencing. Another role ascribed to the CHV1 gene p29 as suppressor of RNA silencing has been to enable co-infection of *C. parasitica* with RNA viruses belonging to families *Reoviridae*, *Narnaviridae*, *Partitiviridae*, *Totiviridae*, and *Megabirnaviridae*. Such synergistic co-infections frequently result in enhanced hypovirulence and further reduction of blight symptoms on chestnut trees.

Sclerotinia sclerotiorum betaendornavirus 1

Sclerotinia sclerotiorum betaendornavirus 1 (SsEV1) is a nonenveloped, dsRNA virus belonging to family *Endornaviridae*. Endornaviruses persist in low copy numbers in infected cells, generally share a "*symbiotic*" relationship and rarely kill their hosts. SsEV1 was isolated from the white mold fungus, *S. sclerotiorum*, a major pathogen of several crop plants that causes extensive yield loss globally. Next Generation Sequencing techniques have been used successfully to identify diverse viruses that infect *S. sclerotiorum* including putative hypoviruses and endornaviruses. Genome sequences of SsEV1 isolates have been reported from the United States, China, and New Zealand. Based upon features like a small genome size (approximately 10.5 kb) and an absence of a "*nick*" or discontinuity in the 5′ region, SsEV1 is classified as a betaendornavirus. Other betaendornaviruses reported from pathogenic ascomycete fungi include *Alternaria brassicicola betaendornavirus 1*, *Gremmeniella abietina betaendornavirus 1*, and *Tuber aestivum betaendornavirus 1*. In contrast to betaendornaviruses, alphaendornaviruses have been reported from plants, fungi, and oomycetes. Alphaendornaviruses have larger genome sizes than betaendornaviruses and contain a site-specific nick at the 5′ ends. In fact, the type species of endornaviruses is *Oryza sativa alphaendornavirus* (OsEV). Its coding sense strand RNA (13.95 kb) contains the diagnostic nick at position 2525 (which is 1.2–2.7 kb from the 5′ terminus). Interestingly, endornaviruses are believed to have originated in fungi and then moved to plant hosts by **h**orizontal **g**ene **t**ransfer (HGT).

Like all members of *Endornaviridae*, SsEV1 has no true capsid and is nonenveloped. The linear dsRNA genome of SsEV1 contains an ORF flanked by **unt**ranslated **r**egions (UTR) (Fig. 9.4B). Typically, the 5′ UTR is longer than the 3′ UTR, which ends in a stretch of poly C residues. The ORF encodes a single polyprotein containing domains for a MTR, a **c**ysteine-**r**ich **r**egion (CRR), a putative DEXDc, a viral RNA **hel**icase (Hel), phytoreo_S7 (S7), and a **RNA-d**ependent **RNA p**olymerase (RdRp). Conserved sequence motifs within the MTR show homology with counterparts in Sindbis-like, ssRNA viruses, and several mycoviruses. In SsEV1, the MTR at the amino terminus is followed by the CRR domain, which contains two signature motifs CXCC and CXCCG (where "*X*" represents any amino acid). It is proposed that the CRR may encode viral proteases that cleave the polyprotein. However, experimental evidence supporting such proteolytic processing in *Endornaviridae* is still awaited. A DexDC domain follows the CRR and contains conserved motifs similar to DExH box Hels. A second Hel domain contains motifs reminiscent of Hels from superfamilies 1 and 2. The RdRP domain occurs after the Hel domains and is proximal to the carboxyl-terminus of the SsEV1 polyprotein. It contains at least eight conserved motifs typical of virus-encoded polymerases.

Endornaviruses like SsEV1 rarely cause symptoms or kill their hosts. Further work is necessary to understand the mechanisms for persistence and coevolution of endornaviruses with their plant and/or fungal hosts. The occurrence and distribution of SsEV1 corresponds with the prevalence of its cosmopolitan fungal host, *S. sclerotiorum*. SsEV1 isolates are usually detected using standard molecular biology techniques like isolation of dsRNA and RT-PCR as demonstrated for SsEV1-11691. Metagenomic approaches to characterize viromes from cultured pathogenic fungi as well as environmental samples have also revealed novel isolates of SsEV1 with potential to improve current understanding of the biology of these cryptic fungal viruses.

VIRUSES INFECTING PROTOZOA

RNA viruses of *Leishmania*

Many protozoan parasites, like *Trichomonas*, *Leishmania*, *Giardia*, *Plasmodium*, and *Entamoeba*, are infected with viruses classified in the family *Totiviridae*, in any of three genera, *Giardiavirus*, *Leishmaniavirus*, or *Trichomonasvirus*. Two other genera are recognized in the family, *Totivirus* and *Victorivirus*. Viruses in the genus *Totivirus* have been found to infect the yeasts *Saccharomyces cerevisiae*, *Scheffersomyces segobiensis*, and *Xanthophyllomyces dendrorhous*, the smut fungus *Ustilago maydis*, and the black summer truffle, *Tuber aestivum*, and more recently, mosquitoes, shrimp, and plants. Members of *Victorivirus* infect only filamentous fungi.

Members of the family *Totiviridae* are characterized by isometric virions 40 nm in diameter that contain a nonsegmented, linear dsRNA genome of

4–7 kb. Based on studies in yeast, the virion is composed of more than a hundred capsid protein subunits and one to two capsid-RdRp subunits surrounding a single genomic dsRNA molecule. The genome encodes only two proteins. It contains two overlapping ORFs, *gag* and *pol*, which respectively encode the capsid and the RdRp. The RdRp is expressed as a CP/RdRp fusion protein, either as a consequence of ribosomal frameshifting or as a direct fusion with CP. Some viruses contain two additional small 5′-proximal ORFs. There are no reports that these small ORFs encode gene products.

First described more than two decades ago, ***Leishmania* RNA viruses** (LRV) were reported to exist within many species of the protozoan genus *Leishmania* as a stable infection or endosymbiont. The virus was characterized in *L. braziliensis braziliensis* and *L. braziliensis guyanensis* (both belonging to the subgenus *Viannia*), and *L. major*. High diversity in nucleotide sequence (less than 40% homology) between LRVs of *L. (Viannia)* and *L. major* has categorized *Leishmania* viruses into the groups LRV1 and LRV2, respectively. LRV seems to follow the generic *Totiviridae* conformation described earlier (Fig. 9.5). The 5.3 kb dsRNA genome is never completely uncoated within the host cell. Viral polymerase synthesizes mRNA, which is translocated to the cell cytoplasm. There, transcripts are equipped with cap structures derived from host mRNAs by a cap-snatching mechanism mediated by the virus capsid protein. Plus-strand viral transcripts direct the translation of ORF2, major CP (Gag), and the minor fusion protein of OFR3 CP-RdRP (Gag-Pol) via a ribosomal frameshift. Predicted protein sequences of the other ORFs (1 and 4) have so far shown no evidence of encoding polypeptides. Mature virions are transmitted to new cells during cell division.

While there is no evidence of abnormal effects to *Leishmania* infected with the virus, a particularly curious relationship exists between LRV1 of certain strains of *L.* (*Viannia*) and the human host. A "*Russian doll*" model of hyperpathogenesis is used to describe the relationship. Hyperpathogenesis

***Leishmania RNA virus 1* (*Totiviridae*)**
Monopartite, dsRNA

RNA (5283 bp)

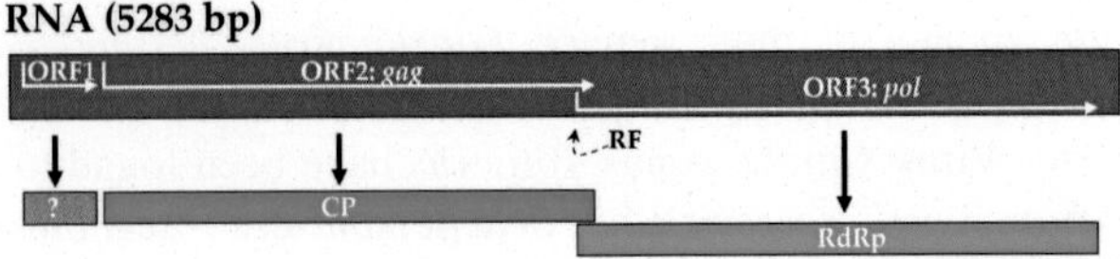

FIGURE 9.5 Genome organization of *Leishmania RNA virus 1*. The monopartite genome of this virus is composed of a single dsRNA molecule that contains three ORFs: ORF1 (of unknown function), ORF2 (*gag*, encodes the capsid protein), and ORF3 (*pol*, the replicase, RdRp). The overlap between ORF2 and ORF3 can produce a CP-RdRp fusion product or alternatively express the RdRp by ribosome frameshifting in the region of the overlap.

refers to the situation whereby a pathogen living endosymbiotically within another pathogen increases disease virulence. In this instance, LRV1 appears to enhance infection by its protozoan host inside its vertebrate. Virus dsRNA is recognized by the host's innate immune system thereby enabling *Leishmania* to subvert the host immune response, resulting in increased severity of leishmaniasis, the disease caused by *Leishmania*.

Leishmaniasis is one of the most important human protozoan parasitic diseases worldwide. Each year, about 2 million people contract leishmaniasis, caused by protozoan parasites from more than 20 *Leishmania* species. The parasite is transmitted to humans by the bite of infected female sand flies. After transmission by sand flies, *Leishmania* parasites infect and form vacuoles in neutrophils, macrophages, and dendritic cells in the vertebrate host. There they multiply, ultimately rupturing the host cell, releasing stages to infect new macrophages, including those that circulate in the blood. There are three main forms of the disease: cutaneous, mucocutaneous, and visceral leishmaniasis. Overall, the mechanisms involved in lesion development are poorly understood. But recent evidence suggests that dsRNA of virus-infected *Leishmania* acts as a virulence factor. It has been shown that a strain of *L. braziliensis guyanensis* infected with LRV1 induced more metastasizing lesions in mice than a virus-free one. It is posited that LRV1 dsRNA behaves as a strong innate immunogen, inducing a hyper-inflammatory immune response by the pathway of toll-like receptor 3. This pathway induces production of a type-1 interferon response resulting in an amplification of inflammatory responses, contributing to the exacerbation of the disease. This fascinating discovery stresses the importance of not considering any pathogen in isolation, but rather in the context of potential biotic partners. These interactions can increase the complexity of host–pathogen relationships and bring into question the concept of "*one microorganism, one disease*," as proposed in Koch's postulates.

Acanthamoeba polyphaga mimivirus

Giant viruses infecting marine or freshwater eukaryotic algae referred to as **N**ucleo**c**ytoplasmic **l**arge **DNA v**iruses (NCLDVs) were described in 2001. Initially the group comprised four families, *Poxviridae*, *Asfarviridae*, *Iridoviridae*, and *Phycodnaviridae*. Four other families, *Ascoviridae*, *Marseilleviridae*, *Mimiviridae*, and *Pandoraviridae* ("*unclassified dsDNA viruses*"), were added after 2007. Diverse eukaryotes, such as protozoa, vertebrate animals, and insects, host these viruses. As their name suggests, NCLDVs replicate in both the nucleus and cytoplasm (phycodnaviruses, asfarvavirus, and iridovirus), or only within the cytoplasm (poxviruses). They possess large genomes (100 kb to up to 2.77 Mb) and a diverse gene repertoire that encodes up to 900 proteins. To put the size of these large viral genomes into perspective, the smallest free-living bacterium, *Mycoplasma genitalium*, encodes 470 proteins. Indeed, they appear somehow closer to a typical cell than any other described virus, encoding proteins that have never been previously identified in viruses but have closely related eukaryotic homologs. NCLDVs belonging to *Mimiviridae* were among the first to be described.

Mimivirus particles are easily visible under the light microscope. They are composed of an icosahedral protein capsid with a diameter of 500 nm that is uniformly covered with a 125 nm thick layer of closely packed fibers. The fibers are morphologically unique among NCLDVs. A prominent five-fold star-shaped structure is located at one icosahedral vertex. This structure, coined the stargate, facilitates entry and infection of host cells. The interior of the particle contains two lipid membranes; an inner membrane encapsidating the nucleic acid, and another that is in direct contact with the interior of the protein capsid. A number of proteins are also packaged in the particles; proteins involved in transcription, DNA repair, and RNA, protein, or lipid modifications. The mimivirus genome, a single linear dsDNA molecule in excess of 1 million bp, encodes most of the proteins essential for virus replication and expression. Genes relevant to mRNA translation are also present: tRNAs and tRNA charging, initiation, elongation, and termination factors, with the exception of ribosome components themselves. Genes that are unique to giant viruses encode proteins that are involved in nucleotide synthesis, amino acid metabolism, protein modification, and lipid or polysaccharide metabolism. The mimivirus genome also encodes a complete set of DNA repair enzymes capable of correcting nucleotide mismatches as well as errors induced by oxidation, UV irradiation, or alkylating agents. Some genes of mimiviruses are homologs in cellular organisms that appear to have been acquired by HGT to the virus, are paralogous genes (duplicated genes), or orphan genes (genes that lack a homolog in any sequence database). Taken together, these viruses do not show strong dependence on the host's systems of replication or transcription for completing their replication, which is typical of viruses.

Acanthamoeba polyphaga mimivirus (APMV) is the prototype of the *Mimiviridae* family. The virus was serendipitously discovered during investigations on the pathogenic microorganisms associated with the 1992 outbreaks of pneumonia in England. Free-living *Acanthamoeba* were isolated from water samples collected from a hospital water cooling system and thought to contain a new type of intracellular Gram-positive bacterium. Ultrastructure studies by electron microscopy in 2003, however, revealed icosahedral-like virus particles with a 750 nm diameter size. Viral identity, and not bacterial, was confirmed by sequence analysis. This new virus was then named based on its ability to infect *A. polyphaga*, and because its structure was very similar to that of parasitic bacteria (i.e., mimics a microbe). Most of the currently identified mimiviruses also infect *Acanthamoeba*. *Acanthamoeba* is a freshwater protozoan and an opportunistic human pathogen. It can be found in a variety of environments, from aquatic environments, soil, and air, and is part of the vertebrates' normal microbiota. The life cycle of the amoebas involves two cellular forms: one that is metabolically active, known as the trophozoite form, and a dormant form, called the cyst. Encystment (i.e., conversion from a trophozoite to a cyst) is triggered by adverse environmental conditions. APMV is unable to infect preformed cysts, only trophozoites. Moreover, APMV infection promotes a reduction in the expression of an amoebal encystment-mediating subtilisin-like serine proteinase, thereby preventing encystment and allowing APMV to complete its replication cycle. Considering the ubiquity of *Acanthamoeba*, giant viruses could hypothetically be found everywhere. Accumulating studies suggest that vertebrates may also be hosts of these viruses. The detection of these viruses of amoebae in humans and the study of their potential pathogenicity are emerging fields.

APMV is unique in its replication cycle. The process involves a rapid takeover of cellular machinery along with extensive membrane biogenesis and a highly coordinated assembly of membrane structures. The virus enters the amoebal host via phagocytosis. Once in the cell, the stargate channel in the virus capsid opens, there is fusion between phagosome and virus membranes, and release of the internal core containing the genome. Replication occurs exclusively in the cytoplasm. The early stages of infection initially take place within the viral core, but later continue in viral factories derived from host nuclear membrane or **e**ndoplasmic **r**eticulum (ER). Described as cytoplasmic inclusions, virus factories consist of an accordion-like mass of membranes, permeated with ribosomes, and the machinery necessary for capsid and DNA synthesis. At any given moment, all stages of replication including capsid assembly and genome encapsidation occur concomitantly. As with other DNA viruses, transcription proceeds in a temporal manner; namely, the early, intermediate, and late stages culminating with the packaging of capsids with viral DNA. Assembly occurs at the surface of the virus factories. ER-like cisternae (flattened membrane discs) are recruited to the periphery of the viral factory. Numerous vesicles bud from these cisternae and eventually fuse to generate open single membrane sheets. It is from this sheet that multiple vesicles emerge and icosahedral capsids self-assemble around these structures. Filling with virus DNA and associated proteins occurs by the transfer of nucleic acid from the interior of the virus factory through a transient pore (other than the stargate). Once filling is complete, the portal is sealed with a protein plug. The remaining structural layers are acquired when the capsid passes in succession through (1) a membrane embedded with integument protein and (2) a membrane, containing on its distal side, a coating of surface fibers. Hundreds of new viruses are released by amoebal lysis.

Since APMV was named the prototype of the family in 2005, over 100 other mimiviruses-like viruses have been isolated: smaller giants of the family *Marseilleviridae*, and then pandoraviruses, which are the current record holders for virion and genome size. These viruses have been found in locations where amoebae normally thrive; seawater, soil, aerosols, and man-made aquatic environments such as sewage, fountains and air conditioners, in addition to harsh, unexpected ecosystems such as permafrost. The most recent, in 2017, are the Klosneuviruses isolated from wastewater in Austria. Klosneuviruses encode an even larger collection of genes needed for protein manufacture and are able to incorporate 14 of the 20 different amino acids into proteins without any help from the host. Analysis of their protein-manufacturing genes reveals close relationships to cellular organisms, rather than viruses.

VIRUSES INFECTING CHROMISTA

Chromista is an assemblage of disparate living organisms. Among its members are the stramenopilous organisms (those with a unique, composite, tubular, flagellar hairs) and plastid-related groups, like the haptophytes and the cryptomonads. In addition, there is the pseudofungus lineage (the Oomycetes) and the chromophytous algal lineage, and so diatoms, golden algae, giant kelps, slime-nets, water molds and the downy mildew pathogen are all chromists. Despite their diversity, these organisms share some or all of the following characteristics: presence of tubular mitochondrial cristae, chloroplast ER (complex plastids, with extra envelope membranes), and tubular mastigonemes (tubular "*hairs*") on at least one flagellum.

Viruses that infect chromists are characterized by almost the same variability in terms of genome organization, virion shape and replication strategies. The genomes can be comprised of dsDNA, ssDNA, dsRNA or ssRNA (+) molecules; the virions (when present) are almost always, icosahedral, which may be surrounded by a lipid-protein envelope. These viruses belong to the families *Endornaviridae* (some members also infect plants or fungi), *Marnaviridae* (one species), *Narnaviridae* (family members infect fungal and protozoan hosts) and *Phycodnaviridae* (with virus members also present in species of the kingdoms Animalia and Plantae). Other Chromista viruses that are not assigned to defined families include those belonging to the genera *Bacilladnavirus*, *Bacillarnavirus*, *Didnodnavirus*, *Labyrnavirus* and *Rhizidiovirus* (also capable of infecting fungi). In total, viruses belonging to the aforementioned families and genera do not surpass 70 different species, from which less than 30 are associated with chromists (Table 9.1). Viral metagenomics has barely opened the window on the complexity of virus community associated with these organisms.

The viruses described below infect members of the oomycetes. Oomycetes include important biotrophic and hemibiotrophic pathogens of plants and animals. Among them, and probably the best known and most studied is *Phytophthora infestans*, the causal agent of late blight of potato and tomato. There is also *Plasmopara halstedii* which is a quarantine pest of many ornamentals of the family Asteraceae. Most, if not all viruses of chromists, induce no discernable disease phenotype in their hosts. In some instances decreased

TABLE 9.1 Viruses Infecting Species Belonging to the Kingdom Chromista (as of March 11, 2017, ICTV)

Virus Species	Genome	Chromista Species
Family *Endornaviridae*		
Phytophthora alphaendornavirus 1	dsRNA	*Phytophthora* (Douglas fir isolate), *P. ramorum* (oomycetes)
Family *Marnaviridae*		
Heterosigma akashiwo RNA virus	ssRNA(+)	*Heterosigma akashiwo* (alga)
Family *Narnaviridae*		
Phytophthora infestans RNA virus 4	ssRNA(+)	*Phytophthora infestans* (oomycete)
Family *Phycodnaviridae*[a]		
Chrysochromulina brevifilum virus PW1	dsDNA	*Chrysochromulina brevifilum* (alga)
Ectocarpus fasciculatus virus a	dsDNA	*Ectocarpus fasciculatus* (alga)
Ectocarpus siliculosus virus 1	dsDNA	*Ectocarpus siliculosus* (alga)
Ectocarpus siliculosus virus a	dsDNA	*Ectocarpus siliculosus* (alga)
Emiliania huxleyi virus 86	dsDNA	*Emiliania huxleyi* (alga)
Feldmannia irregularis virus a	dsDNA	*Feldmannia irregularis* (alga)
Feldmannia species virus	dsDNA	*Feldmannia* spp. (algae)
Feldmannia species virus a	dsDNA	*Feldmannia* spp. (algae)
Heterosigma akashiwo virus 01	dsDNA	*Heterosigma akashiwo* (alga)
Hincksia hinckiae virus a	dsDNA	*Hincksia hinckiae* (alga)
Myriotrichia clavaeformis virus a	dsDNA	*Myriotrichia clavaeformis* (alga)
Pilayella littoralis virus 1	dsDNA	*Pilaiella* sp. (algae)
Unassigned Families		
Genus *Bacilladnavirus*		
Chaetoceros salsugineum DNA virus 01	ssDNA (+/−)	*Chaetoceros salsugineus* (diatom)
Genus *Bacillarnavirus*		
Chaetoceros socialis f. radians RNA virus 01	ssRNA(+)	*Chaetoceros socialis* (diatom)
Chaetoceros tenuissimus RNA virus 01	ssRNA(+)	*Chaetoceros tenuissimus* (diatom)
Rhizosolenia setigera RNA virus 01	ssRNA(+)	*Rhizosolenia setigera* (diatom)

(*Continued*)

TABLE 9.1 (Continued)

Virus Species	Genome	Chromista Species
Genus *Dinodnavirus*		
Heterocapsa circularisquama DNA virus 01	dsDNA	*Heterocapsa circularisquama* (dinoflagellate)
Genus *Labyrnavirus*		
Aurantiochytrium single-stranded RNA virus 01	ssRNA(+)	*Aurantiochytrium* sp. (heterotrophic algae)
Genus *Rhizidiovirus*		
Rhizidiomyces virus	dsDNA	*Rhizidiomyces* sp. (water molds)
Genus Tombunodavirus		
Plasmopora halstedii virus A	ssRNA(+)	*Plasmopora halstedii* (oomycete)
Sclerophthora macrospora virus A	ssRNA(+)	*Sclerophthora macrospora* (oomycete)
Tombunodavirus UC1	ssRNA(+)	Unknown (metagenomic study)
Unclassified		
Phytophthora infestans RNA virus 1	ssRNA(+)	*Phytophthora infestans* (oomycete)
Phytophthora infestans RNA virus 2	dsRNA	*Phytophthora infestans* (oomycete)
Phytophthora infestans RNA virus 3	dsRNA	*Phytophthora infestans* (oomycete)

[a]Not including viruses of the green algae *Chlorella* (Kingdom Plantae), that can be found as a symbiont of some Chromista species (e.g., *P. bursaria*). That is, the primary host is a plant and not a chromist organism.

virulence (hypovirulence) of the pathogenic host is the only phenotype associated with the presence of the virus. The study of oomycete viruses has attracted much attention recently for several reasons, including the potential applications of hypovirulence-associated viruses that might help in controlling crop diseases as well as the potential for engineering oomycete viruses as vectors useful in exploring various molecular pathways and interactions and their potential for the control of important plant pathogens.

Plasmopara halstedii virus A

P. halstedii is the causal agent of downy mildew in various hosts, including sunflowers. First described in the United States, *P. halstedii* is presently found worldwide wherever sunflowers are cultivated, and has been submitted to quarantine regulation in the European Union since 1992. The pathogen causes damping-off disease in seedlings. In adult plants, it leads to dwarfism, leaf bleaching and may severely impact seed yield. Control of *P. hasltedii* typically involves chemical treatment of seeds and deployment of major resistance genes. However, constant fungicide application in the field to control this pathogen has led to the development of resistance against various chemicals, thereby reducing the effectiveness of these treatments. In addition, the pathogen has overcome some plant resistance genes.

Almost 30 years ago, 37 nm-diameter isometric virions, with peplomers (glycoprotein spikes), were discovered in a single North American pathotype of *P. halstedii*. The virus was subsequently named *Plasmopara halstedii virus* (PhV), and is one of the few characterized viruses of the oomycetes (Fig. 9.6). Its genome is composed of two ssRNA(+) molecules. RNA1 is 2793 nucleotides long, contains a poly(A) tract, and carries a single ORF with motifs of an RdRp similar to those found in the oomycete virus, Sclerophthora macrospora virus A (SmV-A). RNA 2 is composed of 1526 nucleotides, has a poly(A) tail, and codes the **c**apsid **p**rotein (CP) of the virus (ORF2). Interestingly, the RdRp proteins of both viruses show high similarity with those of viruses belonging to the family *Nodaviridae* [ssRNA(+) viruses]. However, this is not the case with the CPs. CPs of PhV and SmV-A show greater similarity to ssRNA(+) viruses belonging to the family *Tombusviridae*, ssDNA(+/−) viruses of *Circoviridae* and the new group of hybrid DNA-RNA viruses, the so-called cruciviruses. The CPs seem to be completely different from those of other viruses of *Nodaviridae*. Given these

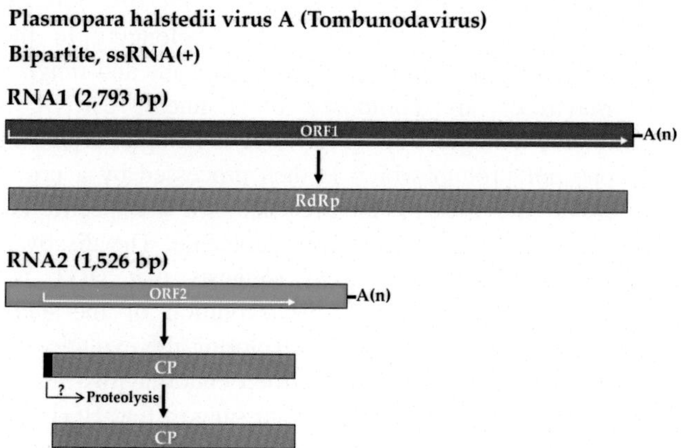

FIGURE 9.6 Genome organization of Plasmopara halstedii virus A of the genus Tombunodavirus. The genome of this virus is comprised of two ssRNA(+) molecules that possess a genomically encoded poly(A) tail. Both molecules contain a single ORF; RNA1 encodes the viral replicase, while RNA2 encodes the coat protein. Upon translation, an unidentified protease removes 21 amino acids from the N terminus of the coat protein to generate the mature product (*black box*).

peculiar characteristics, PhV and SmV-A are apparently closer to Tombunodavirus UC1 and form a new, independent group between the families *Nodaviridae* and *Tombusviridae*.

PhV seems to have a very narrow host range, limited to *P. hasltedii*. Genetic diversity of PhV in sunflower downy mildew populations is of low variability (0.3% or less), regardless of the host pathotype or geographic population of the oomycete or susceptibility to the fungicide metalaxyl. Recent studies have reported that the presence of PhV leads to hypovirulence effects with less sporangia and lower virulence of *P. halstedii* on sunflower. These results indicate that PhV has the potential for the development of biological agents for virocontrol of this plant disease.

Phytophthora infestans RNA viruses

P. infestans is a hemibiotrophic pathogen of worldwide distribution. The oomycete infects numerous members of the plant family Solanaceae as well as species in the families Convolvulaceae and Polygonaceae. *P. infestans* is best known for causing potato late blight, the disease that led to the Irish potato famine in the 1840s. Losses caused by *P. infestans* infection within the genus *Solanum* alone, and in particular in potato and tomato, amount to hundreds of millions of dollars annually. *P. infestans* is a natural host of viruses with dsRNA or ssRNA(+) genomes.

P. infestans Viruses With a dsRNA Genome

Phytophthora alphaendornavirus 1, or PEV-1, was the first nonplant endornavirus to be described. Up to 2005, viruses belonging to the family *Endornaviridae* were known mainly as viruses of plants and fungi. The genomes of endornaviruses are composed of a linear, dsRNA molecule 9.8−17.6 kb and have a single ORF (Fig. 9.7A). A single large transcript is translated into one polyprotein which is then processed by a proteinase. A unique feature of endornaviruses is the presence of a site-specific nick in the 5′ region of the coding (plus) strand RNA molecule. The discovery of the *Grapevine endophyte alphaendornavirus* suggests that HGT is one of the important keys to understanding the evolution of the endornavirus genome. Moreover, phylogenetic analyses exploring the evolutionary origins of protein domains in endornaviruses identified endornavirus-like sequences in plant, fungal, and bacterial genomes, which suggest that HGT might have occurred from an endornavirus to the host or from the host to an endornavirus. The virus is transmitted horizontally (through mating) or vertically (from mother to daughter cells). Endornaviruses do not form virus particles.

PEV-1 was originally isolated from a *Phytophthora* species infecting the evergreen conifer, Douglas fir. PEV-1 has since been found in *P. ramorum*, and other *Phytophthora* species. PEV-1 has a dsRNA genome 13.9 kb long

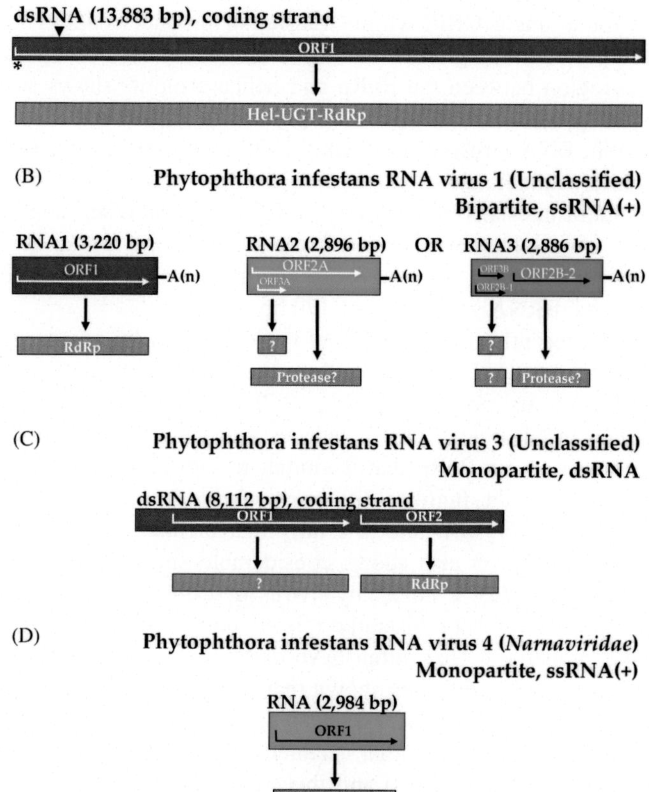

FIGURE 9.7 Genome organization of *Phytophthora infestans* RNA viruses. (A) The genome of PEV-1 consists of a dsRNA molecule with a single break (*solid black, inverted triangle* at position 1215) in the coding strand, and contains a single ORF. The deduced translation product is a polyprotein c.4612 aa when translated from the AUG codon at position 5–7 (unfavorable for translation by scanning) or 4548 aa if translated from the AUG codon at position 167–169 (in favorable conditions for translation; is indicated by the *asterisk*). The polyprotein (deduced from the second initiation site; *asterisk* in the figure) includes sequences typical of a viral RNA **hel**icase (Hel), an **UDP-g**lycosyl**t**ransferase (UGT), and an RdRp. The presence (and activity) of a putative virus-encoded protease has not been reported. No virions are produced by members of the family *Endornaviridae*. (B) The monopartite genome of the dsRNA virus Phytophthora infestans RNA virus 3 contains two ORFs. Although they appear in the figure as separate ORFs, they are probably overlapping, and linked by a potential frameshift sequence. These two ORFs are typically expressed as a fused protein. ORF2 codes for the RdRp of the virus, which is closely related to the same enzyme observed with dsRNA viruses of filamentous fungi. The function of ORF1 is not known. The virus is not known to produce virions. (C) The genome of Phytophthora infestans RNA virus 4 is an ssRNA(+) molecule that possesses a single ORF and encodes a RdRp. Although, in general, genomes of members of the family are not known to have a poly(A) tail, the sequence immediately following the ORF of this virus has a very low G + C content. Also, no members of the family are known to produce virions. (D) The bipartite genome of Phytophthora infestans RNA virus 1 is comprised of two ssRNA(+) molecules. However, two distinct forms of the RNA2 molecule (RNA2 or RNA3) have been reported. RNA1 has a single ORF that codes for a putative RdRp typical of ssRNA viruses, while RNA2 or RNA3 codes for a putative trypsin-like serine protease and other products of unknown function. The two genomic RNAs contain genome-encoded poly(A) tails of different lengths.

that codes for a single ORF, whose translation product contains an RdRp motif at the C-terminal region and an RNA helicase motif at the N-terminal region. The region between the RdRp and helicase motifs shows significant amino acid sequence similarity to UDP glycosyltransferases, believed to be present only in DNA viruses. Presumably UDP glycosyltransferases function as pathogenicity modulators.

Phytophthora infestans RNA virus 2, PiRV-2, on the other hand, appears to represent a new member of, as of yet, an unknown group of dsRNA viruses. This little-known virus was shown to possess both RdRp and protease motifs in its dsRNA genome of 11,170 bp.

Finally, the recently described PiRV-3 is composed by a nonsegmented, dsRNA molecule of 8112 nucleotides that seems to code for two long, overlapping ORFs on its plus strand (Fig. 9.7B). These two ORFs may be expressed as a fused protein. ORF1 encodes a protein of unknown function, while ORF2 encodes an RdRp that is similar to the RdRps of some dsRNA viruses belonging to the families *Totiviridae* (whose members infect Fungi and Protozoa) and *Chrysoviridae* (a family that includes viruses capable of infecting Fungi). PiRV-3 also shows considerable similarity to Phlebiopsis gigantea virus 2 (PgV2), a novel mycovirus described a few years ago. It seems that these viruses of distantly related hosts (oomycetes and fungi) share similar properties. They often have dsRNA genomes, lack traditional virus particles and it has not been shown that they have an extracellular component to their infection cycles. The oomycetes and fungi share many lifestyle features, including similar habitats and similar morphologies (filamentous mycelia and spores) and this might explain why some viruses are able to infect species of both groups.

PiRV-3, however, seems to belong to a new family of viruses, yet to be described, consisting of two genera: genus 1 that might include two viruses from plant-feeding insects (Spissistilus festinus virus 1 and Circulifer tenellus virus 1) and Cucurbit yellows-associated virus, and genus 2 composed by PiRV-3, along with PgV2, Fusarium virguliforme RNA viruses 1 and 2, Fusarium graminearum virus 3, and Diplodia scrobiculata RNA virus 1. Members of genus 2 share other features, which include long 5′ UTRs, along with 7–15 AUGs upstream of the putative start codon in ORF1—that seems to be expressed in a CAP independent manner by means of an internal ribosomal entry site. The aforementioned characteristics, along with other differences in ORF1 between members of the two genera, shed some doubts on the existence of only one new family, and two families have also been posited. Further studies will reveal which of these two contrasting hypotheses holds true.

P. infestans Viruses With an ssRNA Genome

PiRV-4 is a new member of the family *Narnaviridae*. Members of the *Narnaviridae* family of viruses are composed of ssRNA(+) genomes that code for a single protein, the RdRp (Fig. 9.7C), which is distantly related to that of bacteriophages in the family *Leviviridae*. Narnaviruses do not encode a protein capsid, and lipid vesicles, rather than virus particles, are associated with their infected hosts. There are two genera of the family, the genus *Narnavirus* and the genus *Mitovirus*. The single feature distinguishing the two genera is the site of translation of the RdRp, and presumably the site of replication of the viruses. Members of the genus *Narnavirus* are found in the cytosol of fungi, and those of the genus *Mitovirus* are confined to the mitochondria of *S. cerevisiae* and the oomycete, *P. infestans*. An unclassified virus of this family that infects the kinetoplastid (Protozoa), *Phytomonas serpens*, has been recently reported. Virus transmission is horizontally through mating, from mother to daughter cells, by cytoplasmic exchange, sporogenesis, or hyphal anastomosis. On the other hand, another ssRNA(+) virus isolated from *P. infestans* and designated as PiRV-1 appears to be related to the *Astroviridae* family of viruses. Members of the latter infect animal hosts (vertebrates), and cause viral gastroenteritis in humans.

PiRV-1 is composed of a bipartite genome. RNA 1 of 160 nucleotides (plus a poly(A) tail) encodes an RdRp typical of ssRNA(+) viruses (Fig. 9.7D), such as the plant-infecting potyviruses and mammal-infecting astroviruses. RNA 2A is 2776 nucleotides long, polyadenylated, and codes for two potential ORFs: one with the potential to encode a polyprotein of 847 aa including a possible trypsin-like serine protease, and a second putative ORF of unknown function. RNA 2B is an alternative form of RNA 2A in which a stretch of 19 nucleotides has been replaced by a sequence that is 9 nucleotides long. Both RNA molecules (1, and 2A or 2B) share extensive similarity at their 5′ and 3′ UTRs, which presumably indicates that both RNAs represent interdependent segments of a single viral genome. As with the previously described viruses, PiRV-1 causes no obvious symptoms in its host and it is not known whether the virus has any impact on the physiology of *P. infestans*. Early attempts to engineer PiRV-1 as virus-based vector have yielded promising results, but the current vectors are still not fully optimized for gene expression studies in *P. infestans*.

FURTHER READING

Abergel, C., Legendre, M., Claverie, J.-M., 2015. The rapidly expanding universe of giant viruses: Mimivirus, Pandoravirus, Pithovirus and Mollivirus. FEMS Microbiol. Rev. 39, 779–796. Available from: https://doi.org/10.1093/femsre/fuv037.

Banik, G.R., Stark, D., Rashid, H., Ellis, J.T., 2014. Recent advances in molecular biology of parasitic viruses. Infect. Disorders Drug Targets 14, 155–167. Available from: https://doi.org/10.2174/1871526514666140713160905.

Casjens, S.R., Hendrix, R.W., 2015. Bacteriophage lambda: early pioneer and still relevant. Virology 479–480, 310–333. Available from: https://doi.org/10.1016/j.virol.2015.02.010.

Eusobio-Cope, A., Sun, L., Tanaka, T., Chiba, S., Kasahara, S., Suzuki, N., 2015. The chestnut-blight fungus for studies on virus/host and virus/virus interactions: from a natural to a model host. Virology 477, 164–175. Available from: https://doi.org/10.1016/j.virol.2014.09.024.

Gómez-Arreaza, A., Haenni, A.-L.-, Avilán, L., 2017. Viruses of parasites as actors in the parasite–host relationship: a "ménage à trois". Acta Tropica 166, 126–132. Available from: https://doi.org/10.1016/j.actatropica.2016.11.028.

Hyman, P., Abedon, S.T., 2012. Viruses of microorganisms. Scientifica 2012, 734023. Available from: https://doi.org/10.6064/2012/734023.

Nuss, D., 2005. Hypovirulence: mycoviruses at the fungal-plant interface. Nat. Rev. Microbiol. 3, 632–642.

Pietilä, M.K., Demina, T.A., Atanasova, N.S., Oksanen, H.M., Bamford, D.H., 2014. Archaeal viruses and bacteriophages: comparisons and contrasts. Trends Microbiol. 22, 334–344. Available from: https://doi.org/10.1016/j.tim.2014.02.007.

Roossinck, M.J., 2015. Metagenomics of plant and fungal viruses reveals an abundance of persistent lifestyles. Front. Microbiol. 5, 757. Available from: https://doi.org/10.3389/fmicb.2014.00767.

Ruggiero, M.A., Gordon, D.P., Orrell, T.M., Bailly, N., Bourgoin, T., Brusca, R.C., et al., 2015. A higher level classification of all living organisms. PLoS One 10, e0119248. Available from: https://doi.org/10.1371/journal.pone.0119248.

Zangger, H., Ronet, C., Desponds, C., Kuhlmann, F.M., Robinson, J., Hartley, M.-A., et al., 2013. Detection of *Leishmania* RNA virus in *Leishmania* parasites. PLoS Neglect. Trop. Diseases 7 (1), e2006. Available from: https://doi.org/10.1371/journal.pntd.0002006.

Chapter 10

Host–Virus Interactions: Battles Between Viruses and Their Hosts

Gustavo Fermin[1] and Paula Tennant[2]
[1]Universidad de Los Andes, Mérida, Venezuela, [2]The University of the West Indies, Mona, Jamaica

An inefficient virus kills its host. A clever virus stays with it.

James Lovelock

This chapter is about a war: the one fought by viruses in order to ensure their replication, and by the host cell in its defense of its *status quo* and the restriction of infection by an intruding agent. The outcome is not always certain, and depends on the balance of actions and counteractions between host defense systems and the virus' escape mechanisms. The arms-race between viruses and their host organisms follow more or less the following trend:

1. Host resistance by means of a mutated virus receptor that is unable to recognize the pathogen, and hence infection and virus replication are halted. Conversely, mutations in the virus allow for receptor recognition and entry of the virus, and infection.
2. Host innate immunity that by diverse mechanisms actively prevents virus replication. As we will see later, viruses counteract host immunity mostly by avoiding or exploiting weak points in the mechanisms of innate immunity.
3. Adaptive (acquired) immunity by means of which host cells discriminate self from nonself and collect molecular information on the virus in order to impede or reduce, in a highly selective and effective manner, its replication. Viruses, on the other hand, can avoid recognition and/or replication impairment.

Viruses. DOI: https://doi.org/10.1016/B978-0-12-811257-1.00010-3

4. Finally, and as a last resource, **p**rogrammed **c**ell **d**eath (PCD) in an infected cell prompts self-destruction in order to avoid infection of other cells. Of note, the same system that induces PCD can induce dormancy (stasis) in bacteria.

In this chapter, we examine the interactions between a virus and its host that are critical for the establishment and maintenance of infection, including the array of host antiviral mechanisms and the equally impressive diversity in virus evasion mechanisms. Historically, prokaryotic viruses are better known so emphasis is given to eukaryotic viruses and their hosts since they present more challenges in terms of the devastation they cause. The chapter, while devoted to virus–host interactions that lead to disease, does not cover the multitude of phenotypic manifestations of virus diseases—the reader is directed to Chapters 6–9: Viruses as Pathogens: Plant Viruses, Viruses as Pathogens: Animal Viruses, With Emphasis on Human Viruses, Viruses as Pathogens: Animal Viruses Affecting Wild and Domesticated Species, and Viruses of Prokaryotes, Protozoa, Fungi, and Chromista, and to specialized literature for detailed study, including that listed at the end of the chapter.

TYPES OF VIRUS–HOST RELATIONSHIPS: TERRORISM VS GUERILLA WARFARE

Viruses establish different types of relationships with their hosts, ranging from terrorism to guerilla warfare, or more specifically, (1) the acute or "*hit-and-run*" where there is a rapid burst of virus replication but the virus is quickly cleared from the host (e.g., influenza viruses, Ebola virus, rotaviruses); (2) the persistent or "*slow-and-low*" characterized by continuous virus replication at sufficiently low level, not to seriously damage the host or to provoke a defense response in the host of sufficient magnitude to risk clearance (e.g., hepatitis viruses); and finally (3) "*latency*" where the virus remains largely dormant with occasional reemergence (e.g., herpes viruses).

Indeed, the virosphere includes many serious pathogens, but it is important not to lose sight of the fact that virus–host interactions leading to the development of disease is an exception rather than a common outcome and, in most cases, hosts are capable of counteracting the harmful effects of viruses. Mounting evidence suggests that viruses that have a long-term evolutionary relationship with their host are normally relatively benign. These viruses are dependent on their hosts for their survival, thus evolution has engineered the interaction to run smoothly. In some instances, viruses interact with their hosts independently of their role as pathogens and establish relationships that benefit their hosts. These virus–host interactions, namely the mutualistic partnerships viruses establish in almost all holobionts, are addressed in Chapter 11: Beneficial Interactions with Viruses. That being said, we address the interactions between a virus and its host that are critical for the establishment and maintenance of infection.

WEAPONS AND DEFENSES I: PROKARYOTIC VIRUS–HOST INTERACTIONS

The initial interaction between a bacteriophage and its susceptible bacterial host involves attachment of the virion to receptors, such as host pili, proteins, oligosaccharides or lipopolysaccharides, or digestion of the bacterial cell wall by polysaccharide-degrading enzymes. After entry, the cell's metabolism is hijacked by the virus' early expression gene products. Virus genome replication, viral protein production, and virion assembly follow, ending in lysis and the so-called lytic cycle of the virus. On the contrary, the virus may undergo a lysogenic cycle in which lysis is precluded; the cell remains alive and transmits the virus genome passively with the host DNA for many generations. Lysogenic status is acquired and maintained in different ways: in the case of *Escherichia virus Lambda*, a short conserved sequence allows for site-specific recombination between the virus and the bacterial genome leading to the insertion of the whole virus genome in the bacterial chromosome. Virus genome expression is effectively impeded by a virus-encoded protein called cI. The presence of the "*prophage*" prevents phages of the same type or closely related from becoming established in the same cell. That is, lysogenic immunity is conferred. In other phages, like phage P1, lysogenic cell status is not attained by integration into the host chromosome, but by the residence of the virus as a plasmid in the host's cytoplasm. In both cases (with or without phage insertion), however, lysogeny derives from the very presence of the prophage in the host cell. A switch to a lytic cycle is possible and can be initiated when stressful conditions (i.e., DNA damage) induce the excision of the phage genome, which is followed by the expression of lytic genes, genome replication, packaging into progeny phage particles, and bacterial lysis. To survive phage-infections, bacteria have evolved a diverse array of antiphage systems. Similarly, phages have coevolved to overcome these bacterial defense mechanisms.

One strategy of resisting phage attack involves the prevention of phage adsorption by concealing receptors with an additional physical barrier, namely an extracellular matrix, competitive inhibition of the receptor by a ligand other than the phage virions themselves, or modifying receptor structures through mutation. The latter method, the most complex, is used, for example, by *Enterococcus faecalis*. Phage infection of *E. faecalis*, an opportunistic Gram positive bacterium that resides in the human intestine, is facilitated by an integral membrane protein (PIP_{EF}, for **p**hage **i**nfection **p**rotein from ***E****. faecalis*). A 160 aa hypervariable region determines phage tropism to different strains of the bacterium. Phage resistance in *E. faecalis* is acquired by mutations in the gene encoding $PIP_{EF;}$ that is, by increased mutations, the bacterium can develop resistance. However, high rates of mutation in virus structural protein genes can overcome this resistance, and result in specificity to defined strains of the host. In other hosts, resistance determinants are not encoded in the bacterial genome, but on autonomous replicons, for example plasmids: the resistance is transmissible once the plasmid harboring resistance determinants is transferred to the susceptible host.

Other sophisticated mechanisms used to inhibit phage infections include Sie and RM systems. The Sie (**S**uper**i**nfection **e**xclusion systems) are membrane-anchored or membrane-associated proteins that act to block phage DNA injection into host cells, while RM systems (**R**estriction-**M**odification) of self/nonself discrimination destroy invading DNA. RM systems are composed of a restriction endonuclease and a cognate methyltransferase. The methyltransferases methylate self-DNA at specific recognition sites, whereas nonmethylated, nonself, foreign DNA is recognized and cleaved by host restriction endonucleases into harmless fragments. Viruses circumvent RM systems by various means: avoidance by mutation of the restriction sites recognized by host restriction endonucleases, use of different nucleotides (for example, hydroxymethyl cytosine instead of cytosine in the genome of *Escherichia virus T4*), use of genome-coded virus proteins that inhibit host endonucleases or increase the activity of host modifying enzymes (e.g., methylases), or the use of proteins that specifically block restriction sites recognized by the host's RM system.

Other mechanisms widely present in prokaryotes are the so-called **t**oxin/**a**ntitoxin (TA) systems that function through induced cell suicide or dormancy. These systems, at least 33 in *Escherichia coli*, are composed of a stable toxin and an unstable antitoxin. Under normal conditions, the toxin, which is either a holin-like protein or an endonuclease cleaving ribosome-associated mRNA (RNA interferase), is maintained in an inactive state via interaction with the antitoxin gene product. The latter consists of either a small RNA that prevents translation of the toxin gene or a protein that forms an inactive complex with the toxin. Virus infection can inactivate the antitoxin and so release the toxin, which kills the affected cell or induces dormancy, and restricts the impact of the infection. Similarly, **ab**ortive **i**nfection (Abi) systems are forms of "*bacterial apoptosis*" or PCD that lead to death of the infected cell as a sacrifice to protect the surrounding population. Abi systems, effective variants of the TA systems, encompass postinfection mechanisms that hinder all stages of phage replication.

Known mostly from bacterial hosts, the aforementioned mechanisms are expected to be widespread in archaeal hosts too. Little is known about virus entry into archaeal cells but it seems that their formidable morphology plays a role in the initial interactions with host cells. Adsorption and entry of the virus into the cell vary depending on the virions' appendages. Upon binding to archaeal glycoproteins, recognition ensues and the virion can change from a closed conformation to an open conformation that facilitates movement of the virus DNA genome toward the tip of the appendage and contact with the cell surface, resulting in virion disassembly and entry of the viral DNA via a mechanism that resembles entry of filamentous phages into their bacterial hosts. Archaeaviral genomes, similar to bacterial viral genomes, coopt the cell's transcriptional apparatus for viral transcription in a highly controlled fashion. The outcome of the lytic cycle resembles that of observed in bacteria. In some archaeal viruses, however, virus extrusion—like in the case of *Sulfolobus islandicus rod-shaped virus 2* (*Rudiviridae*)—follows after massive degradation of host chromosomal DNA and virion assembly. About 50 virions align side-by-side in the cytoplasm with the concomitant appearance of pyramidal structures (VAPs, for **v**iral-**a**ssociated **p**yramids) at the cell surface. VAPs disrupt the S layer (part of the cell envelope) and point outward; virions and cytoplasm are released upon their opening, leaving behind an empty perforated cell. Finally, acquisition of lysogeny has also been observed in archaea infected by some archaea viruses. These viruses encode an integrase that mediates virus genome integration by site-specific recombination.

As mentioned earlier, a variety of antiviral mechanisms to resist virus infection are also widespread in archaeal viruses. A recent discovery is the adaptive CRISPR-Cas (**C**lustered **R**egularly **I**nterspaced **P**alindromic **R**epeats and **C**RISPR-**as**sociated genes) immune system in both archaeal and bacterial organisms. The system functions via a distinct self/nonself-recognition mechanism that is partially analogous to the mechanism of eukaryotic **RNA i**nterference (RNAi). CRISPR-Cas systems consist of two modules: a locus for memory storage (the CRISPR array of palindromic repeats), and the *cas* genes encoding the machinery that drives immunity. The sequence-based information stored in the former allows sequence-specific degradation of invading pathogens, including viruses, by the latter. This way, the system is not only specific, but heritable and adaptable also by virtue of the stored memory of the CRISPR array component, thanks to the generation of new spacers (short specific sequences) created and donated (inserted) by Cas1-Cas2 complexes. Not surprisingly, some phages evade this exquisite adaptive immune system by coding their own CRISPR-Cas system, which targets genomic islands of the host in a way that allows the lytic infection cycle to proceed. Chapter 13: Viruses as Targets for Biotechnology: Diagnosis and Detection, Transgenesis, and RNAi and CRISPR/Cas-Engineered Resistance, reviews the application of CRISPR-Cas systems in engineering resistance to eukaryotic viruses in organisms other than bacteria and archaea.

WEAPONS AND DEFENSES II: EUKARYOTIC VIRUS–HOST INTERACTIONS

As with prokaryotic viruses, eukaryotic viruses extensively use the host intracellular machinery for replication of their genomes, expression of viral genes, and establishment of infection. As a consequence, they interact profoundly with the host during their biological cycle.

Virus Recognition, Attachment, and Entry

Virions encounter their hosts either by physical forces (Brownian movement, air delivery by aerosols, etc.) or via biological or inanimate vehicles (refer to Chapter 5: Host Range, Host–Virus Interactions, and Virus Transmission). They exploit fundamental cellular processes to gain entry, largely defined by the interactions between virus particles and their receptors at the cell surface. These interactions between the virus and host cell receptors are highly specific and define tissue tropism, that is, the types of cells in which the virus is able to replicate. In many cases virus binding causes conformational changes in cell receptors that help penetration (and uncoating) or induce conformational changes in the virion itself making it more prone to membrane association, and more competent for penetration and uncoating. Plant viruses rely instead on mechanical breaches via wounding of the plant cell wall to directly introduce virions into host cells. Similarly, cell entry of mycoviruses does not involve the typical virus–host receptor interactions and is limited to intracellular routes, including cytoplasmic exchange during cell division, mating and anastomosis, and by spores. Either way, it is also important to keep in mind that the process of infection of a cell by a virus, and its dispersal from an infected cell to another, are two different processes that most probably require different processes and signaling pathways.

Virus Uncoating

For those viruses that have not delivered their genome directly into the host cell, uncoating is a multistep process involving the release of the virus genome from its capsid or capsid-envelope coats. This built-in capacity for disassembly is referred to as structural metastability. That is, virions are assembled as stable structures within infected producer cells, and the very same capsids must be efficiently disassembled in newly infected cells. Altering the environment of the producer cell during morphogenesis or entry is necessary for viruses that uncoat and assemble in the same compartment and may involve individually or in combination: (1) structural modifications triggered by the host receptor/virion interaction; (2) exposure to reducing conditions in the cytosol; (3) low calcium concentrations; (4) interaction with a cytosolic receptor; and (5) interaction with an uncoating cellular factor (which after infection must be inactivated in order to allow the later step of virion assembly *de novo*). Alternatively, morphogenesis and uncoating may be spatially separated and require distinct organelles. Some viruses that replicate in the nucleus must wait until mitosis in order to gain access and for the uncoating processes to be initiated. In other cases, low pH provides the signal for the virion to fuse with endosomal membranes and release the capsid into the cytosol, sometimes with the assistance of uncoating factors provided by the host. Additionally, limited, selective proteolysis may also play a role in the uncoating process of certain viruses (e.g., some reoviruses and vaccinia viruses).

Virus Factories for Virus Replication and Assembly

The processes of genome replication and virion assembly take place in specific intracellular compartments referred to as virus factories. Virus factories can be envisioned as "*organelles*" created *de novo* from recruited cellular and viral components such as cell membranes, cytoskeletal elements, and mitochondria. Essentially all the structural and energy components required for virus replication and virion assembly are provided in one location, with the added benefit of protection from the host's systems of defense and cellular degradation machinery. A crucial step in the biogenesis of virus factories is the early interaction between the virus polymerase(s) and the host cell endomembranes to build **r**eplication **c**omplexes (RC). Viral genome copies are produced (refer to Chapter 3: Replication and Expression Strategies of Viruses), and then transferred to virion assembly sites. Some viruses first preassemble their RCs at the internal face of the plasma membrane, which are later internalized and transported to lysosomes (*Togaviridae*) where replication continues; or replication occurs in mini-nuclei that are assembled with the virus genome and rough **e**ndoplasmic **r**eticulum (ER) cisternae from the host (*Poxviridae*). But in general, viruses with a cytosolic replication strategy replicate in **s**ingle-**m**embrane **s**pherules (SMS), **t**ubulo-vesicular **c**ubic **m**embranes (TCM) or **p**lanar **o**ligomeric **a**rrays (POA). SMS are built with nonstructural viral proteins, viral RNA(s), and host cofactors and endomembranes, and contain a narrow opening to the cytosol. They form from invaginations derived from the ER (*Bromoviridae*), Golgi apparatus (*Peribunyaviridae*), chloroplasts (*Tymoviridae*), mitochondria (*Nodaviridae*)), lysosomes (*Togaviridae*), or peroxisomes. TCMs, on the other hand, consist of highly curved 3D-folded lipid bilayers, mostly originated from the ER, that build a platform for virus replication and virion assembly (*Coronaviridae*). POAs are planar 2D arrays of viral RNA polymerases that play both structural and enzymatic roles. POAs have been observed with polioviruses.

Virion Assembly and Egress

Virion assembly and egress are highly coordinated processes that lead to virus genome packaging and transmission of newly formed infectious particles. Ordered, reproducible structures comprised of many copies of a few or single coat protein subunits are assembled from the crowded cellular milieu for export from the cell. In some instances, virion assembly is coupled to virus genome replication and involves either packaging of the genome into a pre-formed protein shell (procapsid) or coassembly of the capsid around the viral genome. Export from the infected cell typically involves cell-to-cell and long-distance movement of the viral particles and/or viral genome, and transmission to new hosts. Plant viruses move cell-to-cell via plasmodesmata. In the case of animal viruses, egress from infected cells is accomplished by either lysis of the cell or budding (nonlytic viruses) into a cellular membrane, concomitant with the acquisition of the envelope. If budding occurs on the host cell's surface membrane, the released virions are immediately located outside the cell in the extracellular space, but if budding occurs through the membrane of an internal organelle (ER, for example), virions are released in the lumen and exit the cell via the conventional secretion pathway or exosomes. DNA viruses that replicate in the nucleus have to cross the nuclear membrane prior to budding at cellular membranes. Members of the *Herpesviridae*, for example, follow a two-step mechanism of exit: first from the nucleus to the cytoplasm, and then from the cytosol to the external milieu.

Virus-Induced Changes in Selected Metabolic Pathways

Viruses have seemingly evolved strategies to manipulate host metabolic pathways for multiple ends; an increased pool of free nucleotides and amino acids for a sustained and rapid generation of new virus genomic molecules, messenger, and other nucleic acids indispensable for replication and expression cycles, and finally, provision of ATP for the high energy costs associated with genome replication and packaging. Successful virus infection ensures that these alterations provide optimal environments for virus replication and spread, but do not cause host cells to lose their autonomy (the cells are transformed into virocells). Below, general trends of metabolic changes are addressed, but the reader must keep in mind that they are not universal changes since every virus has particular molecular needs and use different ways to meet its material and energy requirements.

Glycolysis

Upon animal virus infection, the glycolytic pathway is induced, including increased glucose intake, glucose degradation (glycolysis), and lactic acid generation. Induced glycolysis can be accompanied with apoptosis, but this is not always the case. Increased glycolysis leads not only to augmented generation of ATP, but also to the generation of metabolites that feed into other metabolic pathways (notably, fatty acid synthesis) that are required, for example, in virion assembly. In virus-infected plants, a common biochemical change is an increase in respiration—accompanied by reduced photosynthesis, accumulation of nitrogenous compounds, and increased oxidase activities.

Fatty Acid Synthesis

Many viruses require fatty acids at some stage of their life cycles; concomitantly, they are able to induce fatty acid synthesis in infected cells by redirecting glucose-derived carbon into the former pathway. To attain this effect, some enzymes involved in fatty acid synthesis are induced; some lipids, on the other hand, are stored as lipid droplets that can lead to steatosis (or fatty liver disease, as caused by *Hepacivirus C*, HCV). Besides disturbances in synthesis and accumulation, other viruses induce rearrangements in existing cellular membranes to form vesicles around viral replication sites or virus factories.

Glutaminolysis

Some viruses require glutamine for their replication. Poliovirus replication, for example, is more efficient in the presence of glutamine than glucose. Glutamine is absolutely required for the replication of *Vaccinia virus* and can be found in their replication factories.

Photosynthesis

Chloroplasts are common **t**argets of **m**any plant **v**iruses [(e.g., TMV, *Plum pox virus* (PPV), and *Potato virus Y* (PVY)] that are exploited for virus propagation and replication. To support the latter activities, these viruses downregulate many chloroplast and photosynthesis-related genes, detrimentally affecting both chloroplast structure as well as the process of photosynthesis itself. Transcriptome and metabolome analyses have documented reduced photosynthesis, increased lipid peroxidation, and accumulation of **r**eactive **o**xygen **s**pecies (ROS). Importantly, ROS can (1) elicit pathogen restriction and often localized death of host plant cells at infection sites, and (2) act as a diffusible signal that induces antioxidant and pathogenesis-related defense responses in adjacent plant cells. Chloroplast–virus interactions also impair chloroplast division, altering consequently the normal number of chloroplasts per cell and affecting chloroplast relocation movement within the cell.

Hormone Synthesis and Signaling

Finally, viral infections also result in hormonal disruption and temporal changes in hormone signaling in plants. It appears that these disruptions at very early steps of viral infection orchestrate and execute the response in host–virus interactions to either induce resistance or to favor virus multiplication, spread and symptom development. Connections have been noted between interactions of specific virus factors with cellular components, and alterations in hormone synthesis and signaling. Plants typically respond to virus infection with a complex scenario of sequential, antagonistic, or synergistic action of different hormone signals leading to defense gene expression. Hormones such as salicylic acid and jasmonic acid/ethylene are regarded as the backbone of the defense response. Growth-promoting hormones (auxin, cytokinins, gibberellic acid, and abscisic acid) either inhibit or potentiate the balance in mediating resistance or susceptibility against the invading pathogen.

Virus-Induced Changes in Selected Structures and Functions of the Cell

Structural changes induced in host cells upon viral infection are called cytopathic or cytopathogenic effects, and the responsible virus is said to be cytopathogenic. Common examples are rounding of the infected cell, loss of contact inhibition, fusion with adjacent cells to form syncytia (or giant cells), and the formation of inclusion bodies within the nucleus or cytoplasm, that represent either altered host cell structures or accumulations of viral components. Diverse cell organelles are not exempt and are targets for different virus-induced damage.

Endoplasmic Reticulum

During the processes of virus entry, replication, or egress, viruses either deform, rearrange the ER membrane (invaginations and/or exvaginations) or generate ER-derived structures (virus factories). However, the ER is at the core of the **r**egulated **i**ntramembrane **p**roteolysis (RIP) signaling mechanism. Intracellular signaling pathways thus connect to the nucleus, triggering antiviral gene expression in response to virus-induced changes to the ER. Viruses cleverly hijack the cell intramembrane proteases involved in RIP to facilitate their proliferation.

Cytoskeleton

Protein filaments of the cytoskeleton (filamentous actin, intermediate filaments, and microtubules) have been shown to play a role in the process of infection, replication, virion assembly, and budding in several viruses across different families. The extent of cytoskeletal reorganization varies among different viral infections and although the end result may not be cell death, there are important consequences (cytopathic effects) for cell properties as shape, movement, and intracellular transport that are regulated by the cytoskeleton.

Nucleus

The nucleus is often used as the site of DNA virus replication. Viruses derive several advantages from processes occurring in this organelle, notably the proximity to polymerases and replication factors. However, viruses other than DNA viruses also exploit the nucleus to complete their replication, and as with other organelles, can induce morphological changes in its structure thus affecting its normal functioning. A few examples of virus–nucleus interactions and induced effects include (1) disruption or exploitation of host cell nucleocytoplasmic trafficking pathways in order to access nuclear functions (e.g., nuclear proteins required for replication); (2) disruption of the nuclear envelope (Parvoviruses cause transient disruptions in the nuclear envelope when they deliver their genomes into the nucleus for replication using a mechanism that has yet to be elucidated). The nucleus can also be a part of the egress mechanism of certain viruses (e.g., herpesviruses). In other cases of nuclear egress, DNA viruses exit the nucleus through the nuclear pore complex or rupture the nuclear envelope); and (3) disruption of host-cell transcription and innate antiviral responses. *Influenza A virus* (IAV) replication occurs at viral ribonucleocapsid complexes in the nucleus. Since IAV lacks the enzyme for adding 5′ caps to its RNAs, it snatches the 5′ ends of host capped RNAs to prime its transcription. Apparently noncoding RNAs, especially small noncoding RNAs such as snRNA and the recently identified class of promoter-associated **c**apped **s**mall **RNAs** (csRNAs) serve as the major cap donors to IAV mRNAs, thus raising the possibility that IAV uses snatching preferences to modulate host-mRNA splicing and transcription. The host cell nucleus is also essential to retrovirus replication as the reverse transcribed DNA is integrated into the host genome where it is transcribed by the host machinery. Additionally, the Gag protein may coopt cellular splicing signals to ensure encapsidation of the unspliced genomic viral RNA. Picornavirus proteins are known to enter the nucleus of infected cells to limit host-cell transcription and downregulate innate antiviral responses.

Mitochondria

Mitochondria are used as sites of virus replication or they are either directly targeted by viral proteins or influenced by the physiological alterations to the cellular environment during viral pathogenesis (like deregulated calcium homeostasis, endoplasmic reticulum stress, oxidative stress, and hypoxia). Specific changes induced by viruses can include (1) membrane rearrangements (invaginations or spherules) that are generally associated with RNA replication and presumably sequestration viral RNA replication intermediates from dsRNA-triggered innate immune responses (see section under *Innate immunity*); (2) microtubule-mediated clustering of mitochondria around virus factories in the cell to meet energy requirements during viral replication (HCV); (3) hijacking of mitochondrial proteins for use as primers in genome replication (HIV-1) or for the transportation of virus genomes to the nucleus in initiate replication (adenoviruses); (4) degradation of mitochondrial DNA (which likely contributes to cell death and tissue damage) in an attempt to evade mitochondrial antiviral responses (herpes simplex viruses); and (5) modulation of mitochondrial membrane permeability to effectively control cell death or apoptosis. Mitochondrial control of apoptosis during viral infection is an important aspect of innate immunity. Anti- or proapoptotic viral proteins coopt host mitochondria to modulate mitochondrial membrane permeability, affecting mitochondrial bioenergetics in favor of replication (HIV-1, HCV, IAV). Altering apoptosis during early stages of virus infection prevents the host immune response and promotes genome replication, whereas induction of apoptosis during later stages of virus infection is used for the release of progeny virions to surrounding cells.

Chloroplasts

Chloroplasts create an optimal niche for the replication of some plant viruses that are also safe from the host's RNA-mediated defense response. These viruses establish specific interactions between chloroplast membranes and virus-encoded proteins associated with virion uncoating, assembly (tombus-viruses), and movement (*Cauliflower mosaic virus*, CaMV). However, viral influence on chloroplast structures and functions usually leads to decreased photosynthetic activity and symptom development. Virus-infected plants manifest striking mosaic symptoms associated with altered chloroplast number and structure (swollen chloroplasts because of invaginations at the outer membranes, large amounts of starch and plastoglobulin accumulation, and disintegrated grana stacks). Impaired division and internal movement of chloroplasts have also been reported. On the other hand, chloroplast factors seem to play active roles in plant defense against viruses; they elicit the effector-triggered immune response against viral pathogens and generate defense signaling molecules such as ROS, nitric oxide, salicylic acid, and jasmonic acid. The chloroplast also serves as a sensor for detecting perturbations in the subcellular environment, and actively communicates these signals to other organelles.

Ribosomes

The most conspicuous interaction viruses establish during infection is that with cellular ribosomes. Viruses must engage ribosomes and the host translation machinery to produce polypeptides. To this end, they have evolved a wealth of mechanisms to customize the translation apparatus to meet their specific needs and effectively maintain viral protein synthesis. This is particularly true for RNA viruses. Some of the mechanisms employed include the use of internal ribosome entry sites, leaky scanning, non-AUG initiation, reinitiation, ribosomal frameshifting, and read through. However, in the process, these mechanisms stifle innate host defenses that are aimed at inhibiting protein synthesis in the infected cell; so much so that translation repression is now being proposed as an alternative RNA silencing mechanism operating against viruses and is mediated by imperfect base-pairing of miRNAs to target mRNAs (see the following section *RNA silencing*). But some viruses do more. Animal viruses, for example, target host translation initiation factors for degradation in order to promote their own translation to the detriment of the host (e.g., Poliovirus, *Indiana vesiculovirus*), while others recruit host translational factors to the sites of viral replication (poxviruses, reoviruses). Plant viruses also recruit cellular translation factors to translate their viral RNAs as well as to regulate their replication and potentiate local and systemic movement through the host. It is not surprising that many natural plant recessive resistance genes have been mapped to mutations of translation initiation factors (eIF4E and eIF4G or their isoforms).

Ubiquitin-Proteasome Machinery

The **u**biquitin–**p**roteasome **s**ystem (UPS) is widely conserved among eukaryotes and is the principal mechanism of protein catabolism in the cytosol and nucleus. Since many viruses use the ER for replication, translation, and/or virion egress, and the ER is finely regulated by UPS, UPS malfunctioning is critical in the manifestation of virus-induced diseases and in virus resistance. In vertebrates, the proteasome 26S plays a fundamental role in the adaptive immune response involving the presentation of viral antigens displayed by **m**ajor **h**istocompatibility **c**omplex class I proteins (MHC class I proteins, see below *Adaptive immunity*). Certain viruses can stall the proteasome, helping the virus to propagate since antigen presentation by MHC class I proteins has been abrogated. For some plant viruses (e.g., some members of the *Tombusviridae* family) ubiquitination is essential for infection and replication. In *Turnip yellow mosaic virus* (*Tymoviridae*), phosphorylation and ubiquitination modulate the pool of viral RdRp available for replication, thus affecting RNA accumulation and recombination, and providing temporal regulation of (-) and (+) RNA strand synthesis. In other cases, the UPS machinery also controls coat protein accumulation and movement thereby reducing its potential cytotoxic effect. On the other hand, some virus proteins disarm host antiviral responses by targeting cell proteins involved in immunity pathways for degradation.

Other Host–Virus Interactions and Immune Evasion

Most of the interactions we have described up to this point are related to the virus subverting the host cell in order to satisfy its energy requirements and sequester building materials for virus replication and virion assembly. It would be erroneous to believe, however, that these are the only interactions that elicit symptom development and disease. Indeed, when a virus makes contact with a host cell a dynamic cascade of events is initiated that culminates in altered gene expression patterns in both the host and the pathogen. These changes lead to the adaptation and persistence of the pathogen or to its clearance from the host. Unbiased and global understanding of the transcriptomes (dual RNAseq) of both host and pathogen is providing new insights into these host–pathogen interactions. It has been demonstrated that: (1) host genes may be up- or down-regulated upon viral infection, including defense-related host genes; (2) RNA silencing appears to be one of the most important biochemical pathways related to plant (and animal?) defense against viral infection; (3) counter silencing, on the other hand, is put in place by the virus in an attempt to evade or subvert host defense systems; (4) multiple virus infections are common, not always related to a synergistic effect but to a facilitating mechanism yet to be fully described, and most probably related to shutting down RNA silencing; and (5) besides cryptic (asymptomatic) infections, some viruses in one host cause no discernable symptoms but in another host are known to cause severe damage, regardless of virus titer. Taken together, viral genomes and viral proteins contribute not only to the development of disease, but to the elicitation of host responses that are designed to protect the host and eradicate the infectious agent.

We will now broaden our review to take account of this complex interplay between viruses and their eukaryotic hosts, and the dynamics of the host's antiviral response. Resistance, defined as the host's ability to limit virus infections, relies on the concerted action of innate and adaptive components of an immune system. The first line of defense across taxa is formed by the innate immune system and encompasses a collection of host defenses ranging from the nonspecific barrier function of epithelia in animals, or waxy cuticle and cell wall in plants, to the highly selective recognition of pathogens through the use of receptors. In contrast, the adaptive immune system, though slower and dependent upon elements of the innate immune system for the initiation, creates receptors for virtually any antigen. The adaptive immune system relies on a diversity of specialized immune cells that roam around the organism. These cells provide an extraordinary virus-specific immune capacity and memory, while minimizing self-reactivity. The system in plants is less complex. Plants lack specialized mobile immune cells. Yet, every plant cell is capable of launching an effective immune response. Although the mechanisms of surveillance and defense differ between host organisms, there are structural and strategic similarities that seem to be universal in the ongoing battle against an old and tricky enemy, functioning on the principle of self/nonself-discrimination. Since resistance mechanisms act by directly limiting the pathogen, once a host evolves resistance to a particular pathogen, the pathogen evolves a method to subvert the resistance. This is the never-ending molecular arms race that characterizes nearly all host–parasite interactions. A few of these virus/host strategies will be examined.

Innate Immunity

The innate immune system, mediated by phagocytes, including macrophages and dendritic cells in animal hosts, utilizes germline-encoded receptors called **p**attern **r**ecognition **r**eceptors (PRRs). PRRs recognize general

pathogen-**a**ssociated **m**olecular **p**atterns (PAMPs) and specifically use two criteria to detect viral nucleic acids: unusual biochemical features present in viral, but not host, nucleic acids and the ability of the host to restrict its own nucleic acids to specific locations within the cell. Nucleic acid sensing PRRs include the **T**oll-**l**ike **r**eceptors (TLRs) and **C**-type **l**ectin-like **r**eceptors (CLRs) located at the plasma membrane or in endosomes, and the intracellular PRRs found in the cytoplasm or nucleus of mammalian cells (**r**etinoic acid-**i**nducible **g**ene **I** (RIG-I)-like receptors (RLRs), **c**ytosolic **DNA r**eceptors (CDRs), and **NOD**-**l**ike **r**eceptors (NLRs), **AIM2**-**l**ike **r**eceptors (ALRs), and **c**yclic **GMP**-**AMP s**ynthase (cGAS).

Cell sensing of RNA and DNA viruses differs. In RNA sensing pathways, a cytosolic family of RNA sensors (RLRs) distinguishes between virus-associated RNAs from cellular RNAs. A downstream signaling pathway is activated by **mi**tochondrial **a**nti**v**iral **s**ignaling proteins (MAVs), which then form filamentous aggregates. These aggregates serve as platforms for interacting with other proteins involved in the signaling pathway, the end result being the expression of specific transcription factors and the transcription of the interferons, IFN-α, and IFN-β. Type I IFNs, in turn, activate autocrine (same cell) and paracrine (cell-to-cell) responses that ultimately induce the transcriptional activity of hundreds of IFN-induced genes, among which are antiviral factors aimed at inhibiting virus replication. RIG-I, for example, recognizes PAMPs in a wide range of RNA viruses. Besides short dsRNAs with a 5' mono-, di- or triphosphate moieties, RIG-I can also recognize self from non-self RNAs by the nature of cap modifications in the viral RNA—except in those cases where this structure is acquired by the virus via a cap-snatching mechanism. In the latter case, the virus evades detection, and hence infection ensues. Other RLRs are also able to recognize double and single stranded RNAs of viral origin, or to regulate the activity of different components of the antiviral sensing activity of the cell. Additionally, mammalian cells detect cytosolic viral RNAs, including dsRNAs, by means of the **p**rotein **k**inase **R** (PKR) activity. dsRNA-activated PKR autophosphorylates as the first step of the signaling pathway, which leads to the complete shutdown of host protein translation, impeding thus viral protein biosynthesis. Viruses avoid this translational arrest by preventing PKR activation, inducing its degradation or bypassing it by displaying RNA structures impervious to PKR interaction.

Cytosolic DNA viruses are detected by a set of different receptors, cyclic GMP-AMP synthase [cGAS], IFI16, DHX9, and DDX41. Distinct aspects of DNA sequence and or structure are recognized by different receptors; viral DNA may be sensed by conformational changes experienced by cyclic GMP-AMP synthases upon DNA binding, or by other features that are-sequence independent but length-dependent. After a DNA recognition signal is received, signaling pathways initiated by receptors activate the **st**imulator of **in**terferon **g**enes (STING), a transmembrane protein in the endoplasmic reticulum. The pathway is collectively known as the **I**FN-**s**timulatory **DNA** pathway (ISD). Once activated, STING translocated from the ER to other cellular compartments (Golgi, endosomes, autophagy-related compartments), is palmitoylated, and after few steps of signaling for activation of other factors, it ultimately leads to MAVs signaling. Although cell sensing of RNA and DNA viruses differ, the outcome is the same in terms of the antiviral response elicited.

Binding of PRRs to their specific ligands evokes intracellular immune signaling cascades that culminate in the robust expression and secretion of antiviral cytokines, such as type I interferons. Some of the roles of IFNs as mediators of cell transcription regulation have been covered earlier in this section. They are small proteins that not only directly inhibit virus replication inside an infected-cell, but are also able to inform nearby cells of the presence of a virus. In doing so, IFNs help neighboring cells heighten their defense responses, and attract immune cells from elsewhere in the body to the site of infection. Cytokines further stimulate the recruitment of antigen-presenting or cytotoxic leukocytes to the site of viral infection. Altogether, the cytokine-coordinated response is very rapid and is usually successful in deterring virus infection (otherwise, adaptive immune response kicks in with all the complex paraphernalia that characterizes it). More on PRRs and the innate immune responses to viruses and their induction pathways follows.

1. **T**oll-**l**ike **r**eceptors (TLRs). TLR3 and TLR7 sensors detect endosomal dsRNAs and ssRNAs, respectively; by means of cellular adaptors, they activate antiviral transcription programs. Additionally, TLRs and RLRs (below) upregulate the IFN-induced 2'-5' **o**ligo**a**denylate**s**ynthetase (OAS), which upon sensing of viral RNAs synthesize 2'-5' adenylate, an activator of the latent cytoplasmic ribonuclease RNASEL. This RNAse cleaves viral RNAs preventing thus their replication.
2. **R**IG-I **l**ike **r**eceptors (RLRs). Different RLRs detect cytosolic viral RNAs with different structures. While some detect short dsRNAs and RNAs with 5' triphosphates, others sense longer dsRNAs and highly structured RNAs. They primarily act by inducing IFN-dependent pathways.
3. **NOD**-**l**ike **r**eceptors (NLRs). Although NLRs are mainly used for the detection of bacterial pathogens, NLRP3 recognizes the cytosolic dsRNA/ssRNA of influenza viruses and Sendai virus, activating the inflammosome, a component of the vertebrates immune system that induces interleukins 1 (regulator of immune responses, inflammatory reactions and hematopoiesis) and 2 (inducer for the production of IFN gamma and of natural killer cells, cytotoxic lymphocytes that induce apoptosis of the target cell). NOD2, on the other hand, seems to play an important role in the restriction of syncytial respiratory virus, *Influenza A virus,* and Human cytomegalovirus by IFN induction.
4. **C**-type **l**ectin **r**eceptors (CLRs). Unlike previous examples, CLRs tends to favor virus transmission, infection and inflammation (e.g., HIV).

Nonetheless, in few reported cases CLRs do induce protective responses by detecting dead cells and promoting presentation to $CD8^+$ T cells (*Vaccinia virus*, and *Human alphaherpesvirus 1* and *Human alphaherpesvirus 2*) and in the case of *Chikungunya virus*, CLRs suppress the pathological inflammatory response induced by the virus.

5. **AIM2**-**l**ike **r**eceptors (ALRs). The role played by ALRs in the activation of the ISD pathway (triggered by the detection of DNA viruses, as explained above) is under debate. However, most ALRs are IFN-stimulated genes, which is consistent with an antiviral function.
6. **C**yclic **GMP-AMP s**ynthase (cGAS). The cGAS-STING pathway plays an important role in type I interferon responses against DNA viruses, including herpesviruses.

Another component of the innate immune response is the complement. Complement consists of an interacting set of enzymes (e.g., proteases), which, upon activation, give rise to a cascade of reactions that result in a membrane attack complex disrupting the lipid bilayers of the infectious agent. Most, if not all, enveloped viruses are susceptible to complement-mediated lysis. The complement forms part of a surveillance system that is independent of antibodies and immune cells. Its unique capacity to sense intruders relies on the interplay between pattern recognition molecules and protein scaffolds (and cell surface receptors). The complement also facilities the elimination of immune complexes and plays a major role in mounting an adaptive immune response, involving antigen presenting cells, T-, and B-lymphocytes. It has become increasingly appreciated that complement function in host defenses extend beyond innate immune responses.

All successful viruses possess effective strategies to evade or inhibit the activation of intracellular PRRs. *Dengue virus*, for example, prevents RLRs from accessing viral RNA by inducing the formation of specific replication compartments that are confined by cellular membranes. Others, replicate on organelles, such as the endoplasmic reticulum, the Golgi apparatus, and mitochondria. *Mammalian 1 bornavirus*, in the *Bornaviridae* family, encodes phosphatases that process the 5′-triphosphate group on their genomes to 5′-monophosphate in order to escape surveillance by RIG-I. Viruses resist IFN activity by blocking downstream IFN-dependent signaling, preventing further IFN synthesis or by inhibiting the activity of the proteins produced following IFN induction. Several poxviruses encode for IFN analogs that bind to host IFNs preventing thus their action and efficacy. On the other hand, some viruses are refractory to IFN action simply by mutation of the gene coding the viral protein targets. Complement evasion via virus mimicry is found with poxviruses. *Vaccinia virus* encodes for, and vaccinia virus-infected cells secrete, a soluble protein that is very similar to host complement regulators in order to inhibit complement-mediated lysis.

Similar to animals, plants mediate early steps of the innate immune response through PRRs. However, studies on plant PRRs for viral recognition and the underlying signaling pathways have just started and details remain largely unknown. One study with *Arabidopsis*, revealed a dsRNA PAMP that induced a signaling cascade involving **S**OMATIC **E**MBRYOGENESIS **R**ECEPTOR-LIKE **K**INASE **1** (SERK1) dependent antiviral resistance. It is clear that upon PAMP perception, a complex set of responses that limit pathogen infection and establishment typically ensue, involving salicylic acid accumulation, ROS production, ion fluxes, defense gene activation, and callose deposition.

In addition to pattern-triggered immunity, there is a second tier to detection in the plant's innate immune system that responds to pathogen virulence factors or avirulence (*avr*) factors, namely **e**ffector-**t**riggered **i**mmunity (ETI). Intracellular ETI receptors, encoded by resistance (*R*) genes, recognize pathogen virulence factors. Recognition triggers signaling cascades and transcriptional reprogramming overlapping with those induced by pattern-triggered immunity, but with the added advantage of a rapid but sustained and robust response. Most ETI receptors are members of a family of **n**ucleotide-**b**inding **s**ite **l**eucine-**r**ich **r**epeat (NBS-LRR) proteins that share domains with mammalian **NOD**-**l**ike **r**eceptors (NLRs). Plant NBS-LRRs can be subdivided into two subfamilies: one with an N-terminal **T**oll/**i**nterleukin-1 **r**eceptor (TIR) domain, which shows homology to *Drosophila* Toll and human interleukin 1 receptor signaling domains (the TNLs), and one with a **c**oiled-**c**oil (CC) domain (CNLs). These two families utilize distinct signaling pathways to drive immune responses, which can include an oxidative burst, hormonal changes, upregulation of immune genes, and a type of rapid cell death that restricts the virus, termed the hypersensitive response. Any protein component of a virus (coat protein, movement protein, replicase) can function as the specific *avr* determinant to elicit resistance mediated by a given *R* gene. Tobacco N protein represents a well-characterized example of the TIR-NBS-LRR class of R proteins in plant−virus interactions. The N resistance protein directly interacts with the helicase domain of the TMV replicase to trigger the resistance response. In addition to dominant *R* genes, recessive *R* genes have also been reported to confer resistance against plant viruses. The recessive gene encoded products are involved in compatibility functions; they are not immune receptors and are not associated with the ETI but rather act as essential translation initiation factors required for the virus to complete its biological cycle. Recessive resistance is conferred by a recessive gene mutation that encodes a host factor critical for viral infection.

To counteract this dual defense and establish infection in susceptible hosts, viruses have evolved effectors that suppress pattern/ETI innate responses. For example, PVY's HC-Pro and PPV's capsid protein function as pattern-triggered immunity repressors, and CaMV-TAV (**tra**n**s**a**ctivator/ **v**iroplasmin) functions as an effector protein suppressing both pattern-triggered and ETI-based responses.

Hormone Action in Many Facets of Plant Virus Resistance

Resistance to plant pathogens, including viruses, is mediated in part by hormone action and is mainly mediated by the action of **s**alicylic **a**cid (SA), **j**asmonic **a**cid (JA), and **et**hylene (Et). Other plant hormones, namely **aux**ins (AUXs), **br**assinosteroids (BRs), **c**yto**k**inins (CKs), **g**ibberellic **a**cid (GA), and **ab**scisic **a**cid (ABA) may play a role in modulating (positively or negatively) plant–pathogen interactions involving viruses. Hormone action in plants is an intricate physiological world in which each hormone has its own signaling pathway; nonetheless, these pathways can establish synergistic or antagonistic interactions for certain responses. For instance, in the same response (namely, resistance to biotic stress), SA, JA, and Et can interact antagonistically; the SA signaling pathway (aimed at counteracting biotrophic and hemibiotrophic pathogens) can repress the JA/Et pathway (mainly devoted to deterring herbivory and necrotrophic infections), but the opposite is also true.

SA is essential for the establishment of local and systemic resistance and mediates *R*-gene resistance, basal immune responses, and the general antiviral resistance mechanism driven by small interfering RNAs known as RNAi (**RNA** **i**nterference or **RNA** **s**ilencing; see below *RNA silencing*). Biosynthesis of and signaling mediated by SA are dependent on the action of the so-called *R* genes in plants whose products, upon interaction with *avr* products of the pathogen, determine the outcome of the host/pathogen interaction. Any mutation that impairs SA biosynthesis and signaling leads to increased virus susceptibility, even if *R* genes are functional. Conversely, mutated *R* genes can lead to compatible interactions with corresponding *avr* products since the SA pathway is not induced. Extreme resistance (resistance that does not implicate local necrosis) is also mediated by SA, and involves the almost total elimination of the virus.

JA, on the other hand, is mainly involved in resistance to necrotrophic pathogens and insect infestation; in conjunction with Et, it modulates the general response known as **i**nduced **s**ystemic **r**esistance (ISR), sometimes triggered by nonpathogenic bacteria. Surprisingly, ISR induced by the latter leads to a reduction of virus symptom expression in certain plants. JA may also modulate early participants of the SA pathway.

ABA's contribution to viral resistance is pleiotropic, that is, it encompasses many different mechanisms and effects simultaneously. The main action of ABA in virus resistance is the indirect accumulation of callose in plasmodesmata with the concomitant hindering of virus cell-to-cell movement, preventing thus local and systemic infection. Although ABA is a strong antagonist of SA, it participates in other defense responses against viruses, including miRNA and siRNA biogenesis, essential components of the RNAi pathway. ABA-mediated induction of RNA silencing has been reported for *Bamboo mosaic virus* (*Potexviridae*); but in this case the effect seems to be mediated by the action of various, but not all, protein-coding genes of the *ARGONAUTE (AGO)* gene family (see below *RNA silencing*).

CKs, which normally promote cell proliferation and elongation, among other physiological roles in plants, seem to also play a role in plant defense responses against viruses. For instance, it appears that levels of CKs help in defining the amplitude of the virus immunity related to the SA response.

Finally, BRs, a hormone family involved in cell elongation, proliferation, and differentiation, apparently act positively in virus resistance, independent of the SA-dependent pathway. Interestingly though, BRs can also enhance SAR, one of the hallmarks of the general SA plant defense.

Viruses, of course, can also counteract these interactions and signaling pathways, and cause disease. In susceptible hosts, plant viruses often manipulate biochemical events and molecular interactions required for their replication and movement, leading to misregulation and disruption of hormone signaling.

RNA Silencing

Another major plant defense response against viruses is based on RNA interference, in which host cellular machinery targets virus-derived nucleic acids. Up until the mid-1990s, it was generally accepted that immunity in eukaryotes is based on the recognition of nonself proteins. Work in the plant science community, however, proved otherwise and demonstrated the existence of an additional defense system based on the recognition of a pathogen's nucleic acids. The defense system referred to as **RNA i**nterference (RNAi), or post-transcriptional gene silencing in plants or quelling in fungi, is generally thought of as an innate immunity mechanism. All forms of cellular life appear to have evolved this sophisticated mechanism, which involves not only innate immunity mechanisms, but in some cases, epigenetic immune memory (i.e., carryover of small interfering RNAs across generations), as well as a distinct type of adaptive immunity, the piwi RNA mechanism.

The canonical inducer of RNAi is long, linear, base-paired dsRNAs, introduced directly into the cytoplasm or taken up from the environment. These dsRNAs are processed by dicer-like nucleases into **s**mall **RNAs** (sRNAs). There are two major classes of sRNAs: **mi**cro**RNA** (miRNAs) and the **s**mall **i**nterfering **RNAs** (siRNAs). miRNAs are encoded by endogenous miRNA genes, transcribed by RNA polymerase II into **pri**mary **miRNAs** (pri-miRNA) and then processed by Dicer-like nucleases. Most processed miRNAs are 21–24 nucleotides derived partly from dsRNA precursors that form imperfectly base-paired hairpin structures. On the contrary, siRNAs are products of foreign nucleic acid, but can also arise from endogenous genomic loci. Following on from processing, a single guide strand from both siRNA and miRNA in association with Argonaute proteins eventually form a macromolecular complex called RISC (**R**NA **i**nduced **s**ilencing **c**omplex). RISC directs sequence-specific cleavage of complementary target RNAs or interferes with their translation. Guide RNA confers sequence specificity to the RNA silencing reaction; whereas, the precise nature of silencing is determined by properties of the Argonaute proteins, including ribonuclease activity and subcellular localization. The primary dsRNA trigger also induces synthesis of secondary siRNAs, through the action of RNA-dependent RNA polymerase enzymes. This secondary pool of siRNAs amplifies and sustains the silencing response, and in some organisms, such as plants and nematodes, leads to systemic silencing that spreads throughout the organism. In other instances, guide siRNA and other Argonaute proteins induce chromatin modifications and transcriptional silencing of complementary genes by methylation in the nucleus.

While limiting virus infection, RNAi orchestrates sequence homology-dependent degradation of viruses (including virus satellites and viroids). That is, the dsRNA viral molecule represented by the virus genome itself, or by any other ds molecule derived from its replication, is targeted for degradation. Experimental evidence indicates that, at least in plants, the RNAi pathway also modulates the immune pathways (presented earlier) by two different mechanisms: fine tuning of hormonal action or silencing of viral genes involved in virulence. For instance, in certain cases RNAi against viruses is enhanced in plants by SA, most probably downstream in the pathway and this leads to the generation of specific siRNAs. To evade RNA silencing, viruses have evolved a variety of counter-defense mechanisms such as RNA-silencing suppressors that interfere with the generation of siRNAs, their amplification or their translocation thus preventing local and systemic resistance. Moreover, some virus-derived siRNAs share sufficient complementarity with host transcripts to promote their silencing (i.e., host mRNA silencing). To complicate things further, certain plants have evolved specific defenses against virus silencing suppressors, providing yet another example of the never-ending arms race between viruses and their hosts.

While plant siRNA sequences defend against viral infections, mechanisms of translational repression of host miRNA are predominantly found in vertebrate hosts. Various animal virus families encode miRNAs that participate in autoregulation of viral gene expression as well as target host mRNAs; together these activities maintain latent and persistent virus infections. Virus-encoded miRNAs are generated by Epstein–Barr virus (*Human herpesvirus 4*), as well as by other members of the families *Herpesviridae*, *Polyomaviridae,* and *Retroviridae*. So far, few virus-encoded miRNA genes have been described in plant viruses. Undoubtedly, further research on miRNAs of cellular and viral origin is expected to widen our understanding of the complex exchange and interactions that are established between organisms and viruses in all kingdoms of life.

Along with miRNAs, other noncoding RNAs, such as **l**ong no**nc**oding **RNAs** (lncRNAs) and **v**ault **RNAs** (vtRNAs), seem to play a role in the arsenal eukaryotes possess to fight against viral diseases. lncRNAs are transcribed by RNA polymerase II, the same protein that transcribes protein coding genes, or III, to generate 200 nucleotides transcripts (or longer) with a 5'-end cap and with or without a poly(A) at the 3' end. They fold into specific 3D structures that interact with genomic DNA, miRNAs, mRNAs, and proteins and are found in the nucleus, cytoplasm, or exosomes, where they play a role in signaling, decoying, guiding, and scaffolding. In vertebrates, lncRNAs are involved in the induction of interferon production, the activation or translocation of transcription factors or the regulation of antiviral proteins or pathways. In plants, lncRNAs have been shown to participate in RNA-directed DNA methylation (i.e., transcriptional gene silencing mediated by siRNAs)

vtRNAs, on the other hand, are small noncoding RNAs of 80–150 nucleotides that are produced by RNA polymerase III and are mostly found free in the cytoplasm or associated with cytoplasmic particle vaults (large ribonucleoprotein particles with barrel shaped morphology). Among other functions, they act as novel suppressors of tumors, but also seem to play a key role in inhibiting innate immune responses against certain viruses, like IAV.

Other Innate Immune Responses Against Viruses

Nonsense-Mediated mRNA Decay

Recent evidence indicates that cellular RNA quality control systems, such as the **n**onsense-**m**ediated mRNA **d**ecay (NMD) pathway, can restrict viral infections. The NMD pathway is a surveillance pathway that safeguards the quality of mRNA transcripts in eukaryotic cells by targeting faulty cellular transcripts for degradation by multiple nucleases or decapping enzymes. NMD depends on protein translation and is triggered by ribosomes, terminating translation in an "*unusual*" position along the mRNA. A number of features can predispose mRNA to NMD, including long or intron-containing 3' **u**n**t**ranslated **r**egions (UTRs), the presence of short open reading frames in the 5' UTR, or alternative splice events that introduce a premature translation termination codons; features that are commonly associated with virus mRNAs because of their inherent compact genome structures (refer to Chapter 3: Replication and Expression Strategies of Viruses). In fact, NMD seems to play a role in deterring virus infection with members of the families *Alphaflexiviridae* and *Tombusviridae* in plants and fungi, and *Togaviridae* in animals (all with ssRNA(+) genomes). In the latter, virus genome expression requires the synthesis of a subgenomic mRNA for translation of the second ORF. This genomic arrangement is perceived as a very long 3' UTR and triggers the NMD degradation pathway. Complementary replication-derived sRNA(-) intermediates in all ssRNA(+) viruses also function as triggers.

Whether mRNAs from DNA viruses are subject to NMD, or if NMD plays a role in limiting virus replication in mammalian hosts is not yet known.

Not surprisingly, viruses have evolved mechanisms that interfere with or modulate NMD. Unspliced *Rous sarcoma virus* RNA, for example, contains a stability element downstream of an internal stop codon that inhibits recognition of the normal *gag* termination codon. *Human immunodeficiency virus 1* (HIV-1), on the other hand, recruits UPF1, a crucial component of the host's NMD pathway, in its biological cycle. UPF1 acts as a positive regulator of HIV-1 mRNA translation and is involved in the nuclear export of viral RNA. HIV-1 virions containing UPF1 are more infectious than particles lacking it; additionally, reverse transcription is more efficient and less costly.

RNA Exosome

Similarly, the exosome-mediated decay of unadenylated mRNAs is considered an innate immune response against viral infections. The nuclear RNA exosome is a multi-subunit complex that controls RNA homeostasis, mainly by 3' to 5' catalytic degradation of RNA substrates. It plays a key role in RNA processing and surveillance pathways such as NMD. Current evidence suggests that impaired RNA exosome activity suppresses IAV infections, which means that some viruses have the capacity of hijacking these mechanisms in order to establish infection successfully. Viruses protect the 3'-ends of their mRNAs by either carrying a poly(A) tail to mimic the cellular mRNAs or a stable 3' stem loop structure that inhibits the action of exosome-associated 3'-5' exonucleases.

Additionally, antiviral responses may be profoundly influenced by exosome-mediated transfer. It is a common phenomenon in eukaryotic cells to release membrane vesicles: small-sized lipid-bilayer enclosed bodies (exosomes) upon the exocytosis of multi-vesicular bodies. An extensive body of evidence has shown that exosomes mediate intercellular communication and also play crucial roles in antiviral innate immune responses. They deliver host miRNAs to regulate the innate immune response. Some viruses utilize exosome-mediated transfer. They use exosomes to transfer their viral RNA to neighboring cells. Work with retroviruses demonstrated this previously unsuspected strategy of evading the innate response, by simply not inducing it.

Adaptive Immunity

Plants, as do insects, rely entirely on innate immune responses to recognize and mount resistance responses against pathogenic invaders. Vertebrates, too, depend on innate immune responses as a first line of defense, but they also mount much more sophisticated defenses, called adaptive immune responses, in instances where viral replication outpaces innate defenses and the host response is overwhelmed or evaded. As the name suggests, "*adaptive*" refers to the differentiation of self from nonself, and tailoring of the response to the particular foreign invader.

The adaptive immune system utilizes two responses: (1) the humoral response that leads to the production of specific, circulating antibodies by B lymphocytes and (2) the cell-mediated response that involves the activation of cytotoxic T lymphocytes that destroy virus-infected cells. In both instances, these cells possess the ability to learn and remember. Memory is maintained by a subset of B and T lymphocytes called memory cells, which survive for years in the body and respond rapidly and efficiently to any subsequent encounters with the virus. Although the adaptive immune response is very specific and robust, its strength can also become its weakness. That is, specific recognition of a particular strain of the virus leads to the perpetuation of clonal cell lineages that are capable of recognizing only slight variants of the region of the virus (i.e., antigenic epitope) that triggered the initial response. If a new variant of the same virus changes enough, it may not be recognized.

Both the humoral and cell-mediated arms of the adaptive immune system work in concert to contain virus infections. If an organism is infected by a virus, initially only a limited number of cells perceive its presence (as explained in previous sections). If the infected cells cannot deter virus replication and virus gene translation, other cells, namely the T cells of the adaptive immune system are alerted of its presence and the cell-mediated response is called into action. The T-cell response typically consists of two major populations of cells making up the circulating surveillance: $CD4^+$ T cells and $CD8^+$ T cells. The latter recognize fragments of virus proteins that are presented by MHC I molecules (**m**ajor **h**istocompatibility **c**omplex class I, MHC I) on the surface of infected cells. MHC class I molecules are expressed on the cell surface of all cells and present peptide fragments derived from the cell's own "*house-keeping*" ubiquitin/proteasome pathways. In a virally infected cell, peptides derived from viral proteins may be presented along with endogenous peptides. The cytotoxic T cells ($CD8^+$) display surface proteins (**T c**ell **r**eceptors, TCRs) capable of recognizing the combination of the MHC molecule and virus fragments, and are activated to multiply rapidly into an army of T cells. Once there is recognition between both cells, $CD8^+$ T cells are activated and destroy virus infected cells via a perforin/granzyme induced apoptosis pathway, thereby hindering the spread of the virus. Cytotoxic cells possess a pool of cytotoxic granules, perforin and granzyme. Perforin, a calcium-dependent cytolytic protein, forms "*pores*" in the membrane of the presenting cell through which co-secreted granzymes (pro-apoptotic proteases) enter and induce apoptosis. Other cytotoxic factors, such as granulysin released by cytotoxic $CD8^+$ T cells, have cytolytic and proinflammatory activities, and cytokines—including γ-interferon and tumor necrosis-α, enhance the killing mechanisms mediated by cytotoxic cells. Another population of T cells known as helper $CD4^+$ T cells recognize viral peptides, (derived from lysosome pathways) that are presented by MHC II molecules on a limited number of cells (dendritic cells and monocytes/macrophages and B cells). Helper T cells are arguably the most important cells in adaptive immunity, as they not only drive the initial expansion of virus-specific $CD8^+$ T cells and the secretion of anti-inflammatory cytokines but they also help activate B cells to secrete antibodies (B cell responses may also be triggered directly in the absence of T cell help).

The humoral arm of the adaptive immune response is tasked with protecting the extracellular spaces. Once activated, B cells secrete antibodies (IgG, IgM, and IgA). Antibodies cannot enter cells, they circulate in the bloodstream and other body fluids and bind directly to the virus (that stimulated their production) before replication begins inside the cell. Antibodies, collectively referred to as immunoglobins, are Y-shaped proteins that by action of their variable regions containing paratopes (the lock), recognize and interact with epitopes (the key) presented by an infecting virus. Successful parotope-epitope (right lock-key pair) interaction may lead to: (1) neutralization of the virus so that it is no longer capable of interacting with and infecting the target cell; (2) agglutination, whereby many virus particles interact with many antibodies rendering them incapable of infection and more "*accessible*" to responses mediated by cells of the immune system; (3) Fc receptor (FcR)-mediated phagocytosis triggered by the binding of phagocyte Fc receptors to antibodies in antibody-virus complexes; and (4) complement-mediated phagocytosis in response to antibodies binding to virus at the surface of a cell (opsonization). The presence of the antibodies and complement proteins facilitates their interaction with immunoglobulin receptors (Fc receptors) or complement receptors on the phagocyte surface leading to a potent defense via ingestion by phagocytic cells.

Once the threat of virus infection is controlled, the immune responses is suppressed, but a small percentage of cells transition into quiescent memory cells, and confer expedited immunity over extended periods of time. Apparently natural killer cells, thought to be a part of the innate immune response, can also exert immunological memory after encounters with viruses, blurring the line between the two arms of the immune response.

Finally, viruses have developed several escape mechanisms that thwart host responses, resulting in uncontrolled or chronic infection associated with pathologic damage or in other instances, productive infection without rapidly killing or even producing excessive damage to host cells. Among these escape mechanisms are viral proteins that interfere with antigen presentation and target both MHC-I (Human cytomegalovirus, Bovine herpesvirus type 1) and MHC-II (*Human herpesvirus 4*) processing. Glycoproteins of Human cytomegalovirus and Murine γ-herpesvirus degrade MHC class I molecules via the **e**ndoplasmic-**r**eticulum-**a**ssociated protein **d**egradation (ERAD) pathway thereby preventing infected cells from properly presenting viral peptides on their surface. Other viruses (*Herpesviridae* and *Coronaviridae*) interfere with the complement activation pathway by avoiding complement binding to antibody-antigen complexes, either by removing (shedding or internalization) these antibody–antigen complexes from the cell surface of infected cells or by expressing Fc receptors. Some poxviruses and herpesviruses encode and express proteins with functional similarities to complement regulators or they incorporate these host proteins into their viral envelopes. Influenza viruses acquire point mutations in genes encoding the surface glycoproteins hemagglutinin (HA; antigenic drift) and neuraminidase (NA) or forms of new virus subtypes with mixed HA and NA from different subtypes by reassortment (antigenic shift). The resulting virus is poorly recognized, if at all, by antibodies and by T cells directed against the previous variant. *Lymphocytic choriomeningitis mammarenavirus* (LCMV) escapes the immune system mainly by two strategies: "*speed*" (the virus is present in high load and in widespread distribution) and "*shape change*" (mutations). The most extreme case of immune suppression is seen with HIV. HIV infection leads to a gradual loss of immune competence. This is in addition to making its surface components difficult to access by neutralizing antibodies, to creating cellular hideouts and destroying immune effectors. On the other hand, the major strategy of Murine cytomegalovirus (MCMV) is to hide from the immune system by latency. After the productive primary infection has been cleared by the immune system, the viral genome is not eliminated, but persists as episomal DNA. Viral proteins, which are the major target for the immune system, are not produced or only to a very limited extent. Productive replication and virus shedding can be reinitiated from latent genomes at some point to allow virus transmission to a new host. Only a few virus families are known to be capable of latency: herpesviruses (of which MCMV belongs) and retroviruses. Latency is a very favorable means for viral survival and represents an example of an interaction engineered to run smoothly.

FURTHER READING

Anand, S.K., Tikoo, S.K., 2013. Viruses as modulators of mitochondrial functions. Advances in Virology. Available from: https://doi.org/10.1155/2013/738794. Article ID 738794.

Chen, N., Xia, P., Li, S., Zhang, T., Wang, T.T., Zhu, J., 2017. RNA sensors of the innate immune system and their detection of pathogens. IUBMB Life 69, 297–304. Available from: https://doi.org/10.1002/iub.1625.

Koonin, E.V., Zhang, F., 2017. Coupling immunity and programmed cell suicide in prokaryotes: Life-or-death choices. Bioessays 39, 1–9. Available from: https://doi.org/10.1002/bies.201600186.

Labrie, S.J., Samson, J.E., Moineau, S., 2010. Bacteriophage resistance mechanisms. Nature Reviews Microbiology 8, 317–327. Available from: https://doi.org/10.1038/nrmicro2315.

Martinez-Martin, N., 2017. Technologies for proteome-wide discovery of extracellular host-pathogen interactions. Journal of Immunological Research 2017. Available from: https://doi.org/10.1155/2017/21976152197615.

Roosinck, M.J., 2010. Lifestyles of plant viruses. Philosophical transactions of the Royal Society B 365, 1899–1905. Available from: https://doi.org/10.1098/rstb.2010.0057.

Sanchez, E.L., Lagunoff, M., 2015. Viral activation of cell metabolism. Virology 479-480, 609–618. Available from: https://doi.org/10.1016/j.virol.2015.02.038.

Schorey, J.S., Cheng, Y., Singh, P.P., Smith, V.L., 2015. Exosomes and other extracellular vesicles in host–pathogen interactions. EMBO Reports 16, 24–43. Available from: https://doi.org/10.15252/embr.201439363.

Wang, H., Peng, N., Shah, S.A., Huang, L., She, Q., 2015. Archaeal extrachromosomal genetic elements. Microbiology and Molecular Biology Reviews 79, 117–152. Available from: https://doi.org/10.1128/MMBR.00042-14.

Chapter 11

Beneficial Interactions with Viruses

Paula Tennant[1] and Gustavo Fermin[2]
[1]*The University of the West Indies, Mona, Jamaica,* [2]*Universidad de Los Andes, Mérida, Venezuela*

When life gets tough only the cooperators survive

Marylin Roossinck

While it cannot be denied that viruses cause catastrophic disease among humans and domesticated plants and animals, there are also viruses that are beneficial to their hosts. These viruses can be essential for the survival of the cellular organism that harbors them, and in some cases, they confer a competitive advantage on the organism. The interaction is best described as mutualistic symbiosis.

The term symbiosis was first coined in the 1800s in an attempt to describe the relationship in lichens—entities made up of two organisms, a fungus and an alga. The term was broadly defined as "*the living together of different species.*" Hitherto it was realized that symbiotic relationships can be obligate, the relationship required for the survival of one or both partners, or nonobligate. The definition is still generally accepted today, but now encompasses several different relationships, including commensalism, antagonism, and mutualism. Commensalism refers to an association where one partner benefits, but the other is not adversely affected; while antagonism is a costly interaction in which one benefits at the expense of the other. Mutualism refers to an association where both partners benefit from the association. The interacting partners in a symbiotic relationship are called "*symbionts*" and the partnership that can involve two or more symbionts, is called the "*holobiont.*" Relationships among symbiotic organisms may change over time and frequently respond to the availability of resources.

Although viruses are often thought of as purely antagonistic, there are examples of viruses that have mutualistic relationships with their hosts. That

Viruses. DOI: https://doi.org/10.1016/B978-0-12-811257-1.00011-5

is, the viruses are beneficial and their net effects on the host are positive, compared with uninfected hosts. Typically, these viruses have a long association with the host and the relationship is essential for the survival of the host; the viruses attenuate diseases caused by other viruses or other pathogens; the viruses are useful to their hosts because they help in warding off competitors; or the viruses help their hosts adapt to extreme environmental conditions. There is also growing evidence that virus–host interactions have many of the features of symbiotic evolution. Where symbiosis gives rise to evolutionary change, it is defined as symbiogenesis. Symbiogenesis, or endosymbiosis, is the extremely rare, but permanent merger of two organisms from phylogenetically distant lineages into one more complex organism. The theory was first recognized in the early 1960s as an important driver of evolution and speciation. Perhaps the best known examples of symbiogenesis include: (1) the intracellular acquisition by an early eukaryote of an α-proteobacterium to give mitochondria; (2) the conversion of a cyanobacterium into the first chloroplast, thereby forming kingdom Plantae; and (3) the acquisition of a red alga to yield more complex membrane topology in the kingdom Chromista. There is growing evidence that virus–host interactions have many of the features of symbiotic evolution and that viruses contributed to host evolution early in the evolution of life and continue to contribute. The most important level of virus symbiotic interaction is at the genetic level.

In symbioses at the genetic level, symbionts contribute whole preevolved genes or regulatory sequences to their partners. As in the case of the acquisition of mitochondria and plastids, this involved the union of whole preevolved genomes of symbionts to form a new holobiontic genome. From the symbiological perspective, the endogenization of retroviruses into vertebrate lineages might be seen as the viral equivalent of the symbiogenetic union of genomes with the ancestral microbes of mitochondria and plastids. Viruses, notably retroviruses, as well as bacteriophages, are capable of endogenization, in which the virus inserts its genome into the chromosomes of the host germ line. The process not only results in the introduction of a preevolved genome that is novel to the host genome, but there is also the introduction of a functional unit of genes that are preevolved to interact with and regulate key aspects of host genetics, immunity, and physiology. Some 8% of the human genome is derived from **h**uman **e**ndogenous **r**etro**v**iruses (HERVs). This number increases to 50% if HERV fragments and virus-dependent entities are included. It appears that these sequences are not "*junk DNA*," but they have made a major contribution to human evolution, and are playing important roles in human embryology, reproduction, and physiology. Another potential outcome of endogenization surrounds host competitiveness and/or the induction of resistance against the same exogenous virus, or a related strain.

A similar complex, multifaceted contribution of viral symbiosis is increasingly found when symbiotic viruses exchange genomic elements. Viruses in mixed infections frequently reassort and recombine, giving rise to new virus variants—some of which could become new species. This gives viruses an enormous level of flexibility and a capacity for very rapid evolutionary change. An example of new RNA virus species likely formed through symbiogenesis includes *Poinsettia latent virus* (genus *Polemovirus* of an unassigned virus family), which appears to be a recombinant between a polerovirus and a sobemovirus. Symbiogenesis is also involved in the evolution of DNA viruses. Recently described geminiviruses have been attributed to recombination.

Our new understanding of the role of viruses in mutualistic symbiotic relationships is expanding as our knowledge of the virosphere increases. This chapter explores the nature of these interactions with bacterial, fungal, insect, mammalian, and plant hosts. We highlight some of the dramatic examples of beneficial viruses, demonstrating why viruses need to be taken seriously not just as pathogens, but as integrated members of holobionts (that is, all living organisms they associate with). Finally, beyond the richness of biotic interactions they establish, viruses are crucial players in the cycling of nutrients and the regulation of biogeochemical cycles. Taken together, viruses seem to also have a say in the homeostasis of ecosystems.

PHAGE–HOST INTERACTIONS

The abundance (and diversity) of prokaryotes is astonishingly high; they are present in all environments, provide essential ecological services, and are required to sustain almost every other form of life. But "*their*" viruses (phages) surpass them, both in terms of number and types. A conservative estimate calculated over 10^{31} phages in the biosphere, with the vast majority in the oceans. It is essential, thus, to briefly summarize the roles played by phages in the life and ecology of their hosts in this rich and complex environment. Here, viruses interact with cellular organisms, bacteria, archaea, and eukaryotes. While some forms of lysogeny have been described as mutualism, lytic phages are predators of prokaryotes, but mortality and the release of cell lysis products into the environment can strongly influence microbial food web processes and biogeochemical cycles. It is evident from recent studies that the environment where bacteria, archaea, and their viruses coexist is a highly dynamic network of genetic exchanges with input from other vehicles of genetic transmission, including plasmids and **g**ene **t**ransfer **a**gents (GTAs). Only viruses will be covered here, but the reader is encouraged to delve further into the dynamics of the mobilome.

Phages Affect Host Population Sizes and Biogeochemical Cycles

One of the most important roles of phages in planktonic communities is likely the maintenance of diversity, but the contribution of lytic marine phages to the conversion of **d**issolved **o**rganic **c**arbon (DOC) into **d**issolved **o**rganic **m**atter (DOM) should not be underestimated. Lytic viruses represent one of the main causes of microbial mortality in these communities. Cell debris derived from lysis along with released viruses (and excreta from primary producers and grazers) are sources of DOM. This in turn is taken up by bacteria, and other organisms, to keep the carbon cycle going. Of course, other inhabitants of the ocean participate in this cycling of carbon since not only bacteria are infected with viruses. Of equal importance are the increases in prokaryotic respiration and production during virus infections, and thus the transformation of organic matter and the regeneration of inorganic N and P. Upon host lysis, virus-induced influx of organic matter is consumed by organisms, adding to the cycling of nutrients. It is estimated that between 6% and 26% of the organic carbon generated by photosynthesis is recycled back to DOM by viral cell lysis.

Phages Manipulate Host Metabolism

In deep-sea hydrothermal vents, viral infections play a pivotal role in the ecology, population dynamics, and evolution of mesophilic and thermophilic communities consisting of thermotolerant bacteria and archaea. Ever since lysogeny was shown to frequently occur in these environments, it has been speculated that an increased potential for horizontal gene transfer under these extreme conditions may have enhanced the metabolic flexibility of prokaryotic inhabitants via key gene exchanges (particularly those related with energy metabolism). Apart from these compensation effects on microbial metabolism, phages regulate microbial diversity and abundance and impact global biogeochemical cycles by lysing their hosts. It is estimated that in deep sea 2.2% of bacteria and 2.3–4.3% of archaea die each day as a consequence of viral infections. This means a respectable amount of material is being recycled daily due to the action of these marine phages.

Not surprisingly, the same outcomes have been observed in freshwater environments (including surface water and groundwater). And in extreme environments, such as Antarctic soils, the structure of viral communities has demonstrated that (1) viruses are highly diverse (members of 38 virus families were found); (2) there is both competition and cooperation between viruses (positive correlations between the families *Myoviridae* and *Mimiviridae*, and negative among these families and *Siphoviridae*); and (3) viruses in this cold, desertic soil interact with each other, their hosts, and the environment in an intricate manner (for example, soil pH and altitude were the dominant abiotic drivers of the composition of viral communities, while soil pH and calcium content acted as the main abiotic drivers of the most significant viral genotypes). All combined, viruses are key players in maintaining host diversity and functioning of microbial communities of this particular environment. If we extrapolate these very features to other environments richer in flora and fauna, it is very likely that viruses not only form part of the communities to which their hosts belong; they also regulate their own existence and the existence of those not directly involved in the partnership.

Phages Contribute to Microbial–Metazoan Symbioses That Affect Metazoan Fitness

Scientists have only recently begun to appreciate the role of phages in microbial–metazoan symbioses that occur in complex natural environments. One environment is that of metazoan mucosal surfaces, the main points of entry for pathogenic microorganisms. In metazoans (multicellular animals with differentiated tissues), a protective layer of mucus lines body surfaces that interact with the environment (fishes and other animals) or internal organs (respiratory tract). Mucus is a slimy, aqueous secretion that is produced by epithelia. Invariably, it is a good substrate for bacterial attachment and growth. Evidence indicates that lytic phages are present in mucus in concentrations that are at least four times higher than those found in surrounding tissues. It was subsequently realized that these phages act as predators and control bacterial abundance via cell lysis. This benefits the metazoan host by limiting mucosal bacteria. Even more, interaction of virions with mucin glycoproteins slows their diffusion through the mucus layer creating thus a protective barrier. This derived immunity guards animals from bacteria in the environment, and when present on internal epithelia, it also protects against pathogenic bacterial infections. The phage-bacterial population dynamics creates a highly complex, and yet efficient, ecosystem where the nonhost interacts symbiotically with phages and in return, the phages are exposed to a steady supply of bacterial hosts. Similar relationships are observed with mucus of sea anemones, fishes, and corals.

The human gastrointestinal tract also hosts a complex microbial ecosystem and an intricate group of viruses. Here, the phage community (dubbed as the phageome) obeys more or less the "*rules*" followed by the human bacteriome; that is, some phages (or bacteria) are common to all individuals, while some others are individual-specific (and even body-location specific). Moreover, both communities change depending on age, sex, physiology, nutrition, and health status of the carrier, and impact each other based on composition. Work is only beginning on gut viral ecology, but it is becoming clear that phages contribute to the structure and function of the human gut microbial community, ultimately influencing states of health and disease of the animal host.

Phages have also been found in the human blood (phagemia). How they translocate from the gut to the blood is still a matter of debate, but they are found in sera, and more interestingly, in both diseased and healthy humans. *In vitro* studies attribute permeation through the body to transcytosis across epithelial cell layers in a unidirectional fashion. Either way, phage circulation in humans, and other mammals, implies protective potential against invading bacteria and perhaps other viruses and the development of some forms of cancer. Accumulating evidence indicates that phages may play a role as immunomodulators that help maintain local immune tolerance to foreign antigens derived from, say, gut microbiota. Additionally, they may help in inhibiting local immune and inflammatory reactions (e.g., inhibition of skin allograft rejection).

VIRUS–FUNGUS INTERACTIONS

Mycoviruses Increase Their Hosts' Fitness

Like bacteria, fungi have evolved systems that confer a marked advantage in the competition with other organisms for available nutrients. The killer phenomenon in *Saccharomyces cerevisiae* is one such example. In the early 1960s, scientists found that certain strains of *Saccharomyces cerevisiae* (commonly known as wine or baker's yeast) secrete protein toxins that are highly lethal to sensitive *S. cerevisiae* strains. They became known as "*killer yeasts*," and the toxins that they produced were called "*killer toxins*." The killer strains themselves were found immune to their own toxin, but they were susceptible to toxins secreted by other killer yeasts. Further work showed that most killer toxins act as pore forming agents (ionophores) that disrupt membrane function; others cause G1/S cell cycle arrest and/or rapid inhibition of DNA synthesis. It seems likely that yeast species evolved differently to produce their killer toxins, because the killer phenotype can be attributed to chromosome genes, DNA plasmids, or to the presence of dsRNA viruses. The latter presents an interesting mutualistic symbiotic association involving the fungal host and two dsRNA viruses, a killer virus, and its helper virus. These viruses, called **S**accharomyces **c**erevisiae **v**iruses (ScVs), belong to the family *Totiviridae*. The M-virus, also called killer virus or satellite virus, contains only one ORF that encodes a preprotoxin, the unprocessed precursor of the mature secreted killer toxin. M viruses depend on a second dsRNA helper virus (L-A) for replication and encapsidation. L-A provides the capsids in which both L-A and M dsRNAs are separately packaged. The M-virus is totally dependent on the presence of the helper L-A virus. Yeasts need to harbor both species of viruses in order to maintain the killer phenotype. Of note, the L-A virus alone does not confer a phenotype upon its host nor does it lead to cell lysis or slow cell growth. Similarly, persistent infection of yeast cells with both viruses is symptomless. Transmission is strictly vertical and the viruses are inherited either after cell division or through mating with a donor cell, not by horizontal infection. Thus, the viruses provide the host with a toxin directed against other yeasts strains, while the yeast cells enable their propagation and the viruses gain a more efficient means of transmission. A recent report suggested that killer systems are so beneficial to their hosts that they have resulted in the loss of host RNAi systems (aimed at controlling virus infection, replication, and movement). As is true in many symbioses, the nature of the relationship between viruses and hosts is dependent on the environment: at high pH, the toxin is much less effective and the advantage is lost. After the initial discovery of the killer phenomenon in *S. cerevisiae* in the 1960s, it soon became evident that killer strains are not restricted to members of the genus *Saccharomyces*, but can be found among other fungal species of the genera *Hanseniaspora, Ustilago*, and *Zygosaccharomyces*.

Mycoviruses as Modulators of Symbiotic Associations

Perhaps a more intriguing association is that between an endophytic fungus and a plant that is modulated by a virus. In 2002, it was observed that a type of grass growing in the geothermal zones of Yellowstone National Park—panic grass, *Dichanthelium lanuginosum*- was able to survive high temperatures of up to 65°C due to its association with an endophytic fungus, *Curvularia protuberata.* When grown separately, neither organism could tolerate such high temperatures. Probing revealed that thermal tolerance conferred by the endophytic fungus was actually due to its infection with an unassigned dsRNA virus, aptly named Curvularia thermal tolerance virus (CThTV). It is hypothesized that the mechanism of thermal tolerance involves the manipulation of plant and or fungal gene products that are involved in stress tolerance. This three-way mutualistic symbiosis, involving a virus that infects a fungus which infects a plant, is reminiscent of the interrelationships between bacteriophages, *Wolbachia* bacteria, and insects.

Interactions Between Mycoviruses and Nonhosts

An even more complex relationship involving a fungal virus that can infect an insect and use it as its vector was recently revealed. Sclerotinia sclerotiorum hypovirulence-associated DNA virus 1 (SsHADV-1) is a circular ssDNA geminivirus-related virus that was isolated from a hypovirulent strain of *S. sclerotiorum.* It is known now as *Sclerotinia gemycircularvirus 1*, and is classified in the newly described *Genomoviridae* family of viruses. *S. sclerotiorum* is a notorious plant fungal pathogen with a broad host range including important crops, such as oilseed rape and soybean. In 2016, scientists noticed insects of the *Lycoriella* genus consuming SsHADV-1-infected mycelia of *S. sclerotiorum* in the laboratory. They investigated further, curious to determine whether the insects were capable of acquiring, and subsequently transmiting, the virus harbored by the fungus. Indeed, it was discovered that SsHADV-1 could infect insect larvae after the consumption of infected mycelia, and that viruliferous adults transmitted the virus transovarially (parent to offspring transmission). SsHADV-1–infected fungus could also suppress the production of repellant volatile substances by their plant hosts to attract adult insects and thus facilitate colonization and egg-laying, increasing the chances of virus acquisition and transmission. Virus infection even stimulated female adults to produce more eggs. These findings represent a unique example of virus interactions spanning both fungal and animal kingdoms.

In other virus–fungus associations, the virus can act as a mutualist of the plant by altering the ability of plant pathogenic fungi to cause disease. These virus infections generally result in reduced virulence of the fungus host, termed hypovirulence, and offer the potential for development of biological control strategies for fungal diseases of plants. The best studied example is the interaction between *Cryphonectria hypovirus 1* (CHV1; *Hypoviridae*) and the chestnut blight fungus, *Cryphonectria parasitica*. Chestnut blight, caused by the fungus *C. parasitica*, is a serious tree disease in forests and orchards. The pathogen infects the bark and cambium of trees through wounds, and causes lethal bark cankers that eventually girdle the branch or stem, leading to death of the entire tree. Infection of *C. parasitica* with CHV1 results in nonlethal and inactive cankers. CHV1 is an unencapsidated RNA virus of the genus *Hypovirus*. The virus not only reduces the pathogenic potential of the fungus; it further inhibits sexual reproduction, strongly attenuates growth and asexual sporulation, and reduces pigmentation of *C. parasitica* cankers. The virus, although not a mutualist of its fungal host, is beneficial for the plants that harbor the fungal hosts. Chestnut blight in Europe has been maintained at low severity in most regions with CHV1 for over 40 years. The *Cryphonectria*–hypovirus pathosystem has led to the discovery of hypovirulent strains in other fungi: *Ophiostoma ulmi* (the causative agent of Dutch elm disease), *Bipolaris victoriae* (the causative agent of Victoria blight of oats), and *Sclerotinia sclerotiorum* (the causative agent of white mold).

VIRUS–ARTHROPOD INTERACTIONS

Arthropod Viruses Protect Their Hosts

Mutualistic interactions between viruses and some insects (and other arthropods) illustrate well cases in which viruses exert a positive effect on host development. In nature, for example, many wasp species parasitize a wide range or arthropods within which they lay their eggs and larvae develop, sometimes feeding on the living, parasitized host. Although the unwilling host possesses the immunological capacity to eliminate the eggs of the invading wasp, a virus has intervened and suppressed the host's immune response to the advantage of the parasitizing wasp. That is, virus-infected wasps are not eliminated by their arthropod host, leading to successful parasitism. Phylogenomic studies suggest that these viruses actually contributed to wasp diversification and adaptation to their hosts. Members of the *Polydnaviridae* family of viruses are integrated into the genome of the wasp hosts, and thus their genomes are inherited as proviral forms that replicate as the host chromosomes do. The virus ancestor of the members of the family integrated into the genome of parasitoid wasps (families Hymenoptera,

Ichneumonoidea, and Braconidae) then experienced extensive rearrangements after which wasps produced their "*own*" functional viruses as enclosed fragmented dsDNA molecules (encoding only virulence genes). This domestication, which allowed the virus to be efficiently transmitted at lower rate of mutation, happened at least twice in the evolutionary history of wasps. Wasps, in turn, gained viral genes that allowed them to avoid or overcome host defenses of the arthropods they parasitize. Not surprisingly, some wasps are assisted further by *another* virus. The aphid *Acyrthosiphon pisum*, for example, is host to the parasitic wasp *Aphidius ervi*. The aphid is able to eliminate wasp larvae due to the symbiotic association with the bacterium *Hamiltonella defensa*. It turns out that the protective role played by the bacterium is due to the presence of a bacteriophage called APSE (***A***. ***pisum*** **S**econdary **E**ndosymbiont, like *Hamiltonella virus APSE1*, *Podoviridae*), which harbors a gene that encodes a toxin potentially effective against eukaryotes. This protective "*system*" seems to be present in other aphids, and involves the aforementioned APSE and other APSE phages. Of note, the phage present in *H. defensa* was apparently acquired laterally from another, unrelated bacterium of the genus *Arsenophonus*. Truly, these represent remarkable examples of complex and intricate evolutionary relationships.

But, there are also much simpler interactions that achieve the same endpoint of successful interactions with insect hosts. Cotton bollworm (*Helicoverpa armigera*), a migratory moth pest of cotton and other economically important crops of the old world and Australasia, is host to *Lepidopteran iteradensovirus 5* (*Parvoviridae*), an icosahedral, nonenveloped ssDNA virus. The virus can be horizontally and vertically transmitted by the moth, and viral infection of the host leads to no detrimental effect on fitness-related traits; that is, the virus and the moth exist in mutual tolerance. From the point of view of the moth (not the cotton plant, though), this interaction brings certain advantages: virus-infected *H. armigera* experiences increased larval and pupal developmental rates and extended lifespan and fecundity of females. Notably, the virus confers resistance to *Helicoverpa armigera nucleopolyhedrovirus* (*Baculoviridae*), a baculovirus used as a biocontrol agent against the cotton bollworm, as well as to low doses of Bt toxin.

In another instance of beneficial interactions between a virus and its host (but not the plant that is parasitized by the arthropod), it has been observed that the parasitic mite *Varroa destructor* is advantaged while transmitting the *Deformed wing virus* (DWV; *Iflaviridae*). Because of the immunosuppressive effect of DWV in bees, feeding and reproduction of the parasitic mite are enhanced. That is, the mite vectors the virus to bees, and as a consequence of this interaction, the mite is protected from the potential defense response of the third party within the ménage-a-trois. In other three-way interactions, it has been demonstrated that viruses are capable of modifying the feeding behavior of their insect vectors in a way that enhances their transmission.

Arthropod Viruses Mediate Host Responses

Worth mentioning, too, is the case in which a virus promotes developmental changes in its insect host. In some aphids, it has been demonstrated that infection of nymphs by an ambidensovirus (*Hemipteran ambidensovirus 2)* promotes their development to winged morphs; otherwise, they remain wingless. These infected, winged aphids subsequently colonize their plant hosts and establish colonies of wingless aphids. As population density increases, the odds of horizontal transmission of the virus between nymphs and the acquisition of the virus increases leading to virus-induced wing formation and winged aphids. These virus-infected winged aphids are likely to colonize neighboring hosts and facilitate dispersal of the virus. The induction of winged morph development as a result of virus infection offers some advantages to the host: phenotypic variation in an otherwise clonal population because of virus-induced epigenetic changes may modify the aphids' response to environmental cues.

UNEXPECTED OUTCOMES IN VIRUS–MAMMAL INTERACTIONS

Viruses are always with mammals. In the case of humans, it has been estimated that between 8 and 12 different viruses (not including the phages associated with members of our bacteriome which are, by the way, the most numerous in our bodies) are carried persistently by any individual. Considering the numbers of all viruses combined just residing at body barrier sites (gut, oropharynx, and skin), there are 10^9 virions per gram, while in blood and urine, this number varies between 10^5 and 10^7 particles per mL. These persistent viruses include mainly herpesviruses, retroviruses, and members of the *Anelloviridae* and *Circoviridae* families of viruses, and in some other individuals, papillomaviruses and hepatitis viruses. In other cases, HIV can also be present with no clinical manifestation of AIDS. Viruses in mammals (and in other animals) can be present as endogenous insertions that can amount to 80% of their genomes (8% in humans). We are dealing here with those viruses that coexist for most of the hosts' life causing no obvious clinical disease, and that, for some, can have a beneficial effect on their carriers instead.

Animal Viruses Can Protect Against Other Viruses

A virus in mammals can cause disease, as explained in other chapters; but some viruses present in persistent infections may lead to a different outcome. For one thing, these persistent viruses (sometimes with subclinical manifestations) might indirectly promote host health by eliminating other viral pathogens by cross-immunity (vaccine-like protection), or indirectly by interfering with the virus infective/replicative cycle. For example, individuals with HIV coinfected with *Pegivirus C* (*Flaviviridae*) show delayed progress to AIDS. Modulation mechanisms of HIV infection by pegiviruses include direct interference with HIV entry and replication, and indirectly by regulation of host factors that ameliorate the progression of the disease. Of note, human pegiviruses are the most prevalent RNA virus worldwide; nonetheless, no pathogenicity has been associated with their infection. Pegiviruses have also been detected in other primates, as well as in bats, rodents, horses, and pigs.

Surprisingly, some viral infections can protect against infections caused by nonviral pathogens. In mice, for instance, latent infection with murine gammaherpesvirus 68 or murine cytomegalovirus confers protection against the pathogenic bacteria, *Listeria monocytogenes* and *Yersinia pestis*. Patients with higher *Hepatitis B virus* (*Hepadnaviridae*) viremia show reduced *Plasmodium* spp. parasitaemia, which explains why malaria is asymptomatic (inflammatory cytokine production decreases) in these cases. Evidence suggests that persistent latent (but not lytic) infection with gammaherpesviruses can confer resistance to lethal bacterial infection and lethal lymphoma challenge through increased macrophage activation and the production of γ-interferon (an antiviral cytokine) or augmented natural killer cell activity. In mice, these latent infections seem to delay the onset of diabetes type-1 as well.

Retroviruses and Emergence of Placental Mammals

Mutualistic interactions with viruses can be analyzed from an evolutionary perspective. They are probably the best analyzed (and accepted) of all purported mutualistic interactions among viruses and their hosts so far. Although viral insertions in hosts' genomes can be better explained in cases where a virus employs a retrotranscription strategy of replication, we know now that other viruses with different replication strategies and genomic molecules are able to insert into their hosts' genomes as a whole genome or as fragments of their genomes. Nonetheless, probably the most well-known case of genomic insertion and ultimate protein exaptation is that related with the origin of the placenta in mammals. Retroviruses use their own genome-encoded reverse transcriptase to synthesize a DNA molecule from its RNA genome; this DNA, in turn, can insert into the host genome. Once inserted, the virus becomes an **e**ndogenous **r**etrovirus (ER) *if* it passes to the germ line and remains stable; the inserted ER is, of course, inherited as part of its host genome. In the course of evolution, one such retroviral sequence was inherited and led to the development of the so-called placental mammals (Eutheria). It was first demonstrated in sheep a decade ago, and later in mice and other mammals, including humans. Amazingly, the integration events that later allowed for the development of a placenta in these animals happened more than once in the evolutionary history of placental mammals. In the case of humans, more than 25 million of years ago an ER element, derived from an ancient retroviral infection, integrated into the evolutionary lineage of primates of the Old World. The newly acquired gene (ERVW-1, for **e**ndogenous **r**etro**v**irus group **W** envelope member 1), located on chromosome 7, codes for syncytin-1 (or enverin), a protein that mediates trophoblast fusion and is essential for the development of the placenta. Another syncytin, syncytin-2, is encoded by the gene ERVFRD-1 (**e**ndogenous **r**etro**v**irus group **FRD** member **1**), located on chromosome 6. In this case, the protein plays a key role in the implantation of human embryos in the womb. Other syncytins have experienced exaptation from several integration events in other mammals. Syncytins are the exaptation version of the *env* gene product of retroviruses. The exapted *env* as a modified, inserted gene for syncytin-1 lies between two LTRs and adjacent to other (nonfunctional) retroviral genes, *gag* and *pol*; it codes for an envelope glycoprotein. Presumably, over the course of evolution, an ancestor of placental mammals captured a viral gene that was involved in the generation of an envelope in the virus that ended up contributing to the evolution of hairy animals. As they are old, the *Syncytin-Car1* gene, for example, was domesticated (exapted) some 60 million years ago before the radiation of members of the order Carnivora, and it is the oldest syncytin known. It seems thus that syncytins in eutherians have appeared on multiple occasions: those from humans (Order Primates), for example, are not the same than those found in mice (Order Rodentia), and these differ from the one found in rabbits (Family Leporidae), and ruminants and Carnivora. As for the noneutherian mammals, which possess a placenta, it has been speculated that they also experienced an integration event of an ancestral *env* gene and that, in the course of evolution, was probably exchanged by newly arrived *env* genes.

To our surprise, ERs have brought other evolutionary novelties with them. It is thought, for example, that in the evolutionary history of humans some other ER elements were operationally incorporated into the development and maintenance of pluripotent specific states. Additionally, specific ERs have been found to contribute to the maintenance of the male germline stability, while others seem to play a role in the innate immune response (e.g., interferon inducible enhancers assisted by ERs are dispersed genomically near to interferon-inducible genes creating thus a network capable of working in a concerted manner due to the very presence of these ERs).

Animal Viruses as Drivers of Protein Adaptation

In analyzing the adaptation of some 1,300 **v**irus-**i**nteracting **p**roteins (VIPs) curated from a study set of 9,000 proteins that are conserved among all mammal genomes sequenced so far, researchers concluded that VIPs account for a high proportion of all protein adaptations (no less than 30% in humans) in these animals. That is, viruses represent, evolutionarily, one of the most efficient drivers of the mammals' proteomes. Viruses infuse, metaphorically speaking, into the realm of mammals' (and surely other living organisms) proteomes obliging them to adapt, suggesting the possibility that viruses (and pathogens in general) might have driven pleiotropic effects on diverse biological functions.

VIRUS–PLANT INTERACTIONS

Viruses Help Plants Cope With Abiotic Stress

More and more evidence is unravelling ways in which plant viruses benefit their hosts through mutualistic interactions. The simplest associations involve viruses that have a persistent lifestyle. As seen in Chapter 6: Viruses as Pathogens: Plant Viruses, these viruses remain with their plant hosts for many generations, and are strictly transmitted vertically at near 100% rates. They are the so called cryptic viruses. Although they are predominant in wild plants, species of the family *Partitiviridae* are also common in crop plants. Various lines of research suggest that cryptic viruses have developed intricate and complex relationships with their plant hosts that are influenced by abiotic factors, such as nutrient availability, water resources, heat and cold stress, and adverse soil conditions. That is, the benefits are conditional. A study with beets, for example, reported that *Beet cryptic virus 1* (BCV; *Partitiviridae*) infection appears to have no substantial effects on sugar beet plants grown in different locations under drought stress. In contrast, root yield and sugar yield were reduced when BCV-infected sugar beet plants were not grown under water deficit conditions. In other words, the interaction is mutualistic under specific environmental conditions; the virus is a pathogen under normal conditions, but it can be beneficial to the plant host under stress conditions. Such environmental dependence is referred to as conditional mutualism.

Similar relationships of conditional mutualism have been reported with plant viruses characterized by an acute lifestyle. These viruses can also enhance the ability of their plant hosts to counteract abiotic stress. *Nicotiana benthamiana* plants infected with *Tobacco mosaic virus* (TMV; *Virgaviridae*), *Cucumber mosaic virus* (CMV; *Bromoviridae*), *Brome mosaic virus* (BMV; *Bromoviridae*) or *Tobacco rattle virus* (TRV; *Virgaviridae*) exhibit better tolerance and survival to drought stress than their uninfected counterparts. The same is true for rice infected with BMV, tobacco infected with TMV, and beet, cucumber, pepper, watermelon, squash, tomato, *Chenopodium amaranticolor,* and *Solanum habrochaites* (a wild relative of tomato) infected with CMV. Beet plants infected with CMV also exhibit improved tolerance to cold temperatures. Tolerance to these abiotic stresses is also correlated with increased levels of osmoprotectants and antioxidants in infected plants. In another case, increased tolerance to water stress in wheat plants infected with *Barley yellow dwarf virus-PAV* (*Luteoviridae*) was attributed to increased production of phytohormones and salicylic acid. Clearly, these plant–virus interactions have widened our perspectives on the role viruses play in natural and food crop systems.

Viruses Help Plants Cope With Biotic Stress

Viruses can ameliorate the effects of biotic stresses as well. There is support for siRNA generated by a tomato endogenous pararetrovirus sequence (LycePRV) in antiviral defenses against exogenous LycePRV and other related viruses. A slightly different interaction occurs in petunia. Very little siRNA is produced from an endogenous virus, *Petunia vein clearing virus* (*Caulimoviridae*), because of chromatin effects and methylation. It seems that, in this case, siRNA does not contribute to immunity, but rather is involved in preventing infectious viruses from entering the meristem. These examples bear striking similarities with cross-protection, a phenomenon discovered in the 1920s, whereby infection with a mild strain of a virus excludes subsequent infection with related severe strains. Presumably, the primary virus, which is the mild strain, triggers RNA silencing that targets both primary and challenge viruses. Mild strains of *Citrus tristeza virus*, for example, protect Pera sweet orange and grapefruit against debilitating stem pitting strains in the field. Protection apparently extends to other pathogens. It has been found that *Banana bunchy top virus* (BBTV)-infected bananas are more resistant to *Fusarium oxysporum* f. sp. *cubense*. Although fungicidal properties are attributed to the BBTV suppressor of RNA silencing, protein B4, there is also the possibility that BBTV infection boosts the plant's defenses against the fungus.

Plant Viruses as Modulators of Symbioses

Other plants harbor viruses that have a persistent lifestyle as integrated viruses. These are the endogenous viruses. Recent data show that plant genomes contain an abundance of endogenous viral elements. Geminivirus sequences were discovered in several *Nicotiana* genomes many years ago. More recently, cytoplasmic RNA viruses have been found integrated in plant genomes. Overall, the flow of genes occurs from viruses to plant hosts, and not the other way around. These integrated elements can be transcriptionally active, implying a possible function for these genes in their plant hosts. *White clover cryptic virus 1* (WCCV-1; *Partitiviridae*) and its host white clover, an important pasture legume, is one example. WCCV-1 is a persistent virus, ubiquitous in white clover, which apparently suppresses the formation of nitrogen-fixing nodules when adequate nitrogen is present in the soil. Rhizobia bacteria provide the plant with nitrogen in the form of ammonium and the plant provides the bacteria with carbohydrates as an energy source. Despite the beneficial aspects of this symbiosis, too many root nodules can unbalance demand and supply and disrupt growth of the host plant. To suppress excessive root nodulation, plants use autoregulation. Transcriptome analysis of *Rhizobium* colonization in white clover identified a gene that appears to suppress nodulation when sufficient nitrogen is present in soils, and thereby save the plant from producing a costly organ when it is not needed. The gene was later realized to be the coat protein gene of WCCV-1; white clover plants at some point in their life history apparently coopted (i.e., exapted) the gene of the virus for their own purposes.

Interactions Between Vectors, Nonvectors, and Plant Viruses

Viruses may induce changes in plant hosts as well as vectors to enhance their transmission. In these cases, virus-induced changes in the host plant usually result in the attraction of the vectoring insect herbivore. Case in point involves two members of the family *Reoviridae* [*Southern rice black-streaked dwarf virus* (SRBSDV) and *Rice ragged stunt virus* (RRSV)], planthoppers and rice plants. SRBSDV and RRSV, that are the etiological agents of serious rice diseases, were found to alter the attractiveness of rice plants to planthopper vectors. The attractiveness was mainly mediated by rice volatiles and was positively correlated with virus titers. In other examples, the impact on the insect vector was examined. Positive or negative effects on vector performance have been demonstrated. One study illustrated positive effects on the thrips insect vector (*Frankliniella occidentalis*) feeding on *Tomato spotted wilt orthotospovirus* (TWSV; *Tospoviridae*) infected pepper plants. Tomato spotted wilt disease is caused by TSWV, a virus that severely damages and reduces the yield of many economically important plants worldwide including tomato and pepper. The virus is transmitted by thrips (Thysanoptera: Thripidae). Apart from the transmission of TSWV, thrips create major damage on plants by causing reduction in plant growth, deformation of plant organs, and cosmetic damage in the form of scarring on leaves and flowers. Generally, pepper plants respond to thrips feeding and damage by producing antifeeding compounds. In one study, juvenile thrips fed on thrips-damaged pepper plants were negatively impacted; that is, survival and developmental rate of the thrips were lower on these pepper plants compared to juveniles fed on healthy peppers. Such negative effects were not observed when juveniles fed on plants that were damaged and inoculated by infected thrips or were mechanically inoculated with the virus. Hence, potential vectors benefitted from seeking infected plants, since virus-infected plants were of higher quality for the vector's offspring; the virus ameliorated the negative effects of thrips damage. It was further shown that TSWV-infected plants produce **s**alicylic **a**cid (SA), making them better hosts for thrips. SA counteracts the **j**asmonic **a**cid (JA) response that is involved in the plants' resistance against pests. It was also reported that thrips infected with TSWV changed their feeding behavior in comparison to uninfected thrips: males feed more, and more importantly, probed more in a way that induced damage to plant cells, the type of probing that is required for virus transmission. On the contrary, negative effects were demonstrated for two aphid vectors, *Myzus persicae* and *Aphis gossypii.* This study explored the effects of *Cucumber mosaic virus* (CMV; *Bromoviridae*) on traits of its squash host plant which affected all subsequent interactions with aphid vectors. CMV was found to increase the attractiveness of infected plants to aphids by inducing elevated emissions of a plant volatile blend very similar to that emitted by healthy plants, but the virus apparently reduced host-plant quality such that aphids performed poorly on infected plants and rapidly migrated from them. Thus, CMV appears to attract vectors deceptively to infected plants from which they then disperse rapidly without colonizing the plant or initiating long-term feeding; a pattern highly conducive to the nonpersistent transmission mechanism employed by CMV.

While most studies have focused on documenting plant-mediated interactions between viruses and vectoring herbivores, more investigations are examining the effects on nonvectoring herbivores. Plant pathogens and insect herbivores are likely to share hosts in their natural habitats. Consequently, virus-induced changes in the host plant can possibly impact nonvectoring herbivores. A few examples follow. White clover plants infected with the potexvirus *White clover mosaic virus* (WClMV; *Alphaflexiviridae*) were found less attractive to fungal gnats, presumably due to virus-induced changes in volatile emissions. Two spotted spider mites *Tetranychus urticae* showed increased preference for and fecundity on TSWV-infected plants compared to healthy tomato plants. Transcriptome profiles of TSWV-infected plants indicated up-regulation of SA-related genes, but no apparent down-regulation of JA-related genes that could potentially induce resistance against the mites. In wild gourd, the bacterial pathogen, *Erwinia tracheiphila*, causes wilt disease and induces the production of volatiles that attract the beetle vector to the plant, and induce movement to healthy plants. In sum, there is accumulating evidence that virus infected plants can present as more suitable or less suitable hosts for nonvector herbivores and invariably impact the fitness of nonvector herbivores. Two primary mechanisms thought to underlie plant-mediated pathogen–herbivore interactions are the activation of defense-related signaling pathways and the induction of primary and secondary metabolites.

FURTHER READING

Górski, A., Dąbrowska, K., Międzybrodzki, R., Weber-Dąbrowska, B., Łusiak-Szelachowska, M., Jończyk-Matysiak, E., et al., 2017. Phages and immunomodulation. Future Microbiology 12, 905–914. Available from: https://doi.org/10.2217/fmb-2017-0049.

Handley, S.A., 2016. The virome: a missing component of biological interaction networks in health and disease. Genome Medicine 8, 32. Available from: https://doi.org/10.1186/s13073-016-0287-y.

Herniou, E.A., Huguet, E., Thézé, J., Bézier, A., Periquet, G., Drezen, J.-M., 2013. When parasitic wasps hijacked viruses: genomic and functional evolution of polydnaviruses. Philosophical Transactions of the Royal Society B 368, 20130051. Available from: https://doi.org/10.1098/rstb.2013.0051.

Márquez, L.M., Redman, R.S., Rodriguez, R.J., Roossinck, M.J., 2007. A virus in a fungus in a plant: three-way symbiosis required for thermal tolerance. Science 315, 513–515. Available from: https://doi.org/10.1126/science.1136237.

Pradeu, T., 2016. Mutualistic viruses and the heteronomy of life. Studies in History and Philosophy of Biological and Biomedical Sciences 59, 80–88. Available from: https://doi.org/10.1016/j.shpsc.2016.02.007.

Rascovan, N., Duraisamy, R., Desnues, C., 2016. Metagenomics and the human virome in asymptomatic individuals. Annual Review of Microbiology 70, 125–141. Available from: https://doi.org/10.1146/annurev-micro-102215-095431.

Roosinck, M.J., 2017. Symbiosis: viruses as intimate partners. Annual Review of Virology 4. Available from: https://doi.org/10.1146/annurev-virology-110615-042323.

Rosenwasser, S., Ziv, C., van Creveld, S.G. & Vardi, A. Virocell metabolism: metabolic innovations during host-virus interactions in the ocean. *Trends in Microbiology*. https://doi.org/10.1016/j.tim.2016.06.006.

Weinbauer, M.G., 2004. Ecology of prokaryotic viruses. FEMS Microbiology Reviews 28, 128–181. Available from: https://doi.org/10.1016/j.femsre.2003.08.001.

Witzany, G. (Ed.), 2012. Viruses: Essential agents of life. Springer, Dordrecht, The Netherlands. Available from: http://dx.doi.org/10.1007/978-94-007-4899-6.

Chapter 12

Viruses as Tools of Biotechnology: Therapeutic Agents, Carriers of Therapeutic Agents and Genes, Nanomaterials, and More

Gustavo Fermin[1], Sephra Rampersad[2] and Paula Tennant[3]
[1]*Universidad de Los Andes, Mérida, Venezuela,* [2]*The University of the West Indies, St. Augustine, Trinidad and Tobago,* [3]*The University of the West Indies, Mona, Jamaica*

One day in the future, a dose of a "ninja virus" might just save your life.

David Robson

The practice of using viruses as therapeutic agents started in the early 13th century during the outbreak of smallpox and the development of the procedure, inoculation. Inoculation, also called variolation, involved the introduction of small amounts of infectious material from smallpox vesicles into healthy subjects, with the goal of inducing protection against the more severe naturally acquired disease. Typically, the subject developed a milder form of the disease, along with pustules identical to those caused by smallpox. As described in Chapter 1: Introduction: A Short History of Virology, Jenner performed variolation on his patients following the observation that dairymaids who contracted cowpox, a minor infection characterized by a few pustules, would not later contract smallpox. He realized that he could not successfully inoculate such individuals with smallpox. On this basis, Jenner concluded that cowpox not only protected against smallpox, but could also be transmitted from one individual to another as a deliberate mechanism of protection. Jenner's work laid the groundwork for what would become live, attenuated vaccines. Today, some of the most common vaccines against measles, mumps, yellow fever, and others, use a similar approach.

Viruses. DOI: https://doi.org/10.1016/B978-0-12-811257-1.00012-7

Vaccines are prophylactic in the sense that they are administered to healthy individuals to prevent disease. By introducing a part of the virus or an inactive virus, the immune system reacts by producing antibodies. Should the virus enter the organism any time thereafter, these antibodies recognize and trigger the destruction of the pathogen. Conversely, there is a growing trend to use vaccines for alleviating chronic infections. These are therapeutic vaccines. A therapeutic vaccine is administered to individuals afflicted with a chronic infection against which naturally produced antibodies are ineffective. The purpose of these vaccines is to increase the activity of the individual's natural defenses. The idea of using viruses as therapeutic vaccines dates back to the 20th century when physicians noted the regression of malignant tumors during natural virus infections. One of the most cited example in the 1900s describes a 42-year-old woman with leukemia who went into remission (symptoms of the disease were abated) after a presumed influenza virus infection. Similar observations were reported in many other instances prompting analysis on the role played by viruses in the clinical remission of diseases.

Today the emerging field, virotherapy, explores the use of native or engineered viruses as infectious agents with inherent cytotoxic activity (viral oncolysis), as antitumoral immune activators (viral immunotherapy), or as delivery vehicles of heterologous genetic material (gene therapy). This chapter presents on the therapeutic use of viruses in the treatment of animal diseases, along with phage therapy and the therapeutic use of bacteriophages in the treatment of bacterial infections. The manipulation of plant virus genomes has also fostered the development of vectors for the production of vaccines, recombinant therapeutic proteins as well as the development of crop protection and biocontrol agents of pests, including viruses themselves. Finally, we briefly examine the application of viruses to nanotechnology and the development of drugs and gene therapies, in addition to the development of diagnostic reagents, novel imaging technologies, and applications in tissue engineering and electronics.

PHAGE THERAPY AND RELATED TECHNIQUES

Phage biotechnology has paved the way for unimagined uses of phage genomes and virions, including the design of novel materials and phage-based therapies. Phages possess many features that make them perfect allies of biotechnology: they cause little harm in nonhost organisms and can be easily removed from the human body. They can be used as-they-are in therapies aimed at deterring bacterial infections; their virions are amenable to manipulation for the display of functional peptide motifs of interest (for us, that is) that can be produced in large quantities by amplification in bacteria; even more, some can self-assemble into nanofibrous tissue-like matrix structures. In other cases, their modified genomes have been extremely helpful in increasing our knowledge and in manipulating organisms, while some of their proteins have helped in designing new approaches for the generation of enormous amounts of sequence data (some nanopore sequencing platforms). This section highlights the potential for whole phage or phage-based antibacterial agents, while the section at the end of the chapter addresses the use of phages in the design of novel nanoengineered systems.

In some instances, the best way to defeat an enemy is by using a proxy antagonist against the former. Such is the case of using a lytic phage against the bacterium *Klebsiella pneumoniae*, that causes hospital-acquired urinary tract infections, pneumonia, and soft tissue infections. A recently described lytic virus (φBO1E; *Podoviridae*) was found to be specific for KPC carbapenemase-producing strains of the *K. pneumonia* bacterium. This was a very important discovery given the pandemic dissemination of the pathogenic bacterium that not only is ubiquitous, but also shows extensive multidrug resistance profiles. The phage is specific for this multidrug-resistant bacterial strain, it does not lysogenize its host and it is envisioned to be amenable for the development of new agents in the treatment of *Klebsiella* infections, or for the decolonization of patients afflicted by the bacterium. The therapeutic use of bacteriophages is not new, and has a long history of use in Russia, Poland, Georgia, and France with notable success, even before antibiotics were discovered. Cases exist where the strategy has actually been employed for the control of bacteria that are resistant to antibiotics and other chemical agents, or when their use is contraindicated for clinical reasons (like cases involving pathogenic bacteria belonging to the genera *Clostridium*, *Staphylococcus*, *Pseudomonas*, *Stenotrophomonas*, *Campylobacter,* and *Escherichia*, to name a few). Phage therapy is employed in different ways and, as with any antimicrobial agent, the key is for phages to reach bacterial targets in sufficient abundance. In general, both direct application of phages to the infection site or via systemic delivery have achieved variable success: (1) topical application (non-systemic) to skin, wounds, mucous membranes, and urogenital and respiratory tract. When applied in bandages, phages are slowly and continuously released for the treatment of ulcers, and of burn and postsurgical wounds. Phage cocktails treat different pathogenic bacteria affecting a wide range of tissues; (2) parenteral (injection, for systemic spread), and (3) *per os* (oral administration of encapsulated phages followed by systemic spread) or colorectal. Inhalation of *formulated* phages has also been used for the treatment of some respiratory infections. Phage therapy may be used in a far different manner; for example, as a prophylactic measure resembling that described for bacterial probiotics. Phages administered orally in mouth rinse formulations for humans can normalize the microflora at infected sites and induce the production of immunoglobulins.

Apart from the conventional approaches briefly mentioned above, other more sophisticated applications of phages in the fight against bacterial infections have surfaced. Engineered phages, for example, are chemically modified in order to evade the immune system and thus prevent their elimination.

Directed recombination permits the creation of phages with a broader host-range, and hence, extended usability. In order to avoid the release of bacterial toxic compounds upon viral-induced lysis, phages have been modified to become lysis-deficient (inactive endolysin). That is, there is bacterial cell degradation without the liberation of damaging biochemicals. Viral engineered endolysin (or chimeric lysins) itself is being assayed as a measure of controlling bacterial infections. A similar strategy involves the modification of phage proteins that can directly inhibit bacterial proteins that are essential for the organism's survival. Finally, phages can also be used as vehicles of lethal substances (antibiotics or bacteriocins) or genes to infection sites; bacteria are killed directly, or there is the expression of a gene whose product achieves the same goal without detrimental effects on host cells. Another approach based on phage-targeted delivery of photosensitizers, or light-activated antibacterials, allows for the elimination of targeted cells on irradiation without the phage actually infecting the bacterium. Despite advances in the use of phages in well-designed strategies of bacterial control, there are many hurdles that need to be overcome before a wider acceptance of virotherapy is achieved.

Preventative interventions, on the other hand, involve the use of phages to engineer pathogenic bacteria *in situ* and impede their ability to transfer antibiotic resistance determinants to other bacteria. Further, antibiotic-resistant bacteria can be reversed to their sensitive wild-type phenotype by transferring, via phage-delivery, susceptibility-conferring alleles. This was demonstrated with *E. coli*. Resistance to the antibiotics streptomycin and nalidixic acid was reversed by using an engineered temperate phage to insert the dominant-sensitive wild-type genes of *rpsL* (a gene that encodes the large 30 S ribosomal subunit) and *gyrA* (gyrase coding gene) into the bacterial genome.

Other nontherapeutic uses of phages as-they-are include their direct application as food additives as a measure of deterring bacterial proliferation. Similarly, phages have been used as external disinfectants. In agriculture, phages can be applied to plant surfaces in order to prevent infections by pathogenic phytobacteria, like those belonging to the genera *Xanthomonas*, *Ralstonia*, *Pseudomonas,* and *Dickeya*. Along the same vein, phages have been employed in wastewater treatment and sludge processing. Since viruses free in the environment can also be inactivated, as explained in Chapter 5: Host Range, Host Virus Interactions, and Virus Transmission, some strategies have been devised to protect good phages. For instance, it was demonstrated that 1% of the biodegradable polymer poly-γ-glutamic acid protects environment-friendly viruses (T2, T4, Ms2) against high temperatures, acidic pH, and UV radiation.

Phage Display

Phage display is a molecular biology technique by which phage genomes are modified in such a way that the coat proteins of assembled virions are fused to other proteins or peptides of interest (of any origin), displaying them thus to the external milieu. The technique has been widely applied to basic research in order to study protein–protein, peptide–protein, and nucleic acid–protein interactions. Phage display has facilitated the analysis of protein function, along other practical applications like protein engineering, drug discovery and selection, discovery of new enzyme ligands, receptor agonists and antagonists, and tumor antigens. Analysis of nucleic acid–protein interactions have helped uncover the intricacies of gene regulation at transcriptional and posttranscriptional levels. Among the variations of the technique, **a**ntibody **p**hage **d**isplay (APD) has proven most fruitful. APD is based on the genetic manipulation of the display phage genome which, after an antibody-library preparation, is used to ligate PCR products of their heavy and light chain genes. The virus-displayed antibodies are then submitted to a carefully designed selection scheme in order to isolate specific **m**onoclonal **anti**b**odies (mAbs) from the vast repertoire of immunoglobulins produced (displayed) this way. Moreover, the selected and cloned mAb can be subsequently modified to allow for its easy production and purification; sometimes, it can also be modified for specific targets and purposes.

Using a different approach, some researchers have applied this technique to the production and selection of cyclic peptides, which by virtue of their structures, are able to bind protein targets with high affinity and selectivity. This in turn, can be translated to the development of agents for research and therapeutics and adds to the arsenal of engineered materials that are being developed not only for the basic and applied fields already mentioned, but also for regenerative medicine. In this case, phage-engineered biomaterials can be represented by those obtained (or applied) (1) *in vitro* (functionalized synthetic substrates [2D surfaces] for the expansion of stem cells, and functionalized hydrogels [3D environments] for the recreation and/or manipulation of niches of the very stem cells), or (2) *in vivo* (functionalized injectable biomaterials aimed at controlling the release of growth factors in order to promote endogenous tissue repair, or functionalized nanocarriers for the targeted delivery of genetic material to reprogram specific cells).

Phage display has been the technique of choice for carbohydrate display (HIV-related glycans) by means of phage Qβ (*Escherichia virus Qbeta*; *Leviviridae*), and of enzyme-display (like alpha-amylases, xylanases and biofilm degrading enzymes) by *Escherichia virus T7* (*Podoviridae*) for diverse purposes.

VIRUS-BASED VACCINE TECHNOLOGIES

Vaccination has proven to be one of the most, if not the most, efficient method of fighting infectious diseases, including those caused by viruses. As explained in the introduction of this chapter, the virus itself (or its virion components) is used to trigger an immune response in the host. The virus is inactivated or attenuated in order to attain this goal. Essentially, the immune system "*remembers*" the exposure and mounts a vigorous defense in the event of a later encounter with the target virus. In many instances, besides the antigen (virions or virion proteins), an adjuvant is also delivered to the immunized animal (including humans). The purpose of the adjuvant is to help enhance the immune response to the vaccine; this in turn, helps limit the amount of antigen required to elicit the protective response.

Different strategies are employed in the development of vaccines aimed at fighting against diseases of viral etiology: namely, (1) conventional vaccines, including live attenuated vaccines and inactivated vaccines, (2) subunit vaccines, only an antigenic part of the virion is used; more recently, virus genes are inserted into expression plasmids that are delivered to the person, (3) **v**irus-**l**ike **p**articles (VLPs) and **v**irus-**b**ased **n**anoparticles (VNPs), and (4) vector-based vaccines, including *Vaccinia virus*-based vaccines (*Poxviridae*), recombinant *Venezuelan equine encephalitis virus* (*Togaviridae*; VEE virus-like replicon particles) vaccines, recombinant human adenovirus-based vaccines, and recombinant **ve**siculovirus-based **v**accines (VSV).

The conventional vaccine platform is based on the direct use of disarmed or "*killed*" viruses, or their attenuated forms derived by heat, radiation, or chemical inactivation of their virions. Sometimes, however, viruses can retain virulence as demonstrated in different animal model systems. Nonetheless, live-attenuated vaccines can be envisioned as one of the most successful weapons against infectious diseases. They have facilitated, for instance, the eradication of smallpox (1980), and almost the complete suppression of poliomyelitis (and in the near future, measles). *Rinderpest morbillivirus* (*Paramyxoviridae*), and the livestock disease it incites, was declared eradicated in 2011 in what has been considered one of the greatest achievements in veterinary history due to the concerted use of mass vaccination and zoosanitary procedures. Unfortunately, this strategy does not always work well; for example, the efficacy (and safety) of a mumps vaccine is questionable, while a rotavirus vaccine license was withdrawn due to its association with intussusception (a medical condition in which a part of the intestine is folded into a section next to it leading to intestinal obstruction), and although vaccines against yellow fever are known to be effective and extraordinarily safe, some concerns have arisen due to unexpected epidemiological complications.

The so-called subunit vaccines do not consist of the virus itself or virus-based vectors, but rather antigenic viral components capable of eliciting a protective immune response. These vaccines can be time consuming to develop since antigenic properties and efficacy must first be assayed independently for each antigen candidate; besides, immunological memory is not always guaranteed. In more recent times, viral genes cloned into expression plasmids are injected directly into the muscles of the patient; gene gun delivery, as well as electroporation after injection, are used as means of DNA delivery. Once in the target tissue cells, expression of the protein occurs only where the plasmid resides. These vaccines are also known as DNA vaccines, and their major advantage is that they allow for the generation of specific antiviral antibodies as wells as cytotoxic T lymphocytes. Additionally, these vaccines are cost effective, easy to manufacture, stable during storage at room temperature, and shipping does not require sophisticated equipment or conditions.

The concept of viral vector vaccines is slightly different from that of subunit vaccines. Here, a virus (not the virus causing the disease) is used as a vector or carrier for a gene coding the target virus antigen; upon expression, the circulating antigen elicits thus an immune response. These vectors can be replication competent or replication-defective, each with advantages and disadvantages. Replication-competent viral vectors cannot be used in immunocompromised populations; but they can induce a strong and long-lasting immune response after immunization. For example, replication-competent, recombinant VSVs induce a strong humoral and cellular response in humans; additionally, they have demonstrated effective postexposure value. Nonreplicative viral vectors, on the other hand, can be used safely with both populations, but require the administration of multi-doses in order to be effective.

A more recent trend is the development of VNPs (virus-based nanoparticles) and VLPs (**v**irus-**l**ike nano**p**articles) as vaccines. Whereas VNPs consist of a modified capsid encapsulating a virus genome, VLPs are made up of protein components only. One advantage of these vaccines is that they bypass preexisting immunity against virus-derived antigens harbored by virus vector-based vaccines. Additionally, they are safer since there is no risk of virulence and they are capable of inducing vigorous immune responses without the need of multiple doses. Recombinant VLPs and VNPs can be engineered to present almost any nonrelated antigenic epitopes. Additionally, depending on the way they were engineered, they can be manufactured and produced in large quantities in heterologous systems including bacteria, yeasts, plants, and cell lines derived from mammals and insects. They are amenable to chemical modification by simply fusing antigenic peptides of interest to the coat protein of the modified virus-derived VLP or VNP. Several strategies based on these two platforms are currently being investigated for the production of vaccines against HIV, Ebola, influenza and *Marburg marburgvirus* (*Filoviridae*), diverse types of cancers, Alzheimer's disease, and hypertension. The vaccine company Novavax, for instance, has developed a baculovirus-based vaccine containing the GP (**g**lyco**p**rotein) of an Ebolavirus plus a matrix adjuvant. It is highly immunogenic in mice, rabbits, and baboons, and shows 100% protection in mice challenged with lethal doses of the virus. Vaccines based on VLPs are also being developed against addictions to tobacco and cocaine.

Overall production of vaccines, whether "*natural*" or synthetic, requires the massive production of the agent. In the former, the main method of producing vaccines is based on the use of embryonated eggs, which is a relatively slow process. In addition, yields are unpredictable and may vary according to the strain of the virus used, as exemplified by the production of influenza vaccines. For influenza vaccines, and other vaccines, an alternative to the allantoic cavity of embryonated eggs is the use of cell-specific lines. More recently, this method has also been applied to the production of synthetic vaccines. Among the cell lines used, we can find Madin Darby canine kidney cells, Vero cells (isolated from the kidney epithelium of an African green monkey), which have been used for the production of the polio vaccine for many years (and now influenza vaccines) and PER.C6, a human retina-derived cell line. Synthetic influenza vaccines have been produced by a baculovirus expression vector system in which insect cell lines are used instead of mammalian derived lines. The production of vaccines in plants is discussed in a later section (*Plant viral vectors*) in the chapter.

Experience has demonstrated that multiple approaches have to be implemented in the fight against prevalent diseases of viral etiology. In the tropics and subtropics, for example, dengue continues to be the most significant arboviral disease. The ailment can be fatal, and it is estimated that half of the planet's population is at risk of contracting the disease. Ten to hundreds of millions of people are infected annually. Although efforts have been made to control the insect vector (see section on *Viruses as biological control agents*), the development of effective vaccines has also been aggressively pursued for many years. These include inactivated, multivalent (specific for more than one antigen) attenuated, subunit, chimeric, and DNA vaccines, but only an attenuated, tetravalent vaccine is currently being used in countries where dengue is endemic. The vaccine, however, confers only limited protection, and it has been associated with some unexpected, undesirable effects. Other immunization strategies based on the use of nonstructural proteins of the virus are also under study. It appears that a multipronged approach aimed at controlling both the virus itself and its vehicle of dissemination is the safest route to take in order to bring wellbeing to the millions afflicted.

Before closing this section, it is important to mention that the most important impediment to the use of vaccines against viruses is the lack of reliable safety data for some vaccines. Moreover, the wide variety of immune evasion strategies of animal viruses limits the predictive power of alternate animal model systems when trying to forecast the efficacy of vaccines intended for humans. And, as been demonstrated with failed experimental vaccines against Ebola, conventional vaccines based on heat, radiation, or formalin inactivation of the target virus, pose a significant risk of reversion to an active, infectious state. Finally, when vaccines are based on virus vectors other than the target virus, a preexisting immunity against the vector might lead to a decrease in the immune response to the target.

MAMMALIAN VIRUSES WITH INHERENT CYTOTOXIC ACTIVITY AS THERAPEUTIC AGENTS

In 2016, some 40 clinical trials initiated recruitment drives for participants to examine the feasibility of a range of **o**ncolytic **v**iruses (OVs) against multiple cancer types. The year marked the most active period of clinical OV studies in the history of the field. OVs are emerging as important agents in cancer treatment. These viruses are capable of replicating specifically in and destroying tumor cells. In contrast to gene therapy where a virus is used as a carrier for the delivery of therapeutic agents, OV therapy uses the virus itself as the therapeutic agent. The viruses recognize and exploit certain traits that are unique to cancer cells.

Cancer cells differ from their normal counterparts because of sustained growth signals, insensitivity to antigrowth signals, evasion of apoptosis, increased angiogenesis (formation of new blood vessels), cell immortality, and tissue invasion and metastasis (development of secondary malignant growths at a distance from a primary site of cancer). These alterations make cancer cells attractive hosts for viruses. Upon infection, OVs hijack the cancer cell's transcriptional and translational machinery and trigger cell death (apoptosis, necrosis, or autophagy) after available cellular resources have been exploited for the synthesis and assembly of progeny virions. Additionally, OV-mediated cell death is accompanied with the release of cytokines, **t**umor-**a**ssociated **a**ntigens (TAAs), and other danger signals, including **d**amage-**a**ssociated **m**olecular **p**attern molecules (DAMPs) and **p**athogen-**a**ssociated **m**olecular **p**attern (PAMPs) molecules. These signals initiate the so-called "*immune-associated*" bystander effect involving the release of cytotoxic perforins and granzymes that bring about the destruction of adjacent nonvirally infected tumor cells. There is also the release of other cytokines, such as **i**nter**f**ero**n**s (IFNs), **t**umor **n**ecrosis **f**actor-**a**lpha (TNF-α), and **i**nter**l**eukins (IL), that promote further cell-mediated immune responses. Taken together, the very nature of OV infection makes for an effective therapeutic agent; OVs selectively target, infect, and destroy cancer cells. Moreover, OVs possess unique properties over traditional drugs. They can replicate and spread, and they can be easily engineered to accommodate increased tropism to cancer cells, selective replication, lytic capacity, and host antitumor immunity.

As with any other drug therapy, OV therapy begins with the administration of the virus. Intravenous and intratumoral administrations are commonly used methods. While intravenous delivery is more practical for clinical use and can readily target multiple tumor sites, intratumoral delivery ensures viral access to the tumor and avoids potential systemic neutralization, which is explained below. Specific targeting of cancer cells is essential for oncolytic virotherapy in order to avoid off-target effects. Many naturally occurring

OVs exhibit preferential tropism for tumor cells, like certain reoviruses, Newcastle disease virus (*Avian avulavirus 1*, *Paramyxoviridae*), *Mumps rubulavirus* (*Paramyxoviridae*), and *Murine leukemia virus* (*Retroviridae*). Others are engineered to adapt or enhance their tumor selectivity, such as *Measles morbillivirus* (*Paramyxoviridae*), adenoviruses, Vesicular stomatitis virus (*Indiana vesiculovirus*, *Rhabdoviridae*), *Vaccinia virus* (*Poxviridae*), and Herpes simplex virus (*Human alphaherpesvirus 1*, *Herpesviridae*). Specific targeting is achieved in two ways, (1) targeting selectivity of viruses prior to entering a cancer cell (transduction targeting) or (2) controlling selectivity of replication once the virus has infected a cell (viral gene inactivation, transcriptional targeting, microRNA targeting sequences).

Once delivered, local spread of OVs presumably occurs by intercellular fusion, by direct transfer of virus from infected to adjacent cells, or by release and local migration of progeny virions through the interstitial space. Systemic spread as free virus particles or as virus infected migratory cells occurs via lymphatic channels or via the bloodstream. As explained above, the overall antitumor effect induced by oncolytic viral treatment consists of two major components: (1) local cell death of both virally-infected and noninfected cancer cells, and (2) induction of a systemic immune response to virally-induced cell destruction within the tumor. Genetic engineering has been employed to enhance OV antitumoral responses. One major strategy is the expression of immune-stimulating molecules, including (1) cytokines; (2) molecules that enhance immune system cross-priming; (3) T lymphocyte costimulatory molecules; and (4) chemokines. Additionally, nonimmune-directed strategies of increasing OV efficacy include the use of (1) suicide genes to make cells more susceptible to apoptosis or treatment with other drugs, (2) extracellular matrix-targeting molecules, (3) fusogenic membrane protein expression, and (4) strategies to prevent viral clearance or inactivation.

Oncolytic virotherapy has come a long way since its inception. New-generation OVs are highly specific and possess multiple tumoricidal mechanisms. But the feasibility of oncolytic virotherapy depends on not only safety, but efficacy. Initial efforts focused on designing viral oncolytic agents with cancer specificity and low chances of regaining pathogenicity, without the possibility of transmission to healthy individuals, undesired side effects, and preexisting immunity. Now, efforts are mainly directed at improving efficacy, that is, efficient delivery of viruses and on getting around the host's immune response. Although the multimodal immunogenic cell death mediated by OV infection is able to activate the host immune system effectively against tumor cells, the same process can be detrimental to the continued propagation of OVs. OV clearance by antibodies generated against viral PAMPs and/or cytotoxic T cells is possible and NK cells can upregulate natural cytotoxicity receptors in virally infected cells. Repeat administrations further promote the development of adaptive antiviral immunity and reduce oncolytic effects. Moreover, a large proportion of the population has been previously exposed to the naturally occurring viruses that are used in the development of OV agents (e.g., HSV). Combination strategies, like chemotherapeutic drugs, can potentiate the cytotoxic mechanisms and remove the barriers to successful oncolytic virotherapy.

Recent approvals of OV-based therapies (e.g., T-VEC for melanoma and H101 for nasopharyngeal cancer) undoubtedly mark a great milestone for the field of oncolytic virotherapy, and establish OVs as a new class of cancer therapeutics. Over 20 virus platforms are under investigation. Exciting results have been observed in combination strategies with all major modalities of cancer treatment, including the field of immunotherapy.

MAMMALIAN VIRUSES AS GENE DELIVERY VEHICLES

Although therapeutic genome editing technologies are under development, approaches involving the addition of therapeutic genes to cells are currently being investigated as an alternative treatment for a wide range of infectious diseases. The approach, referred to as gene therapy or gene transfer, is defined as a collection of strategies that modify the expression of an individual's genes in order to correct the loss of function caused by mutation, to express a deficient gene product at physiologic levels (monogenic disease), or to introduce an additional function to treat (as in cancer) or to prevent (as in a vaccine) disease. In diseases associated with recessive gene defects, the addition of a functional sequence can often revert the phenotype even though mutated copies remain within the genome. A different approach is taken with pathologies linked to dominant mutant copies of a gene that involves knocking out, knocking down, or replacing the mutated copy. In suicide gene therapy, suicide-inducing transgenes are selectively introduced into cancer cells. The gene product either directly kills cancer cells or is an enzyme that metabolizes a systemically available pro-drug to an active antineoplastic agent. Alternatively, cancer cells of the individual are genetically modified so that they trigger an enhanced immune response and attack on cancer cells. In theory, it is possible to transform either somatic or germ cells. However, jurisdictions in various countries prohibit the use of germline therapy.

The introduction of the therapeutic gene into target cells can be achieved in two ways. The genetic material (transgene) may be introduced directly into cells of a patient (*in vivo* gene therapy). Or, the cells are removed from the patient, the genetic material inserted into these cells *in vitro*, and then the genetically modified cells are transplanted into the patient (*ex vivo* gene therapy). The ultimate objective of therapy is to achieve safe delivery of the transgene such that a cure is achieved. To this end, viral vectors have been designed for gene transfer.

Viruses are the most highly evolved natural vectors for delivering genetic material into cells. This feature has led to extensive strategies to engineer recombinant viral vectors for use as stable gene-delivery vehicles. Virus-based vectors harness the viral infection pathway, in particular virus tropism, but avoid the subsequent expression of viral genes that lead to replication and infection. Typically, viral vectors consist of viral particles with nucleic acids encapsidated in capsid proteins and, in many cases, an envelope layer. Generally, one or several nonessential viral genes are deleted to prevent spread within the host as well as to reduce cytotoxicity and to accommodate the gene of therapeutic interest. Vectors generally contain strong promoters to support high levels of transgene expression and they may also be tissue-specific to exclude expression in tissues other than the target tissue. Other means of targeting involve the introduction of specific recognition ligands in the surface of the virus coat to enable infection of specific cells/tissue. There are two types of virus-based vectors: integrating and nonintegrating. The former, such as retroviral and adeno-associated viral vectors, integrate into the host's genome, whereas the nonintegrating vector (e.g., adenoviral vector) is maintained in the cell's nucleus without integrating into the host's chromosomal DNA and remain in the cell in an episomal form. Expression of the transgene is therefore transient when nonintegrating virus vectors are used. Gene transfer via viral vectors is called transduction, while transfer via nonviral vectors (such as polymer nanoparticles, lipids, calcium phosphate, electroporation/nucleofection, or biolistic delivery of DNA-coated microparticles) is referred to as transfection.

Of the clinical trials conducted so far, 67% have utilized viral vectors. Mammalian virus vectors are the preferred vehicle for gene transfer because of their high efficiency of transduction into human cells. Taken together, the concept of gene therapy seems straightforward. However, some fatal adverse events can be associated with virus-based vectors. There is the danger that the transgene might be inserted in the wrong cell, or in the wrong location in the DNA, possibly causing harmful mutations or cancer. In the event that the transgene is unintentionally introduced into the patient's reproductive cells, transmission to the patient's offspring is a possibility. Other concerns include the overexpression of the transgenes at harmful levels, induction of the immune response (toxicity), and transmission of the virus vector to other individuals or into the environment. Nonetheless, this mode of therapeutic gene introduction shows promise. Examples of viral vectors that have been engineered as gene delivery vehicles and their therapeutic applications are presented.

γ-Retroviral Vectors

The unique nature of the retroviral replication cycle, combined with the simplicity of the retroviral genome, makes these viruses attractive vectors for gene therapy. The retrovirus copies its RNA genome into a double-stranded DNA form which is then efficiently and exactly integrated into the host cell genome. The integrated form, the provirus, is transcribed as a normal cellular gene to produce mRNAs encoding various viral proteins both in nondividing cells and in dividing tissues. Retroviral vectors share the basic architecture of its virus progenitor, but with the therapeutic sequence inserted in the region between the LTRs (i.e., LTR-leader-packaging signal-gene of therapeutic interest-LTR). Genes necessary for gene transduction and integration are provided in *trans* and are expressed from plasmids to make replication-defective vectors. All the elements are introduced into the same packaging cell. Retroviral packaging cells provide the necessary viral proteins, but do not package or transmit the RNAs encoding these functions. The most popular retrovirus vectors used in gene therapy are based on the *Murine leukemia virus* (MLV).

The first gene therapy protocols using MLV-based vectors began in the 1990s. In these protocols, patients with **s**evere **c**ombined **i**mmuno**d**eficiency (SCID) due to **a**denosine **dea**minase (ADA) deficiency were treated with a retroviral vector carrying the ADA coding sequence under the transcriptional control of the promoter/enhancers of MLV's LTRs. ADA disease is characterized by a reduced or absent T-lymphocyte and B-lymphocyte function. This leads to extreme susceptibility to frequent infections. In this pioneer trial, partial corrections of the phenotype were obtained. Some patients recovered cell counts and function and showed no adverse effects, while the response of others was more limited primarily due to lower transduction efficacy—possibly due to immune responses against the retroviral vector; yet another set developed leukemia-like conditions caused by the very treatment that cured SCID. In recent years, three modifications have produced vectors with enhanced abilities and a greater potential for safe applications. First, the regulator regions of the virus were engineered to enhance transgene expression or reduce transcriptional silencing in specific target cells. The vectors combine the enhancer and promoter derived from the **l**ong **t**erminal **r**epeat (LTR) of Friend mink cell focus-forming viruses, with the 5′ untranslated leader of the **M**urine **e**mbryonic **s**tem cell **v**irus (MESV), or contain an engineered LTR derived from the Myeloproliferative sarcoma virus. Second, the use of heterologous **env**elopes (env) from other viruses was a major advance that permitted a very broad spectrum of infectivity (retrovirus vectors with high gene transfer efficiency, only partly applies to applications on the blood-forming system). And third, the development of self-inactivating vectors. Viral enhancer and promoter sequences were deleted from these vectors; instead, another viral or cellular promoter is placed internally, downstream of the RNA packaging signal. Because of transcriptional inactivation of the provirus in the target cell, this type of vector is called a SIN or **s**elf-**i**nactivating **v**ector, the risk of insertional oncogenesis is minimized. More recent applications use retrovirus vectors for siRNA delivery into mammalian cells. **RNA i**nterference (RNAi)-based therapeutics have emerged for the treatment of cancer, infectious diseases, and other diseases associated with specific gene disorders.

Lentivirus Vectors

Lentiviral vectors have been shown to be less genotoxic than their γ-retroviral vector counterparts. Upon appropriate modifications, these vectors target a huge variety of cells including quiescent and difficult-to-transduce cells such as hematopoietic precursors, neurons, lymphoid cells, and macrophages. Additionally, lentiviruses can introduce genes into both dividing and nondividing cells. This is accomplished by two virus proteins, "*matrix*" and "*Vpr,*" which interact with the host nuclear import system to allow entry of the virus into the nucleus whether the cell is actively dividing or not. Like simple retroviruses, lentiviruses have been engineered into separate vector systems to prevent an accidental production of a replication competent virus and to prevent the ability of the virus to recombine with other HIV viruses *in vivo*. The three cassettes include, (1) an envelope cassette that delivers viral envelope protein, which is often pseudotyped with Vesicular stomatitis virus G protein (VSV-G) to restrict entry to only specific cell types, (2) a packaging cassette which is devoid of all of the accessory genes of *Human immunodeficiency virus 1* (HIV-1) excluding *Tat* and *Rev*. Rev protein is crucial for exporting the full length, or partially spliced RNAs from the nucleus into the cytoplasm, while Tat protein is required when the vector is not self-inactivated. *Gag-Pol* genes supply all necessary structural and enzymatic proteins required for replication, and (3) an expression cassette that contains the therapeutic gene flanked by LTRs and the Psi-sequence of HIV. The LTRs are necessary to integrate the therapeutic gene into the genome of the target cell, while the Psi-sequence acts as a signal sequence and is necessary for packaging RNA with the therapeutic gene into viral particles that are unable to continue to infect their host after they have delivered their therapeutic cargo.

Reports on preclinical animal models show that lentiviral vectors carrying therapeutic genes were able to treat or cure β-thalassemia, sickle cell anemia, and hemophilia B. Improvements in other genetic disorders like Parkinson's disease, cystic fibrosis, and spinal muscular atrophy were also observed. The first human clinical trial using lentivirus vectors was initiated in 2003. In this trial, a VSV-G-pseudotyped HIV-based vector was engineered to conditionally express an antisense RNA against a virus envelope glycoprotein in the presence of regulatory proteins provided by wild-type virus. Subjects with chronic HIV infection received a single dose of gene-modified $CD4^+$ T cells. One year later, an increase in $CD4^+$ T cells and decrease in the viral load were obtained. Adverse clinical events were not noted for up to two years after the procedure. Other clinical trials using lentivirus vectors for the treatment of rare diseases, in particular of primary immunodeficiencies and in neurodegenerative storage diseases, have since received approval.

Adenovirus Vectors

Adenovirus vectors are probably the most frequently used viral vectors, mainly because of the high levels of transgene expression that can be obtained in a broad range of host cells. While the early versions of adenovirus vectors caused toxic effects and typically induced strong immune responses, more recent versions (carrying deletions in nonessential viral genes) exhibit reduced toxicity and improved gene expression. The vectors are predominantly nonintegrating, episomal, nonenveloped, double-stranded DNA viruses.

Most adenovirus vectors are genetically modified versions of the serotype Ad5. Two types of vectors are available: **r**eplication-**d**efective (RD) and **r**eplication-**c**ompetent (RC). Replication-defective vectors do not carry the virus *E1A* gene. *E1A* initiates the program of viral gene transcription and reprograms multiple aspects of cell function and behavior. These *E1A*-deleted vectors are usually constructed from plasmids or virus DNA containing the genetically modified genome; the vectors are grown on complementing cell lines that retain and express the *E1A* gene. For oncolytic gene therapy for cancer, where replication-competent viral gene expression is needed, *E1A* has been either mutated or placed under tumor-specific transcriptional control. Today's vectors also have deletions of various *E2* and *E4* genes because viral proteins encoded by these DNA sequences were shown to induce most of the host immune response. In addition to decreased toxicity, there is the advantage of prolonged gene expression *in vivo*. Because adenoviruses are immunogenic and there is much preexisting immunity to the virus, their vectors are not optimal for long-term complementation of faulty genes in monogenic diseases, but rather for the delivery of vaccines and cancer therapy. Replication-competent oncolytic vectors have been engineered to replicate preferentially in cancer cells, causing apoptosis, direct cell death, or increasing the sensitivity of cancer cells to antitumor drugs. The vectors have proven highly safe and effective against gliomas (brain tumors).

Adeno-associated virus Vectors

Adeno-**a**ssociated **v**irus (AAVs)-derived vectors are regarded as ideal vectors for gene therapy due to the lack of pathogenesis associated with the progenitor virus and its ability to infect numerous and nondividing cell types (muscle cells, peripheral and central nervous system cells, hepatocytes). Like retroviruses, AAVs are able to integrate into host DNA, in particular at a specific site in chromosome 19. Recombinant vectors, however, lack this characteristic and the associated risk of insertional mutagenesis. They persist primarily as extrachromosomal elements. The vectors also lack the *cis*-active

IEE, which is required for frequent site-specific integration, and they do not encode for Rep. They contain the ITRs because the *cis* signals are required for packaging. AAV vectors for gene therapy have been based mostly on the serotype AAV-2.

Although no human diseases are associated with AAVs, it is normally a weak immunogen, and recombinant viruses can elicit innate and humoral immune responses producing neutralizing antibodies. The cloning capacity of these viruses is limited, but this can be expanded through the generation of hybrid AAV-HSV vectors. Despite these drawbacks, AAV vectors boast over 100 gene therapy trials, mostly treating monogenic diseases. One product has achieved marketing authorization and is the first approved gene therapy treatment. Glybera™ was approved in the European Union in 2012. Glybera™ contains an AAV1 vector for the treatment of patients with lipoprotein lipase deficiency. Glybera™ is designed to restore the LPL (**li**po**p**ro-tein **l**ipase) enzyme activity required for the processing, or clearance, of fat-carrying chylomicron particles and prevent episodes of pancreatitis. AAVs have also been successfully used as gene delivery vehicles for retinal gene therapy. These advances in gene therapy applied to ocular diseases might allow a near future cure of glaucoma, the second leading cause of blindness worldwide.

PLANT VIRAL VECTORS

Engineered vectors based on different plant virus genomes have been developed for a diversity of purposes either through (1) expression or (2) silencing platforms. In the former, viral vectors are engineered in such a way that virtually any protein-coding gene from any organism can be efficiently delivered to the target plant for massive expression and protein production. In the case of viruses engineered to elicit gene silencing in the recipient plant, the purpose is to down regulate the expression of a homologous, resident gene, leading to a loss of function phenotype. This is also called **v**irus **i**nduced **g**ene **s**ilencing (VIGS). Either way, **t**ransient **g**ene **e**xpression (TGE) systems, based on plant virus vectors, have advantages that surpass those of bacteria-, yeast- and insect-based systems, and that of plant transgenesis. TGE in plants is time efficient and scalable; in terms of the target protein, it allows high levels of expression and accumulation, correct protein folding, and expected modifications. Some other advantages will be mentioned in the presentation of vaccines in plant-based systems. Indeed, a movement away from animal experimentation is underway.

In general, plant-engineered viruses can be introduced to plant tissues by conventional methods of inoculation that include the direct transfer of DNA plasmids harboring the engineered virus, or by challenging plants with infectious cDNA clones or viral RNA synthesized *in vitro*. However, protocols based on the use of *Agrobacterium tumefaciens* and its T-DNA (by vacuum infiltration, agroinfiltration, or magnifection) have become more popular. The latter has been dubbed as a "*Launch vector system.*" These plant vectors have been used *in planta* for the production of a wide range of proteins for different purposes including, but not limited, to vaccine antigens, plantibodies (antibodies produced not by but in plants), nanoparticles (mostly virus-like particles), and other protein and protein-RNA scaffolds. The engineered vector fused to the T-DNA is delivered to plant leaves by infiltration of the bacteria (not true transgenesis though); the T-DNA is transported to the nucleus and expression of the engineered genes ensues. The expression of the viral vector itself, on the other hand, allows for cell-to-cell transmission of the modified virus vector. Viral backbones used to attain magnifection include, among many others, those derived from *Tobacco mosaic virus* (TMV; *Virgaviridae*), *Potato virus X* (*Alphaflexiviridae*), *Cowpea mosaic virus* (*Secoviridae*), *Turnip mosaic virus* (*Potyviridae*), and Bean yellow dwarf virus (*Geminiviridae*). The magnifection strategy has demonstrated to be scalable to industrial production of many recombinants proteins in the range of milligrams per gram of fresh leaf biomass.

Almost any gene can be expressed in plants in a transient manner aimed at semiindustrial or industrial production and purification. Production of vaccines in plants, for instance, has gained momentum due to the many advantages that their manufacture in these living systems possess: (1) safety issues related to the manipulation of highly infectious animal viruses are precluded, (2) mass production is attained with low financial input and high yield (for example, the cost of a plant-derived product is 10–50 times lower than similar production in bioreactors with *E. coli*, or 140 times lower than in systems based on the use of engineered baculoviruses in insect cell lines), (3) the strategy (and materials) can be easily transferred to locations where the application of more complex technologies is not possible, (4) coverage of users (patients) is extended (more people benefit at lower costs), (5) logistic use of cold storage, vaccine distribution, and highly skilled personnel for management and administration is lowered or not required. More importantly, however, plants can perform very sophisticated biochemical modifications to the protein product (correct folding, posttranslational modifications including glycosylation, for example) during manufacture. Additionally, more vaccines can be orally administered rather than parenterally.

At the beginning of the production of plant-derived vaccines, transgenesis was the method of choice. Nonetheless, plant transgenesis will not be covered here since the focus of this chapter is the use of viruses as tools for biotechnology. Besides, transgenesis has its own intrinsic complicating issues (e.g., unintended silencing of the gene whose expression is required, position effects, transgene instability over generations, and much more). Some of these issues have been overcome by using transient expression systems based on engineered viral vectors. To name only one case, vaccination programs for cholera, caused by the pathogenic bacteria *Vibrio cholerae*, have been supported with the use of two edible vaccines, Dukoral^{TM} and Shanchol^{TM}. Given that they are effective for only 2–3 years, some researchers have proposed the use of a scalable bioproduction system in plants aimed at producing an aglycosylated variant of the **c**holera **t**oxin **B** (pCTB) via a virus vector-based, large-scale production scheme. A "*deconstructed*" tobamovirus was engineered in order to fuse the recombinant gene to a secretory signal, and later delivered to plants (*Nicotiana benthamiana*) by the *Agrobacterium* vacuum infiltration method. More than 1 g of the recombinant protein was obtained from 1 kg of fresh leaves within 5 days, which amounts to 1000 doses of Dukoral^{TM}. This might constitute, aided with the addition of killed bacteria, the first step for the design of a mass vaccination program. To finalize this section, it is worth mentioning that virus-based plant vaccines aimed at "*vaccinating*" plants have also been developed. Although the strategy presented here can be envisioned as a modified version of cross-protection (see the section on *Viruses as biological control agents*), it is also true that cross-protection is a virus-based plant version of the antigen-based vaccination in animals. Briefly, *Apple latent spherical virus* (ALSV; *Secoviridae*)-based engineered vectors are capable of infecting a wide range of plants with no detrimental effects to target organisms. It has been used in plants to trigger VIGS in gene silencing-directed protocols. When ALSV vectors carry part of the genomes of unrelated viruses like *Bean yellow mosaic virus* (BYMV: *Potyviridae*), *Zucchini yellow mosaic virus* (ZYMV; *Potyviridae*), and *Cucumber mosaic virus* (CMV; *Bromoviridae*), and they are used as a preinfection carrier, plants later infected independently with BYMV, ZYMV, or CMV show homology-dependent resistance; virus multiplication is inhibited, and so viral spread. Moreover, dual infection with ZYMV and CMV in plants preinoculated with the engineered ALSV proved to be protected too. The strategy worked well thus as a preventive measure. Interestingly, however, it also worked well as a therapeutic measure since ZYMV-infected plants challenged *later* with the engineered ALSV exhibited remission of symptoms.

In other instances, VLPs are used as fused peptides to viral coat protein and thus are presented (displayed) on the particle's surface. VLPs can be used for the presentation (display) of antigenic epitopes on the engineered particle. They have been shown to elicit a protective immune response against the target. For example, a VLP derived from *Alfalfa mosaic virus* (*Bromoviridae*) displaying a 21-mer peptide from *Human respiratory syncytial virus* G protein was able to induce a virus-specific immune response in human dendritic cells (*in vitro*), which generated strong CD4^{+} and CD8^{+} T cell responses, and nonhuman primates (*in vivo*) showed a robust humoral and cellular immune response. Using the Launch Vector System mentioned in previous paragraphs, an ample range of different target epitopes were produced in the nonedible plant, *Nicotiana benthamiana*. Successful cases include the production *in planta* (and later being used) of antigens from

Bacillus anthracis, *Yersinia pestis*, Influenza viruses, *Plasmodium falciparum*, *Trypanosoma brucei,* and *Measles morbillivirus*. Of note, some plant viruses are themselves used as inducing agents of immune responses. Nanoparticles constructed from the coat protein of *Papaya mosaic virus* (PapMV; *Alphafelixiviridae*), for example, proved highly immunogenic and were taken up by dendritic cells. The same nanoparticles, modified to display an influenza epitope on their surface, elicited an increased humoral response to influenza virus infection in immunized animals; that is, PapMV acted as a true adjuvant. TMV also stimulates cellular immunity, which is one reason why it can be used as a carrier of an additional, modified antigen and induce the desired immune response against the intended target.

Plant viruses have also been used for the production of nanomaterials, nanospheres, nanowires, nanoparticles (aimed at antigen presentation), therapeutic agents, and for biomedical imaging. Finally, although not presented in this chapter, viruses also contribute genes for their own control (viral transgenes). Their use in engineering resistance against plant viruses is discussed in detail in Chapter 13: Viruses as Targets for Biotechnology: Diagnosis and Detection, Transgenesis, and RNAiand CRISPR/Cas-Engineered Resistance.

The past decade has seen the development and ever-increasing applications of the technique VIGS. When a viral vector harbors complete or partial sequences with high levels of sequence similarity to resident genes, both end up being silenced by the RNAi machinery of plants (VIGS). More than 35 DNA or RNA plant viruses have been modified as VIGS vectors. Among the RNA viruses modified for VIGS, one of the most versatile of all is *Tobacco rattle virus* (*Virgaviridae*), while for DNA viruses we find the begomovirus *African cassava mosaic virus* (*Geminiviridae*), among many others. Once in the plant cytoplasm, both DNA and RNA viruses lead to the production of dsRNA molecules, the *par excellence* trigger of the RNA-mediated degradation mechanisms that will bring about the "*knockdown*" (i.e., lack of translation) of the resident target gene and of the silencer sequence harbored by the engineered virus. Silencing the expression of genes in this specific, controlled manner has helped researchers to uncover gene function and regulation. This strategy has been applied in schemes in which plant genes are silenced by VIGS on a one-by-one basis. Where the engineered vector has silencer sequences for more than one gene, all genes are silenced in a coordinated way. All plant genetic processes can be analyzed by VIGS. Even more, VIGS is a fundamental tool required for plant metabolic engineering since it can facilitate the dissection of complete anabolic and catabolic pathways. In addition, knowing where to stop a biochemical pathway can help in increasing the production of an intermediary enzyme or metabolite of interest. On the other hand, an engineered virus for VIGS of a specific gene can be used to silence the homologous gene in a heterologous system (that is, the strategy and tools are transferable to different plant systems). VIGS has become an important tool for functional genomics − particularly in poorly known plants, genetically speaking.

To conclude with this short account on plant virus vectors, some mention should be made of genetic engineering plant genomes using modified plant virus vectors. Early technologies that made use of nonviral vectors will not be mentioned here though. **G**enome **e**ngineering (GE) in plants involves ways of delivering GE components (genes, and/or recombinogenic enzymes or systems) with the purpose of, in a very specific targeted way, disrupting, inserting or "*correcting*" genes at particular locations within the plant genome. Virus-based vectors, as autonomously replicating agents, offer the possibility of acting as such vehicles. Among them, *Wheat streak mosaic virus* (*Potyviridae*), *Barley stripe mosaic virus* (*Virgaviridae*), *Tobacco rattle virus* (*Virgaviridae*), and *Cabbage leaf curl virus* (*Geminiviridae*) are already in use. These vehicles act as carriers of GE platforms that rely on the use of meganucleases, zinc-finger nucleases, **t**ranscription **a**ctivator-**l**ike **e**ffector **n**ucleases (TALEN), and clustered regularly interspaced short palindromic repeats (CRISPR)/CRISPR associated9 (CRISPR/Cas9). Each of the systems has its own peculiarities, strengths, and limitations. But they are also extremely flexible. In the case of geminiviruses-based vectors, for example, a strategy has been developed known as "***v**irus based gRNA delivery system for CRISPR/Cas9 mediated plant **g**enome **e**diting*" (VIGE). VIGE allows the generation of plant knockout libraries, circumventing the use of the efficient, and yet time consuming, VIGS. These additional techniques (e.g., the use of virus-based vectors for the delivery of the Cas9 and specific sgRNAs) increase the range of strategies available for engineering plant genomes—and thus imagination is the only limiting factor.

VIRUSES AS BIOLOGICAL CONTROL AGENTS

Viruses have been employed for the control of pests (mainly phytophagous insects), and pathogens like fungi and viruses themselves. Three representative cases are highlighted in this section: (1) cross-protection, (2) insect control by viruses and virus-based strategies, and (3) control of phytopathogenic fungi by hypovirulence-causing mycoviruses.

Cross-protection is understood as the biological phenomenon by which a preceding infection with a virus in a host helps prevent the establishment of infection with another virus in the same host. There must be, of course, a high degree of similarity between both viruses in order to attain such an effect. In animal hosts, cross-protection is due to the recognition of antigens present on the second virus by antibodies raised against the first (or vice versa, or both simultaneously). In plants, cross-protection works differently: prior infection with an attenuated virus triggers an RNA-mediated resistance mechanism leading to resistance to subsequent infections with a highly similar virus. In essence, the working force in both cases is the same: similarity, at the protein sequence level (for antibody-mediated responses) or at the nucleotide sequence level (for RNA-mediated resistance). In both cases, the protection is achieved with wild-type, mutant, or engineered viruses (including chimeric viruses). For animal viruses, this strategy is formally the vaccine-based strategy that was presented earlier, and so we will briefly mention here only the application of cross-protection in the management of plant viruses. Cross-protection is a "*natural*" phenomenon whereby a plant is coinfected by two or more different but related viruses (or strains of the same virus) or when the plant has been superinfected (sequential infections). Here we will address cases where this has been performed *intentionally* for the purpose of virus disease control. This typically begins with surveying for and identifying naturally occurring mild strains of the virus. Some mild variants of a virus may slightly affect plant performance. These variants, only studied so far from crop plants, can be found in the field (or obtained by mutagenesis followed by selection), and their protective cross-protection capacity can be highly variable. To overcome this problem, plant viruses have been engineered in a sequence-oriented fashion in a way that the "*cross-protecting*" virus conserves its replication and movement capabilities, and causes only mild symptoms in the plant host. This strategy of using an induced mutant strain has been successful in controlling *Papaya ringspot virus* (*Potyviridae*), while the latter (custom-tailored strains of the virus) has controlled *Pepino mosaic virus* (*Alphaflexixiviridae*). Recent evidence suggests that RNA-mediated pathways (adaptive immune resistance) may not be solely responsible for mediating cross-protection: salicylic acid-mediated pathways (innate immune resistance) also appear to play some role in this mechanism of resistance against plant viruses.

In a different case of biological control, members of the *Baculoviridae* family of viruses are natural "*controllers*" of insects' populations. They are safe for vertebrates and plants, and have found use as biopesticides with success. Baculoviruses show a defined, narrow range of hosts, and it is for this reason they are excellent candidates for species-specific, narrow insecticidal applications. They have been shown to have no detrimental effects on non-target, beneficial insects. Baculoviruses virions exist as **o**cclusion-**d**erived **v**iruses (ODVs) or as **b**udded **v**iruses (BVs), depending on the way they were produced. ODVs are released by cell lysis and are horizontally transmitted to other hosts, whereas BVs bud out through cellular membranes and disseminate within the insect's body. The rationale behind the use of baculoviruses for the targeted control of insect pests is that virions can be ingested by the insect and produce an infection— which is typically fatal. After the insect dies, more virions are released, and thus more target insects are infected. The majority of baculoviruses used as biopesticides belong to the genera *Alphabaculovirus* (αBV) and *Betabaculovirus* (βBV). Some baculoviruses used for the specific control of insect pests include: *Cydiapomonella granulovirus* (βBV) for the control of the Codling moth (*C. pomonella*) that afflicts apples and pears, *Spodoptera littoralis nucleopolyhedrovirus* (αBV) for the control of *S. littoralis* on tomato, cotton, and corn, *Helicoverpa* *armigera nucleopolyhedrovirus* (αBV) used to protect chickpeas (*Cicer arietinum*) from the pod-borer *H. armigera*, and *Autographa californica multiple nucleopolyhedrovirus* (AcMNPV; αBV) employed as a biopesticide against the alfalfa looper, *A. californica*, on alfalfa (*Medicago sativa*). Interestingly, in a creative turn of events, an engineered AcMNPV was shown to work well as an alternative and safe vehicle to target cancer cells; specifically, prostate-specific gene expression of therapeutic genes cloned into the vector has been attained by using human transcriptional control sequences in the engineered virus genome. In the pursuit of insect control, engineered baculoviruses have also been modified to express insecticidal neurotoxic peptides from scorpions, spiders or sea anemones, diverse regulatory proteins (diuretic hormone, juvenile hormone esterase, eclosion hormone, prothoracicotropic hormone), degradative enzymes (chitinases, gelatinases, stromelysin, cathepsin L, enhancing, keratinase and others), and other insecticidal proteins (for instance, toxins derived from *Bacillus thuringiensis*).

Insects that afflict crop plants are not the only targets of virus biocontrol agents. In the case of mosquitoes, which are able to spread an important number of pathogens of epidemiological importance, including viruses (e.g., *Dengue virus*, *Zika virus*, *Chikungunya virus* and *Yellow fever virus*, among others), *Plasmodium* species (causal agents of malaria), and filarial worms (the causal agents of lymphatic filariasis, or elephantiasis, including species of the genera *Brugia* and *Wuchereria*), current studies propose the use of **m**osquito **d**enso**v**iruses (*Parvoviridae*: *Parvovirinae*), or MDVs, for their management. MDVs replicate in the nuclei of infected mosquito cells, and they eventually result in the death of insect larvae in a dose-dependent fashion; even if larvae survive, they do not pupate or eclose to adults. Additionally, MDVs are transmitted vertically by infected female mosquitoes that could help in maintaining these viruses in the population and location. Overall, the net expected result is a decrease in the vector's population. Nonetheless, results so far are not very encouraging, but the strategy is still under evaluation. Finally, another strategy under investigation for both cases of insect control (agricultural insect pests and insects vectoring arboviruses of human health importance) involves the delivery of protective dsRNAs; that is, the trigger for RNAi. Because RNAi machinery is present in all insects, RNAi-based management strategies that target and disrupt the expression of essential genes and impair insects' performance should be possible. Investigations with insect-specific viruses, along with modified viral-like particles, are ongoing.

To end this section, it is important to mention that fungi (mainly phytopathogenic fungi) are also targets of control by fungal-specific viruses (mycoviruses) that incite the development of the hypovirulence phenotype (Chapter 9: Viruses of Prokaryotes, Protozoa, Fungi, and Chromista). In plant pathology, hypovirulence is known as the reduced ability of some strains of a fungus to infect, colonize, kill, or reproduce on susceptible hosts.

The phenomenon is associated with the presence of viruses and associated dsRNAs in the fungal host. Biological control of some fungi has been demonstrated through applications of hypovirulent (virus-infected samples) strains to diseased plants. Disease severity may be reduced along with reproductive capacity of the fungus. Hypovirulent isolates of *Ophiostoma novo-ulmi* infected with *Ophiostoma mitovirus 3a* (*Narnaviridae*), for example, has been effective and persistent (that is, stable) in the control of the phytopathogenic fungus in Europe. Also in Europe, control of Chestnut blight, by means of using hypovirulent strains of *Cryphonectria parasitica* infected with *Cryphonectria hypovirus 1* (CHV-1; *Hypoviridae*), has been reported successful. CHV-1 infects the fungus reducing its parasitic growth and sporulation capacity, hindering thus its dissemination. Both prophylactic and postexposure therapies are possible. Suppression of the severity and incidence of the rapeseed stem rot disease, caused by *Sclerotinias clerotiorum*, has also been reported following the application of hyphal-fragment suspensions of the corresponding fungal strain infected with *Sclerotinia gemycircularvirus 1* (*Genomoviridae*).

VIRUSES AS SCAFFOLDS AND TEMPLATES FOR NANOMATERIALS

Increasingly, biomolecules and microorganisms, including viruses, are being used as building blocks for nanotechnology. Nanotechnology is the study of manipulating matter at the atomic and molecular scale, and involves the synthesis of materials, devices, and systems with atomic precision. A nanoparticle is the most fundamental component in the fabrication of these nanostructures. In general, the size of a nanoparticle spans the range between 1 and 100 nm. Although their physical and chemical properties depend on their composition, size, and shape, nanoparticles share common properties: they are highly mobile and their high surface areas provide a tremendous driving force for diffusion. About 20 years ago, viruses attracted attention as potential scaffolds or templates for applications in nanotechnology because of their distinct properties: their highly uniform structures, small size, and ability to self-assemble. Since viruses are naturally occurring nanomaterials, they are both biocompatible and biodegradable. Viral nanotechnology and **v**iral **n**ano**p**articles (VNPs) hold promise for the development of next-generation tissue-targeting, imaging devices and therapeutics.

VNPs are self-assembled, homogeneous nanoparticles derived from the coat proteins of viral capsids. Usually a distinction is made between VNPs and VLPs; VLPs are the genome-free counterparts of VNPs and are regarded as a subclass of VNPs. Regardless, these virus-based particles provide an engineering scaffold that is superior to synthetic particles. The internal cavity can be filled with drug molecules, imaging reagents, and other nanoparticles, whereas the external surface can be decorated with targeting ligands to allow cell-specific delivery. There is also the application to inorganic materials synthesis on the exterior and interior surfaces. VNPs can be produced in high titers in their natural hosts. VLPs, on the other hand, are produced in heterologous expression systems like *E. coli* and yeasts. VLPs based on eukaryotic viruses tend to be produced using baculovirus vectors in insect cells or adenovirus vectors in mammalian cells. Alternatively, VLPs can be produced directly from VNPs by pH-induced swelling followed by alkaline hydrolysis of the released nucleic acids. Another option involves the disassembly of intact VNPs into their individual coat protein subunits and their reassembly into VLPs, once the nucleic acids are removed. Therapeutic applications (vaccine technologies, drug delivery) of VNPs/VLPs were discussed earlier; applications to imaging, tissue engineering, and inorganic materials synthesis are briefly introduced in this section.

Imaging Molecules

VNPs have been developed as reagents for noninvasive imaging techniques such as **m**agnetic **r**esonance **i**maging (MRI). MRI contrast reagents are developed by decorating *Cowpea mosaic virus* (CPMV; *Secoviridae*), *Cowpea chlorotic mottle virus* (CCMV) and bacteriophage MS2 (*Escherichia virus MS2*, *Leviviridae*) particles with gadolinium, a paramagnetic metal ion. This approach is useful because the large surface area of VNPs facilitates decoration with hundreds of ions. Densely and specifically aligned imaging agents on viruses have allowed for high-resolution and noninvasive visualization tools to detect and treat diseases earlier than previously possible. Encapsulation techniques can also be used to design VNPs with fluorescent cores for use in cellular imaging applications. Coupled to a fluorescent label, M13 phage functions as a microfluidic reporter. Other applications have been to **n**ext **g**eneration **s**equencing platforms (NGS). Proteins of the DNA packaging motor of phage phi29, for instance, consist of a dodecamer channel of gp10 capable of translocating dsDNA and when embedded in a lipid bilayer generate extremely reliable, precise, and sensitive electric conductance signatures upon passage of a charged molecule through a tunnel. Coupled to appropriate sensors, it allows base-by-base DNA sequencing in some NGS platforms.

Scaffolds for Tissue Engineering

Cells that are cultured on a three-dimensional surface have been shown to undergo more rapid growth when compared to cells grown on flat surfaces. This has important applications in tissue engineering: for example, glass slides coated with either rod-shaped TMV or the icosahedral *Turnip yellow mosaic virus* (TYMV; *Tymoviridae*) encouraged bone marrow stromal cells to undergo significantly faster rates of osteogenic differentiation compared to cells grown under standard conditions. One of the main benefits of using

VNPs as a 3D surface for cell growth is that different spatial arrangements of cells can be manipulated, and different cell types can be grown in close proximity to each other in an orchestrated arrangement of different populations of cells. Engineered M13 phages capable of displaying a peptide called RGD (Arg-Gly-Asp) has been used in conjunction with poly(lactic-coglycolic acid) to create hybrid nanofiber matrices that can be efficiently employed as biomimetic scaffolds with applications in tissue engineering. This is due to the fact that matrices enriched with RGD peptides promote both attachment and cellular behavior. Phage M13 has also been engineered to function as a versatile bioink. That is, in the emerging field of bioprinting, aimed at producing tissue-mimetic 3D structures, the phage has been modified in a way that it displays (see earlier section on *Phage display*) integrin-binding and calcium-binding domains that, after being blended with alginate, are able to form Ca^{2+}-crosslinked hydrogels. After optimizing the printing process, cell-laden scaffolds with high cell viability, possessing enhanced proliferation and differentiation rates, are generated.

Templates for Materials Synthesis

Bio-templating is a process involving the deposition of inorganic structures on the internal or external surface of the VNP that is governed by interactions with specific capsid proteins. Rod-shaped *Tobacco mosaic virus* (TMV; *Virgaviridae*) and filamentous *Escherichia virus M13* (*Inovirida*e) particles have been used as structural templates in these applications for the synthesis of semiconducting tubes and wires for potential applications in electronics such as medical devices, data storage devices, or battery electrodes. Phage virions, being amenable to genetic modification, can also be chemically modified. When combined, chemical modification (bioconjugation) with genetic engineering, phages become biological "*lego blocks,*" so to speak, with endless possibilities.

FURTHER READING

Breitbach, C.J., Lichty, B.D., Bell, J.C., 2016. Oncolytic viruses: therapeutics with an identity crisis. EBioMedicine 9, 31−36. Available from: https://doi.org/10.1016/j.ebiom.2016.06.046.

Filley, A.C., Dey, M., 2017. Immune system, friend or foe of oncolytic virotherapy? Frontiers in Oncology 7, 106. Available from: https://doi.org/10.3389/fonc.2017.00106.

Greber, U.F., Suomalainen, M., Stidwill, R.P., Boucke, K., Ebersold, M.W., Helenius, A., 1997. The role of the nuclear pore complex in adenovirus DNA entry. EMBO Journal. 16, 5998−6007. Available from: https://doi.org/10.1093/emboj/16.19.5998.

Haddad, A., 2017. Genetically engineered vaccinia viruses as agents for cancer treatment, imaging, and transgene delivery. Frontiers in Oncology 7, 96. Available from: https://doi.org/10.3389/fonc.2017.00096.

Henry, M., Debarbieux, L., 2012. Tools from viruses: bacteriophage successes and beyond. Virology 434, 151−161. Available from: https://doi.org/10.1016/j.virol.2012.09.017.

Holay, N., Kim, Y., Le, P., Gujar, S., 2017. Sharpening the edge for precision cancer immunotherapy: targeting tumor antigens through oncolytic vaccines. Frontiers in Immunology 8, 800. Available from: https://doi.org/10.3389/fimmu.2017.00800.

Lee, K.L., Twyman, R.M., Fiering, S., Steinmetz, N.F., 2016. Virus-based nanoparticles as platform technologies for modernvaccines. WIREs NanomedNanobiotechnol 2016. Available from: https://doi.org/10.1002/wnan.1383.

Martins, I.M., Reis, R.L., Azevedo, H.S., 2016. Phage display technology in biomaterials engineering: progress and opportunities for applications in regenerative medicine. ACS Chemical Biology 11, 2962–2980. Available from: https://doi.org/10.1021/acschembio.5b00717.

Minor, P.D., 2015. Live attenuated vaccines: historical successes and current challenges. Virology 479-480, 379–392. Available from: https://doi.org/10.1016/j.virol.2015.03.032.

Pokorski, J.K., Steinmetz, N.F., 2010. The art of engineering viral nanoparticles. Molecular Pharmaceutics 8, 29–43. Available from: https://doi.org/10.1021/mp100225y.

Daya, S., Berns, K.I., 2008. Gene therapy using adeno-associated virus vectors. Clinical Microbiology Reviews. 21, 583–593. Available from: https://doi.org/10.1128/CMR.00008-08.

Steinmetz, N.F., 2010. Viral nanoparticles as platforms for next-generation therapeutics and imaging devices. Nanomedicine 634–641. Available from: https://doi.org/10.1016/j.nano.2010.04.005.

Walther, W., Stein, U., 2000. Viral vectors for gene transfer: a review of their use in the treatment of human diseases. Drugs 60, 249–271. Available from: https://doi.org/10.2165/00003495-200060020-00002.

Xie, J., Jiang, D., 2014. New insights into mycoviruses and exploration for the biological control of crop fungal diseases. Annual Review of Phytopathology 52, 45–68. Available from: https://doi.org/10.1146/annurev-phyto-102313-050222.

Yildiz, I., Shukla, S., Steinmetz, N.F., 2011. Applications of viral nanoparticles in medicine. Current Opinions in Biotechnology 22, 901–908. Available from: https://doi.org/10.1016/j.copbio.2011.04.020.

Zaidi, S.S., Mansoor, S., 2017. Viral vectors for plant genome engineering. Frontiers in Plant Sciences 8, 539. Available from: https://doi.org/10.3389/fpls.2017.00539.

Zimmer, C., 2015. A Planet of Viruses, Second Edition University of Chicago Press, Chicago, USA, p. 128.

Chapter 13

Viruses as Targets for Biotechnology: Diagnosis and Detection, Transgenesis, and RNAi- and CRISPR/Cas-Engineered Resistance

Paula Tennant[1] **and Gustavo Fermin**[2]
[1]*The University of the West Indies, Mona, Jamaica,* [2]*Universidad de Los Andes, Mérida, Venezuela*

When the enemy is relaxed, make them toil. When full, starve them. When settled, make them move.

Sun Tzu

"*Friend or foe?*" broadly summarizes our current understanding of viruses, and the relationships they establish with cellular organisms. And for good reason: viruses are able to cause extreme damage, decimation, disease, and death (Chapter 10: Host-Virus Interactions: Battles Between Viruses and Their Hosts); but viruses have also been a part of the natural history of all organisms, and represent a driving force in their evolution. They are currently recognized as playing a fundamental part in the functioning of the biosphere, including the cycling of many components of the planet.

During the last few decades, genetic and molecular studies have led to an enormous increase in our understanding of viruses. Established for many are the mechanisms of replication, genome expression, and packaging of progeny genomes. This, together with the complete nucleotide sequences of genomes and methods of cloning and manipulation of viral genomes, has made possible the use of viruses as vectors for the expression of foreign proteins

Viruses. DOI: https://doi.org/10.1016/B978-0-12-811257-1.00013-9

and as vectors for gene therapy (Chapter 12: Viruses as Tools of Biotechnology: Therapeutic Agents, Carriers of Therapeutic Agents and Genes, Nanomaterials, and More), with the goal of either curing a disease or alleviating an underlying condition. One of the current goals of modern biotechnology is to further uncover the diversity of viruses, the hosts they interact with, the ways they impact their biological functioning, and how we cope with the damage they inflict upon us and the plants and animals we use for our very existence and survival. As an enemy of our comfort, we have been pursuing, since the use of the first vaccines, the development of strategies to fight against viruses and the diseases they cause.

Viruses represent a therapeutic challenge since their lifecycles occur within the host cells and often utilize cellular proteins. It is therefore difficult to identify therapeutic targets, compared to other pathogens, which have their own cellular metabolism that is quite different from that of the host. Viral proteins, then, present useful targets for therapy, for example, inhibitors of viral polymerases, or prevention, for example, viral coat proteins for vaccination (Chapter 12: Viruses as Tools of Biotechnology: Therapeutic Agents, Carriers of Therapeutic Agents and Genes, Nanomaterials, and More). Since some viruses enter an inactive state of persistence or latency where very few viral proteins are produced, strategies that specifically target viral genome sequences are useful to the antiviral armamentarium. Virus sequences themselves have been employed in the development of these strategies, which include transgenesis and the **c**lustered **r**egularly **i**nterspaced **s**hort **p**alindromic **r**epeats (CRISPR)/ CRISPR-**as**sociated protein **9** (Cas9) systems. Finally, any effective virus management strategy requires accurate diagnosis of the viral etiological agent. Various strategies target viruses or rather detect, diagnose, and monitor viral diseases, using biotechnological advances.

This chapter highlights biotechnological approaches that provide novel methods for the control of virus diseases through (1) the development of rapid diagnostic tests based on serology and nucleic acid technologies and (2) the development of resistance strategies/mechanisms utilizing properties of the viral genome.

UNCOVERING THE VIROSPHERE AND HIDDEN WORLD OF VIRUSES

For good or for better we need, and we want to know, the world of viruses in its most intricate detail. The discovery of viruses, their isolation, and the demonstration of their chemical nature was an exciting achievement with far-reaching consequences for our understanding of the living world (Chapter 1: Introduction: A Short History of Virology). More viruses were discovered and described henceforward, and by the latter half of the last century we became aware of the existence of viruses infecting organisms from all kingdoms of life. We devised a system for their classification and naming, and we were amazed by the viruses that were appearing in our lives (apparently out of nowhere)—with on occasion very unfortunate consequences. Various advances in genomic-based tools have further assisted in our quest for more information regarding an old enemy.

Three main ways of uncovering new viruses and their existing variants, with their own advantages and limitations, include: (1) classic sequence-based methodologies, (2) **n**ext **g**eneration **s**equencing (NGS), with their variants, and (3) metagenomics approaches, successfully helped by NGS.

Classic Sequence-Based Methodologies

If we feel comfortable wandering within the sequence space occupied by a known virus, exploring it to its limits will reveal variants in the same space—and sometimes, new unknown viruses. That is, we can explore the intimacy of this space to uncover intraspecific sequence variations of a known virus guided by previous knowledge; population or disease-oriented studies allow the detection of strains of the same virus in the same host population, and epidemiological studies demonstrate the existence of host-specific strains of the virus or geographically distributed virus strains. Either way, in the classical approach of molecular characterization by sequencing of virus genes or genomes, the key point is previous knowledge. In this approach, we use regular protocols in molecular biology that include reverse transcription, DNA amplification, molecular cloning and sequencing (Sanger's method), assisted many times by primers deduced from a closely related virus sequence, the information provided by public databases, and their analysis capabilities. The method is primarily accomplished by amplification of templates with fluorescent chain-terminating nucleotides, followed by capillary electrophoresis of the amplicons and reading the fluorescence signals. The requirement of prior sequence information has made this technique suitable for the discovery of new genotypes of known viruses, but the technique is not appropriate for the discovery of novel viruses.

Next Generation Sequencing

On the contrary, prior knowledge is not required for NGS. The technique does not assume any knowledge about the organisms being investigated. The strategy, aimed at detecting and sequencing viruses, makes use not of specific primers, but of universal primers that are ligated to all the nucleic acid molecules produced during sample preparation; applications to RNA genomes add a reverse transcription reaction step. Then, a parallel, massive sequencing step is set up in order to sequence all tagged molecules present in the sample. As a result, many virus sequence spaces are simultaneously targeted, along with that of the host. NGS is also known as high-throughput sequencing and deep sequencing. The methods include, but are not limited to, pyrosequencing, single-molecule real time sequencing, ion semiconductor sequencing, sequencing by ligation, nanopore sequencing, Polony sequencing, DNA nanoball sequencing, Heliscope single molecule sequencing, and sequencing by synthesis.

NGS has been used to analyze viruses of crop plants thereby uncovering the presence of viruses previously unknown, as well as the presence of known (and unknown) viruses in wild relatives of crop species. The strategy has also been fruitful in revealing the vast diversity of viruses present in insects (most of which are commensals of their hosts), in evaluating the human virome, and in revealing viruses causing human diseases whose etiology was previously a mystery. There is no doubt that NGS has opened endless possibilities for a better understanding of the planet's virosphere. However, the use of less powerful, and yet equally informative, techniques and methodologies (culture, biophysical, and biochemical methods) are required to fully characterize all new viruses.

Metagenomics Approaches

NGS and **M**etagenomics **a**pproaches (MGA) do not differ much regarding their sequencing method, but rather in the source of the target and purpose of the analysis. Nonetheless, metagenomics refers to the analysis of microbial communities from environmental samples (including clinical samples) by means of DNA sequencing. More or less universal primers targeting ribosomal genes are used when applied to cellular organisms, but this is not a *sine qua non* condition for MGA. Contrary to conventional methods of sequencing in which DNA from identical cells (or same DNA source) are extracted and subjected to sequencing, in MGA many different organisms (most of the times unknown) are taken from a defined environmental location from which total DNA is extracted. That is, a microbial community's DNA, including its virus members is collected. It is estimated that between 60% and 95% of sequences of viral origin recovered from metagenomics studies have no known homolog in public databases and represent viral novelties, including new virus species.

Virus metagenomics analyses have revealed a surprising number of new viruses in the following instances: (1) virus detection and identification from samples of total RNA or DNA from individuals or pooled organisms; (2) nucleic acids extracted from the virions isolated from single or pooled individuals; this limits the amount of host DNA and concentrates only on nucleic acids derived from viral elements; (3) targeted sequencing of dsRNA-enriched samples allows coverage of both dsRNA and RNA viruses since hosts do not normally produce dsRNA molecules; and (4) **s**mall **i**nterfering **RNA** molecules (siRNAs, Chapter 5: Host Range, Host–Virus Interactions, and Virus Transmission) synthesized by the eukaryotic host when mounting an immune adaptive response against viruses in a sequence-specific manner. In the latter, endogenous viral elements (i.e., inserted in, and inherited with the host genome) are also detected, if they are transcribed. MGA has proven the best way to study viruses borne in the air, water, soil, and other solid synthetic surfaces. Finally, metagenomics have had such an impact on our knowledge of the viral world and the ways we classify viruses that in the last release (2017) of the ICTV report on virus taxonomy, some new species were described based solely on metagenomics data.

DETECTION OF VIRUSES AND DIAGNOSIS OF VIRUS INFECTIONS

Historically, diseases of viral etiology were diagnosed based on symptomatology, even before the discovery of the entities themselves and the demonstration of the causal relationship between viruses and the diseases they incited. The latter discussions were first initiated by Koch in the 1880s who put forward a series of requirements to prove that the diseases anthrax and tuberculosis were caused by different types of bacterial microbes (see Chapter 1: Introduction: A Short History of Virology for the Four Criteria). Later in the 1930s, the animal virologist Thomas Rivers revised Koch's postulates for viruses and formulated six criteria that should be met in proving that a virus caused a particular disease: the isolation of a virus from the diseased host, cultivation of the virus in host cells, filterability (to exclude other pathogens), the cultivated virus is able to induce similar disease when transferred to a healthy susceptible animal host, the same virus is recoverable from the experimentally infected host, and an immune response to the virus is detectable in the animal. Another revision by Fredericks and Relman occurred in 1996 to include the use of nucleic acid-based evidence, and this allowed for the proof of causation in instances where viruses could not be propagated in host cells. The established criteria include: the detection of nucleic acid of the pathogen in the infected organism, no pathogen associated nucleic acid sequences should be detected in healthy hosts (or tissues) or in hosts after the resolution of infection, pathogen associated nucleic acid sequences should be consistent with the known biological characteristics of that group of organisms, the sequence-based evidence is reproducible, the pathogen can be detected in diseased tissue that contains the nucleic acid sequences, and nucleic acid sequence copy number that correlates with the severity of disease is more likely to be the cause of disease.

Diagnosis, thus, requires a composite of information, including history, physical examination, and laboratory data. Laboratory methods include, but are not limited to, cell-dependent culture of the virus (and determination of its cytopathic effect, if any), observation of virions by electron microscopy (and even optical microscopes for the newly described giant viruses) or of inclusion bodies (if any), detection of virus-specific antibodies (in vertebrates), detection of virus antigens (with specific antibodies raised in different hosts and nonhost mammals) or viral genomes (including the amplification by PCR of specific segments of the virus genome), and host indexing (mostly used with plant viruses). In some instances, sequencing of the virus can help in elucidating not only virus identity, but specificity (host-specific strains of the virus), and/or geographical origin.

With some diseases, such as Ebola virus disease, diagnosis plays a critical role in outbreak response efforts. However, establishing safe and expeditious testing strategies in resource-limited environments remains extremely challenging. According to the World Health Organization (WHO), detection and diagnostic methods for use in developing nations should be "*ASSURED*" (**a**ffordable, **s**ensitive, **s**pecific, **u**ser-friendly, **r**apid and robust, **e**quipment-free, and **d**eliverable to end users), in order to deliver timely diagnoses and effective therapy. These all-in-all methods do exist: for example, easy-to-perform, single-use diagnostic tests are available for the rapid, visual detection of antibodies to HIV 1 and HIV 2 and more recently, dipstick tests that can rapidly distinguish between Zika and dengue virus infections. Similar ImmunoStrips are available for the detection of plant viruses either independently or in combination. But not in all cases will a single available method meet all the mentioned criteria, but a combination might—sometimes increasing sensitivity and/or specificity. A brief description of traditional and newer methods used in the diagnosis of viral infections and detection of viruses is provided below. Since we focus on those methods that use a biological component to detect a biological target, many useful and powerful techniques are not covered here. The reader is recommended to get acquainted with, for example, the use of nanopore-based technologies aimed at detecting and, literally, counting viruses, and other non-biotechnological methods such as electron microscopy and mass spectrometry and its many variants.

Culture Systems: Whole Organisms- and Cell-Based Methods

Virus culture: This technique is mainly used with animal viruses, particularly those of clinical importance. Isolation of the virus can be specifically accomplished by using different cell lines, which vary according to the targeted virus (organotypical). The most common and widely used cell lines for viral diagnosis are primary **rh**esus **m**onkey **k**idney cells (RhMK), primary rabbit kidney cells, MRC-5, human foreskin fibroblasts, HEp-2, and A549. Although detection of the virus is by the pathology it causes (CPE, **c**yto**p**athic **e**ffect), definitive identification may be performed by immunofluorescence staining and other specific diagnostic tests. In some instances, transgenic cell lines in which virus replication is improved are used; in other cases a reporter gene is expressed in the transgenic line only after infection, which allows for the specific detection of the virus via a rapid and simple enzymatic assay. Organ cultures, although useful in maintaining the differentiated state of the target cell, are not widely employed. Sometimes, inoculation of living animals (suckling mice, for instance) is required for investigations on natural infections; embryonated eggs are also useful for the isolation and identification of animal viruses.

Host indexing or Biological indexing: This method, used in the identification of plant viruses or differentiation between strains, involves the mechanical inoculation of a set of different plants (indicator plants) susceptible to the suspected virus. If prior purification is needed, a host where the virus causes local lesions is used before proceeding with the indexing scheme. Sometimes serial passages are conducted until purification is attained. Host indexing can also be performed by grafting parts of the infected plant onto indicator plants. In both instances, there is the development of characteristic systemic symptoms or local lesions if the candidate plant is virus infected. The determination of the virus—vector relationship may be useful, in some instances, for putative virus identification.

Immunological Methods

Immunological methods allow for the detection and quantification of a virus present in a sample, and in vertebrate hosts, the diagnosis of new, recent, or past infections depending on the antibody response in the organism. Immunological detection systems use specific antibodies developed in animals in response to antigens (immunogenic molecules). Polyclonal and monoclonal antibodies can be developed in a number of different ways. The most straightforward is to inject host animals, such as mice for monoclonal development or rabbits or goats for polyclonal development, with live or inactivated material and then fuse the spleens or collect the blood of those animals that have high titers to the target agent. Antibodies are produced by B cells of the immune system after being exposed to femtomolar concentrations of the antigen, generating up to 10^9 specific antibody molecules in a week. They are not only specific but quite stable too. Additionally, the immunized organism generates memory cells that allow the animal to mount a more rapid response after reexposure to the same antigen. They can be used in single detection experiments, but arrays of multiple antibodies raised against many different targets have also been developed.

Precipitation and agglutination tests: Precipitation tests rely on the formation of a visible precipitate when sufficient amounts of the antigen and their specific antibodies are in contact with each other. There are many variations to the precipitation test that are informative and easy to perform, but these tests lack the sensitivity of more modern methods for the detection of low-titer viruses. In agglutination tests, an inert carrier particle is coated with a specific antibody that, upon recognition of its antigen, clumps or agglutinates, resulting in the formation of a complex that is visible to the naked eye or may require an optical microscope.

Complement fixation test: This test can be used to determine whether antibodies against a virus are present in serum. The test is based on the fact that viral antigen and antibody interaction activates (or "*fixes*") the complement and leads to membrane lysis. Red blood cells are used as targets because lysis of their membranes is readily observed.

Hemagglutination inhibition test: This test is specifically used for the detection and relative quantification of virions of animal arboviruses, influenza, and parainfluenza subtypes. It is based on the binding of sialic acid receptors present on the surface of red blood cells to the hemagglutinin glycoproteins found on the surface of these virions and the formation of a network (lattice) created by the bound cells and the interacting virions. Antibodies against viral proteins with hemagglutination activity can block the ability of virus to bind to red blood cells.

Enzyme-linked immunosorbent assay (ELISA or enzyme immunoassay EIA): ELISAs have been the mainstay of immunoassay technology. They provide economical and sensitive methods for the detection of viral antigens or antibodies and are particularly convenient for routine large-scale testing. Although automation ranges from simple workstations to fully automated ELISA platforms that perform the entire process from start to finish, sample processing continues to be time consuming. Fundamental to ELISA is the concept that an enzyme can be bound to antibodies to form a conjugated molecule that has enzymatic activity and is also serologically active. In one of the most popular ELISA applications (direct ELISA), immobilized antigen is detected by an enzyme-linked primary antibody. A chromogenic molecule is added and is converted by the enzyme (peroxidase or alkaline phosphatase) to an easily detectable product. In the so-called indirect ELISA, an immobilized antigen is bound by the specific antivirus, primary antibody. The bound antibody is then detected by a secondary antibody directed against a specific target (epitope) on the primary antibody. In another application, antigen is trapped between a capture antibody and a detector antibody labeled with an enzyme; both are specific to the antigen. In a positive test, the addition of substrate specific to the enzyme results in a color change. The technique is also known as DAS-ELISA, **d**ouble **a**ntibody **s**andwich ELISA. Other variations of enzyme immunoassays include fluorescence polarization immunoassay, microparticle enzyme immunoassay, and chemiluminescent immunoassay.

Western: In this classical technique, the antigen or antigen source, either purified or from a crude sample, is fractionated by polyacrylamide gel electrophoresis. The samples are subsequently transferred to a solid support (PVDF or nitrocellulose membranes) and all unoccupied surfaces are blocked with an inert substrate (say, albumin). The membrane is first incubated with a primary antibody against the target of interest, unbound primary antibody washed off, and then a secondary anti-antibody linked to an enzyme or a fluorescent compound (a conjugate). Presence of the virus protein, thus, is indirectly reported by the enzymatic activity (or fluorescence) of the conjugate, if the reaction is positive. Alternatively, samples can be blotted onto a membrane (no electrophoresis required) and processed as described above, but size estimation of the target virus protein is not possible. However, immunoblotting allows for the simultaneous evaluation of many more samples than western analysis alone. These techniques are often used as confirmatory evidence of a positive ELISA by, for example, detecting antiHIV antibodies in serum.

Immunoelectron microscopy: This technique allows for virus detection by (1) differential trapping of virions on electron microscope grids coated with specific antibodies against different viruses, (2) antiserum endpoint dilutions that coat or decorate virions, and (3) observation of agglutination or clumping at varying concentrations of the antiserum or monoclonal antibodies. One application, **i**mmuno**h**isto**c**hemistry (IHC), uses antibodies to localize viral proteins in fixed tissue sections or cell cultures.

Biosensors: Biosensors are physical devices that utilize a specific bioreceptor surface to analyze intact virions or particular protein components. The most widely used bioreceptor molecules are specific antibodies raised against particular virus antigens. The analytical process of the captured analyte (intact virions or viral proteins) might be based, for example, on an optical detection platform using a **s**urface **p**lasmon **r**esonance (SPR) system. In SPR, the interacting molecule (in this case, an antibody) is immobilized on the sensor surface (ligand); the partner analyte flows through the detection cell where it can interact with the ligand, depending on the specificity of the interaction (positive). As the positive analyte interacts with the ligand, it accumulates on the surface increasing the refractive index. The changes are detected and measured optically. Other biosensor platforms that have been successfully used in virus detection and diagnostics include nanowire field effect transistor, interferometer, impedance-based, electrochemical, resonator, waveguide-mode, and surface-acoustic sensors. Other ligands include glycans (for the interaction with virus proteins, like hemagglutinin of influenza viruses) and aptamers (small, highly specific nucleic acids molecules that have been previously selected from a library enriched after many rounds of selection by its ligand); the latter are also known as chemical antibodies.

Molecular Methods Based on Detection of Nucleic Acids

***P**olymerase **c**hain **r**eaction (PCR) or Nucleic **a**cid-**a**mplification **t**ests (NAATs)*: The development of methods that amplify target nucleic acids has proven efficient, cost-effective and highly specific for the diagnosis of virus infections. The power of PCR results from its ability to synthesize millions of copies of a specific gene or gene segment *de novo* starting with only a few template copies in a large excess of non-target DNA. The specificity of PCR is based on the use of oligonucleotide primers that are complementary to regions flanking the sequence to be amplified. For diagnostic purposes, amplification should be followed by sequencing in order to definitively establish the identity of the target. Many variations of the basic PCR exist. In **r**everse **t**ranscription **PCR** (RT-PCR), RNA molecules are converted into their **c**omplementary **DNA** (cDNA) sequences by reverse transcriptases, followed by the amplification of the newly synthesized cDNA by standard PCR procedures. In real time PCR (or quantitative PCR), differential detection and quantification of a virus is not only possible, but the technique also allows for monitoring responses following antiviral therapy. Multiplex PCR, on the other hand, permits the simultaneous amplification of different targets—provided that a pair of specific primers for each target is used. This way, in one step many viruses can be detected, lowering thus the cost associated with separate reactions. If multiplex PCR is coupled with the use of fluorescent tags, more than 20 different genetic targets can be detected at the same time. When restriction enzymes are included, products from PCR amplifications may reveal **r**estriction **f**ragment **l**ength **p**olymorphisms (RFLPs) and facilitate the characterization of virus variability in epidemiological studies.

Isothermal nucleic acid amplification: This is a simple process that allows for the amplification of nucleic acids at a single temperature instead of the multiple thermal cycling used in conventional PCR protocols. This method does not require a thermal cycler and reactions are performed in a single tube. LAMP (**l**oop mediated isothermal **amp**lification) is one example that uses two to three sets of primers and a polymerase that, in addition to its replication activity, possesses high strand displacement activity. In NASBA (**n**ucleic **a**cid **s**equence-**b**ased **a**mplification), an RNA substrate is converted to its **c**omplementary **DNA** (cDNA) then to dsDNA, more RNA molecules are produced and the cycle starts again resulting in DNA amplification of the RNA target molecule. RPA (**r**ecombinase **p**olymerase **a**mplification) refers to the method by which a combination of a recombinase, **s**ingle-**s**tranded DNA-**b**inding (SSB) proteins and a strand displacing polymerase synthesize DNA *de novo*, delimited by specific primers. If a reverse transcriptase is used to generate cDNA, RNA can also be used as a substrate for amplification. Another variation is HDA (**h**elicase-**d**ependent **a**mplification). Here helicase generates single-stranded templates for primer hybridization and subsequent primer extension by a DNA polymerase.

Molecular hybridization on solid supports (Southern and Northern): Southern and Northern analyses are mostly used in basic research, but are also commonly used to confirm positive EIA screen tests. In both assays, a labeled probe (radioactively or cold-labeled) hybridizes with target sequences under stringent conditions that preclude unspecific interaction with nontarget sequences. In most cases, the probe is a DNA molecule used to detect sequence-driven hybridization with an immobilized target DNA (Southern) or RNA (Northern). The technique can be used for virus detection in cultured cells, tissue sections or purified, immobilized nucleic acids on solid supports.

The size of the target nucleic acid can be estimated if hybridization is conducted after fractionation of total DNA (that has first been restricted) or RNA and transfer to a solid support.

Microarrays: Microarrays allow for the simultaneous detection of all viruses in a sample given that a single microarray cell consists of millions of discrete oligonucleotide probes specific for previously recognized strains and new or "*variant*" viruses associated with outbreaks. Since the technique relies upon random PCR amplification followed by hybridization of the fluorescently labeled products, it is less specific than multiplex PCR. Nonetheless, the use of improved microarrays, based on detection by changes in electrical conductance instead of hybridization, overcomes this limitation. Additionally, it is only informative for the range of known or closely viruses.

CRISPR-based detection of viruses: Based on CRISPR-Cas13a/C2c2, a recently developed strategy named as **s**pecific **h**igh sensitivity **e**nzymatic **r**eporter un**lock**ing (SHERLOCK), facilitates the detection of very low virus titers (for instance, at 10^{-18} attomolar levels) and enables the distinction between strains and closely related viruses, for example, *Zika virus* and its close flavivirus relative *Dengue virus*. In this revolutionary detection system, the enzyme Cas13a is programmed (by sequence-specific small RNAs) to target RNAs that match the tag in a reaction which releases a fluorescent signal. The method works well with blood, urine, and saliva samples.

***R**eplicating-**C**ompetent **R**eporter-**E**xpressing **V**iruses (RCREVs)*: RCREVs are viruses that have been artificially modified in such a way that they retain their own characteristics, but gain those derived from a reporter gene. RCREVs are useful tools for the analysis of virus replication in cells or whole organisms, viral protein transport, and virus movement. Among the reporter genes used in RCREVs, we find the green fluorescent protein and its variants (that fluoresce blue, yellow, and red), bioluminescent luciferases, and nonfluorescent, nonluminescent reporters as the Cre-recombinase and the neomycin-resistance gene. CREVs have been successfully used in serum/virus neutralization tests for the detection of antivirus neutralizing antibodies, as well as in the screening of the effect of compounds and molecules intended as antivirals, in the modification of live vaccines and in tracking of a disease of viral etiology.

Detection of "old" and "novel" viruses: NGS has not only allowed for cheaper, faster, and more accurate sequencing of whole genomes, but it has also been added to the arsenal of methodologies virologists count upon to detect known viruses and new ones. NGS has also been employed in epidemiological studies to facilitate tracking of outbreaks and pandemics contributing thus to the understanding of emergence and transmission profiles of, for example, influenza. A NGS-based approach facilitated the determination of human papillomavirus types, subtypes, and variants present in cervical samples. An unknown, fatal human disease observed in patients receiving visceral-organ transplants, caused by a previously unknown arenavirus, was also revealed by NGS, when classical PCR, culture, microarray, and serological assays failed to detect the virus. In the world of plant viruses, NGS has played a fundamental role in the discovery and variability analysis of viruses and viroids, their host range, their interactions with susceptible hosts and their vectors, their epidemiology and their evolution. Finally, due to its very nature of providing huge amounts of sequence information, NGS has enriched sequence databases of known and unsuspected members of the virosphere.

Viral Targets for Engineered Resistance

Engineered resistance to viruses has been more widely used, and attained, in plants by transgenesis and more recently by CRISPR/Cas9. The former, transgenic plants expressing the coat protein or replicase genes have, undoubtedly, paved the way for understanding viral engineered resistance, despite the strong objections expressed by some sections of society. Examples demonstrate proof of principle that transgenesis may also improve animal health and aid in infectious disease control in livestock. In goats, for instance, pronuclear embryos were transformed by microinjection with an engineered vector harboring the 3Dpol gene (viral RNA polymerase coding gene) of *Foot-and-mouth disease virus* (*Picornaviridae*) with remarkable success. Tongue epithelium cells exhibited resistance against the virus. Viruses can thus provide the elements required for their own restriction.

Typically this involves the activation of cellular mechanisms of recognition and destruction (or avoidance). The main goal of transgenic approaches is to prime the cell to elicit a response against the intended virus target by maintaining and expressing sequences derived from the very same virus. In theory, any virus-derived sequence might be used to engineer transgenic resistant organisms. In practice, however, resistance has been successfully attained with sequences derived from the **c**oat **p**rotein (CP) or nucleocapsid, replicase or movement protein-coding genes. Other transgenic approaches make use of viral sequences, like noncoding sequences, engineered microRNAs, **t**ranscriptional **a**mplification **s**trategy (TAS)-based engineered sequences, heterologous resistance genes (for example, the mice Mx1 protein-coding gene conferred resistance to influenza virus in pigs), and immune defense activators genes. Of note, other applications are aimed at interrupting vector borne transmission and directed thus toward controlling vector populations (particularly of arboviruses). The end result is the same, whether by transgenesis or genome manipulation; virus resistance ensues as a result of disruption of virus function or a host factor. Applications of both approaches are provided below, but for historical reasons, applications to plants are given greater attention.

Local and Systemic Movement of Viruses

Nonfunctional transgenes derived from movement protein genes of *Tobacco mosaic virus* (TMV) have proven effective in conferring virus resistance against not only the homologous virus, but also to other tobamoviruses. The functional form of the transgene, however, can lead to increased severity of the infection, and in some experiments, resistance was reported temperature-dependent. Either way, it seems that resistance is the result of impaired systemic movement of the virus, most probably due to effects on plasmodesmata rather than from a form of interference (as explained below). In other words, resistance is a byproduct of the presence of the virus protein in the transgenic plant. In another approach aimed at inhibiting local and systemic movement of the virus, researchers worked with the virus replicase gene. Tobacco plants transformed with a truncated form of the replicase-coding gene of *Cucumber mosaic virus* (CMV) were resistant to systemic CMV disease. In this case, resistance was not only at the cellular level and due to suppression of virus replication, but also to disruption of the systemic movement of the virus.

CP-Mediated Resistance in Plants

Some 100 years after Jenner's work on the therapeutic administration of cowpox virus, similar observations were made with plant viruses. That is, a mild virus strain was used to protect plants against economic damage caused by a severe strain of the same virus. The phenomenon, referred to as cross protection, was brought to the fore in the late 1920s (Chapter 12: Viruses as Tools of Biotechnology: Therapeutic Agents, Carriers of Therapeutic Agents and Genes, Nanomaterials, and More). It was observed that tobacco plants infected with a light green strain of TMV never developed infections when challenged with a severe TMV yellow mosaic strain. Later in the 1930s, similar effects were demonstrated with an avirulent strain of *Potato virus X* (PVX) in potato, and disease control was achieved using mild strains of *Citrus tristeza virus* and *Papaya ringspot virus* in the 1980s. Although the exact mechanism by which cross protection worked was not clear at the time, evidence suggested that the CP was involved in some instances. These observations formed the basis for engineering virus-resistant plants using the virus' own CP gene, and the concept of pathogen-derived resistance emerged.

Pathogen-**d**erived **r**esistance (PDR) refers to the use of a pathogen's own genes to confer resistance in a host to that pathogen. Since pathogens generate unique and critical products during their association with a host, it was posited in the mid-1980s that the expression of one of these products in the host could disrupt the infection cycle of the pathogen resulting in a resistant phenotype. The first application of the concept of PDR involved the introduction and expression of a viral CP gene. The strategy, which became known as "*coat protein gene-mediated protection,*" is best understood with TMV and tobacco. Transformed tobacco plants accumulating TMV CP were challenged with differing levels of TMV inoculum and found to either show a delay in the appearance of symptoms or no symptom expression at all. It appeared that (1) the expression of the CP gene in the host inhibited a number of stages in the virus replication cycle, (2) CP expression was necessary for resistance, and (3) the CP-mediated resistance against TMV was not immunity, but similar to tolerance.

Subsequent testing of PDR in other virus CP-host combinations as well as other virus genes revealed additional characteristics of the phenomenon. CMV CP transgenic plants expressing what appeared to be equivalent amounts of CP showed different degrees of resistance. CP PVY (*Potato virus Y*) transgenic plants with high levels of CP were not necessarily resistant, while CP PLRV (*Potato leafroll virus*) plants expressing barely detectable levels of CP were resistant to virus challenge. Testing of other viral genes, like the replicase, further substantiated these observations. Plants transformed with a component of the TMV replicase (the 54 kD), the *Pea early-browning virus* or a truncated replicase gene encoded by RNA-2 of *Cucumber mosaic virus* strain Fny, were also reported resistant to virus challenge. There were no doubts that different mechanisms of resistance existed and depended on the virus and the virus gene/protein expressed by the transgenic plant. Investigations aimed at further understanding the mechanisms associated with PDR are described in the next section.

RNA-Mediated Resistance

Work on tobacco in the early 1990s with the potyvirus, *Tobacco etch virus* (TEV), indicated that genetically engineered virus resistance could be mediated by RNA, and not CP. When transgenic plants expressing a nontranslatable version of the TEV CP gene were challenged with virus, some proportion of plants were reported immune to infection. That is, virus replication was never detected in these plants and the resistance could not be overcome by high levels of TEV inoculum (virion or RNA). Interestingly, resistant plants displayed low or undetectable steady state CP transgene mRNA levels, even though nuclear transcription rates were high. Based on these observations, it was hypothesized that the high levels of CP transcription triggered a cellular response resulting in sequence-specific RNA degradation of both transgene transcripts and genomic TEV RNA, causing the resistant phenotype. It was further suggested that aberrant or overexpressed transgenes were part of the RNA targeting and degradation system and that these transcripts generated small RNAs, the trigger for the response. Aptly, the response was called **p**ost **t**ranscriptional **g**ene **s**ilencing (PTGS) or gene silencing. Needless to say, this model generated much interest within both the plant and animal biology communities and subsequent contributions from both communities helped in defining the mechanism of PTGS or **RNA i**nterference (RNAi), which was coined by animal biologists. It was realized that PTGS is a conserved biological mechanism that most likely evolved as a primitive immune response evoked by the presence of foreign nucleic acids. The RNA inducers or the specificity determinants described in the 1990 model were discovered in plants (and nematodes). They consisted of dsRNA of about 25 nts and were complementary to the sequence of the silenced transgene carried by transformed plants. Later biochemical experiments with fruit flies (*Drosophila melanogaster*) identified the RNA directed nuclease, Dicer, that directed PTGS. Dicer-like homologs were subsequently found in plants. Dicer is a dsRNA endoribonuclease that cleaves dsRNAs into short duplexes, siRNA, one strand of which is incorporated into a RISC complex (**R**NA-**i**nduced **s**ilencing **c**omplex), where it serves as guide to direct RISC to complementary RNAs to degrade them or prevent their translation. **Argo**naute (AGO) proteins make up the RNA cleaving catalytic component of RISC. Some of the examples of resistance exhibited by transgenic plants have since been attributed to the processing of dsRNA transgene transcripts produced by nontranslatable versions of the virus CP, or translatable versions of the CP where an insertion pattern promoted the generation of ds transgene RNA.

PDR using viral genes has been applied to the development of virus-resistant commercial varieties, the first of which was transgenic squash with multivirus disease resistance against *Zucchini yellow mosaic virus, Watermelon mosaic virus,* and *Cucumber mosaic virus*. Expression cassettes consisting of single CP sequences joined to a plant promoter were transferred to squash by *Agrobacterium tumefaciens*-mediated transformation. Under field conditions, the transgenic plants were found resistant to all three viruses and displayed impressive yields compared to their nontransgenic, susceptible counterparts. Even after more than 20 years of continuous use, the variety still performs well. Although natural sources of resistance against these viruses have been identified in different cucurbit species, the resistance is multigenic and not always effective. PDR using virus CP genes offers a simple and fairly rapid solution to breeding for multivirus resistance.

Since the release of these transgenic varieties, the process for developing transgenic virus resistance has improved considerably through the use of synthetic transgenes. In one example, these transgenes contain virus sequences in an inverted orientation. The idea is that transcription will generate RNA capable of forming the dsRNA trigger for RNA silencing. Brazil, in 2011, released transgenic common bean developed with this strategy. The *AC1* viral gene that encodes the rep protein, the protein essential for *Bean golden mosaic virus* genome replication, was the target. Transgenic plants expressing a portion of the *AC1* gene in an inverted orientation were highly resistant to the virus. The method has since been applied to the development of resistance against multiple tosposviruses, which are notorious pathogens of a variety of crops and ornamentals. Here, multivirus resistance was engineered using a few hundred base pairs of sequences of different viruses. Still further improvements developed after the discovery of **micro****RNAs** (miRNAs). miRNAs are another class of small RNAs involved in PTGS. While siRNAs result from the action of dicers on large dsRNAs of diverse origins, miRNAs are transcribed from their own genes. They function as negative regulators of gene expression and play an important role in plant development and response to biotic or abiotic stress. In plants, these primary-miRNAs are processed in the nucleus into **pre**cursor-**miRNA** (pre-miRNA) and then into miRNA:miRNA duplexes before being exported to the cytoplasm and incorporated into RISCs. They inhibit gene expression by inducing target mRNA degradation or translation repression. Following the demonstration that plant miRNA precursors could be engineered to express small RNAs (referred to as **a**rtificial-**miRNA** or amiRNAs) with sequences unrelated to the corresponding mature miRNAs, several studies have successfully generated resistance to plant viruses using amiRNA. Basically amiRNAs are developed using endogenous miRNA precursors to generate new target-virus specific miRNAs for gene silencing in plants. Plants are transformed with an endogenous plant miRNA carrying one or two different 21 nucleotides regions of a viral gene, such as a viral RNA-silencing suppressor gene. Compared with siRNAs, miRNAs cause more accurate gene silencing. Although the strategy has been successfully tested against cucomo-, potex-, begomo- and tospoviruses, transgenic plants using amiRNAs have not been approved for commercial use as of 2017.

As mentioned earlier, the specificity and robustness of PTSG or RNAi generated immense interest within the animal biology community. One immediate application of the technology was its use in therapies that directly silence pathogenic genes, disease causing mutant genes or interrupt viral infections. This endogenous pathway could be coopted for targeted RNAi, either through the delivery of exogenous **s**mall **i**nterfering **RNA** (siRNA) to target cells or, as with plants, by transgenic expression of **s**hort **h**airpin **RNA** (shRNA). Once sequences are chosen correctly (usually within a gene essential for replication), they could be processed by the Dicer nuclease, taken up by RISC machinery and used to disrupt virus replication. An added advantage over classical vaccines and small drug molecules is that siRNAs are highly selective for the pathogen and can be produced at large scale quickly. But a key challenge to realizing the broad potential of siRNA-based therapeutics is the need for safe and effective delivery methods. Two approaches have been examined: genomic integration of shRNA constructs, and exogenously delivered siRNA. Exogenous delivery has several advantages over the transgenic approach. Of these is the short-term effect of delivering the small RNAs, the avoidance of the detrimental effects of overexpression of transgenic shRNA, and more importantly, this approach does not require the manipulation of the genome, making it accessible for human therapeutics and bypassing the issues surrounding genetically modified organisms.

Additionally, newly emerging viruses can be addressed more quickly with exogenously delivered siRNA.

HIV was among the first infectious agents that provided *in vitro* proof-of-concept demonstrating the successful delivery and antiviral potential of RNAi. Both synthetic and expressed siRNAs were used to target several early and late HIV-encoded RNAs. Additionally, cellular cofactors, such as the receptor CD4 and coreceptors CXCR4 and CCR5, were down regulated by RNAi resulting in the inhibition of HIV replication. Similar inhibitory effects occurred when the technique was applied to other viruses, for example, *Hepatitis B virus* (HBV), *Influenza A virus*, and *Dengue virus*. siRNAs against sites in the HBV genome, such as the envelope protein and HBx protein (which promotes transcription of the viral genome), effectively prevented viral infection in mammalian cells. Coinjection with a hepatocyte-targeted **N**-**a**cetyl**g**alactosamine-conjugated **m**elittin-**l**ike **p**eptide (NAG-MLP), resulted in the repression of viral RNA, proteins, and viral DNA with long duration of effect. Anti-influenza siRNAs were reported as target-specific and, in some instances, effective in both prophylaxis and therapy. shRNA targeting the polymerase components, NP, PA, or PB1, was effective in experimental animals treated intravenously or intranasally. RNAi-mediated suppression of *Dengue virus* (DENV) has also been investigated. This application involved the use of dendritic cell targeting peptide mediated delivery of siRNA against a conserved sequence in the DENV envelope. The technique was found to effectively block DENV replication in macrophages and monocytes. In addition to RNAi-mediated suppression of DENV itself, RNAi-mediated suppression of viral dependency factors, or factors required by the virus for productive infection, were shown to inhibit DENV replication. Yet another target was tested, that of vector competence. The aim here was to develop transgenic mosquitoes (*Aedes aegypti)* that exhibit reduced vector competence for DENV by triggering the insect's RNAi pathway. *A. aegypti* mosquitoes were manipulated to express **i**nverted-**r**epeat (IR) sequences derived from the premembrane protein coding region of DENV. It was hypothesized that the expression of IR-RNA in the midgut of female mosquitoes would ensure targeting of the virus in its most vulnerable state; that is, after ingestion of viremic blood, at the onset of replication, and before the establishment of infection within the mosquito. Indeed, transgenic mosquitoes poorly supported virus replication and showed high levels of resistance to DENV.

RNAi has also proven effective against a wide range of significant viruses in domesticated animal, both *in vitro* and *in vivo*; these include the *Foot-and-mouth disease virus*, *African swine fever virus*, *Classical swine fever virus*, *Influenza A virus*, *Chicken anemia virus*, and *Infectious bursal disease virus*. Protection has also been achieved against viruses of aquatic species. Exogenously delivered siRNA protected *Litopenaeus vannamei* shrimp and *Marsupenaeus japonicus* (the kuruma prawn), for example, from *White spot syndrome virus*. The potential value of RNAi technology to interrupt viral infection in fish was demonstrated for the *Redspotted grouper nervous necrosis virus* (RGNNV).

While RNAi promises to be a useful tool for controlling viral diseases, major limitations in RNAi technology present obstacles to its present use; incomplete loss of gene function, specific off-target activity, off-target gene degradation, and delivery of RNAi to specific tissues. Chapter 12: Viruses as Tools of Biotechnology: Therapeutic Agents, Carriers of Therapeutic Agents and Genes, Nanomaterials, and More discusses these and other challenges associated with nucleic-acid-based drugs.

CRISPR/Cas9 and Other Targets for Engineering Virus Resistance

Genome-engineering strategies have recently emerged as promising tools for developing virus resistance in eukaryotic species. Among these genome engineering technologies, the CRISPR (**c**lustered **r**egularly **i**nter**s**paced **p**alindromic **r**epeats)/CRISPR-associated 9 (CRISPR/Cas9) system has received special interest because of its simplicity, efficiency, and reproducibility. CRISPR/Cas9 is a prokaryotic molecular immunity system against invading nucleic acids (through horizontal gene transfer) and bacteriophages (Chapter 10: Host–Virus Interactions: Battles Between Viruses and Their Hosts). Once the underlying mechanisms of CRISPR/Cas9 became clear, biologists realized that the system could provide an efficient means of editing genomes. Only two components—the Cas9 protein and the short **g**uide **RNA** (sgRNA) molecule—are required for this engineered system to function. Cas9 is a nuclease that creates a **d**ouble **s**trand **b**reak (DSB) at a specific genomic location, and the sgRNA, that can be programmed to identify any gene sequence, functions as the homing device for directing the Cas9 nuclease to the particular target sequence. In 2012, the system was developed for use as a gene editing tool. That is, the CRISPRCas9 system, derived from *Streptococcus pyogenes*, was modified to recognize specific sequences in the genome and cut the targeted DNA to generate DSBs, which could then stimulate the host cell's genome editing through recombination mechanisms, either **n**on**h**omologous **e**nd **j**oining (NHEJ) or **h**omology-**d**irected **r**epair (HDR) mechanisms. NHEJ can introduce small insertions or deletions (indels) that cause frameshift mutations and the disruption of gene function, while HDR can be engineered for target gene correction, gene replacement, and gene knock-in effects. Recent studies have used CRISPR/Cas9 to engineer virus resistance in plant and animal hosts, by (1) directly targeting and cleaving the viral genome, (2) disrupting a provirus inserted in the host genome, or (3) editing host genes to introduce viral immunity.

Two strategies have emerged as promising tools to introduce virus resistance into plant species: one targeting the virus genome and another that generates resistance by targeting and modifying plant genes encoding host factors required for virus replication. The former example was demonstrated with geminiviruses. It was shown that *N. benthamiana* plants expressing the CRISPR/Cas9 machinery exhibited resistance against *Tomato yellow leaf curl virus*, *Beet curly top virus*, and *Merremia mosaic virus*. These plants displayed considerably reduced viral titers, which abolished or significantly reduced disease symptoms. It was also shown that one sgRNA targeting the *Bean yellow dwarf virus* (BeYDV) genome could confer plant resistance without cleavage activity, suggesting that catalytically inactive Cas9 could be used to mediate virus interference, thereby eliminating concerns of off-target activities in the plant genome. The second approach, that off targeting a plant gene, was demonstrated in a potyvirus-cucumber host system. Translation initiation like factors, eIF4E and eIF(iso)4E, are known to be directly involved in the infection of RNA viruses. These targets were mutated with the CRISPR/Cas9 system in order to develop cucumber plants with resistance to *Cucumber vein yellowing virus*, *Zucchini yellow mosaic virus* and *Papaya ringspot virus*-type W. Disruption of the eIF4E gene in cucumber by CRISPR/Cas9 and the development of virus-resistant plants proved that translation initiation factors are prime candidates for host genes that can be targeted; but technically any host gene encoding a factor that the virus requires is a potential target for modification. One major advantage of targeting and modifying host factors is that CRISPR/Cas9 can be introduced as transgenes to create the genome edits and subsequently removed in later generations by backcrossing to give virus-resistant plants free of genetic modification. Alternatively, the Cas9 protein and the sgRNA can be introduced directly into cells as a ribonucleoprotein complex to avoid the incorporation of transgenes into the genome. The resulting plants would therefore be indistinguishable from plants carrying naturally occurring alleles or plants generated by random mutagenesis, which may make them exempt from current GMO regulations. The debate on this matter, the regulation of CRISPR/Cas9-derived organisms, continues.

CRISPR/Cas9 has found potential applications to human virus diseases. In HIV, for example, the possibility that CRISPR/Cas9 could be used to inactivate or even delete proviral DNA from HIV-1 infected cells was examined. In some studies, the cofactor CCR5 was targeted in pluripotent stem cells and hematopoietic stem cells. HIV-1's entry is mediated by sequential binding of its surface glycoprotein to the cellular receptor CD4 and then a chemokine receptor, CCR5. A naturally occurring genetic mutation, known as CCR5Δ32, was determined responsible for HIV resistance. The mutation results in a smaller protein that no longer sits on the cell surface and as a result, CCR5Δ32 hampers HIV's ability to infiltrate immune cells. Using CRISPR Cas9, Δ32 mutations precisely matching the naturally occurring homozygous CCR5Δ32 genotype were generated. The monocytes/macrophages derived from CCR5Δ32 mutation pluripotent stem cells were resistant to HIV-1 infection. In other studies, the CRISPR-Cas9 system was reportedly useful for editing the HIV genome integrated into host cell genome. Several labs have designed sgRNAs to program Cas9 to cleave different regions of HIV-1 DNA that include either essential viral genes or the viral **l**ong **t**erminal **r**epeats (LTRs). Suppression of HIV-1 production and infection were observed in different cell types including latently infected $CD4^+$ T cells, primary $CD4^+$ T cells and induced human pluripotent stem cells. The extreme variability and the high evolution rate of HIV-1, however, may warrant programming of Cas9 with multiple sgRNAs that target conserved viral DNA regions in order to avoid virus escape. It is possible that CRISPR/Cas9 could be combined with antiretroviral agents, **H**ighly **A**ctive **A**nti-**R**etroviral **T**herapy (HAART) to clear latently infected cells.

A similar strategy proved successful in generating resistance against the virus *Porcine reproductive and respiratory syndrome virus* (PRRSV). PRRSV, an enveloped, ssRNA(+) virus belonging to the *Arteriviridae* family in the order *Nidovirales*, causes one of the most important infectious diseases of pigs worldwide. **P**orcine **R**eproductive and **R**espiratory **S**yndrome (PRRS) manifests differently in pigs of all ages, but primarily causes late-term abortions and stillbirths in sows and respiratory disease in piglets. As with the HIV-1 CCR5 strategy, the CRISPR-Cas9 system was used to introduce a deletion in a cell surface receptor and thereby hampering the virus' ability to infiltrate immune cells. PRRSV has a highly restricted tropism for cells of the monocyte-macrophage lineage. Its entry via fusion with the host cell membrane is mediated by the receptor CD163, in particular, domain 5 of CD163. Using CRISPR/Cas9 gene editing, pig zygotes with a deletion in the CD163 domain 5 were generated. The deletion rendered cells resistant to PRRSV.

Good efficacy was also reported with studies using CRISPR/Cas technologies to disable replication of DNA herpesviruses like Herpes simplex virus type 1 (*Human alphaherpesvirus 1*, HHV 1), Epstein-Barr virus (*Human gammaherpesvirus 4*, HHV 4), and Human cytomegalovirus (*Human betaherpesvirus 5*, HHV 5). Upon infection, the linear viral DNA of these viruses is delivered to the cell's nucleus, where it circularizes to form a viral episome. Depending on several factors, replication can proceed either to a productive infection or to a state of latency. In either case, the viral genome is maintained as extrachromosomal circular DNA. Using sgRNAs, CRISPR/Cas9 was directed to cut virus DNA, whether actively replicating or latent in the host cell, at one or more sites and induce mutations that would cripple virus replication. Although active replication of all three viruses was abolished, the latent genome was cleared only in host cells challenged with EBV. The potent antiviral activities will need to be tested in animal studies, and eventually humans before any application as future therapy.

Targeting RNA viruses using a similar approach was demonstrated with Hepatitis C virus (*Hepacivirus C*, HCV) and a Cas9 isolated from *Francisella tularensis* subsp. *novicida*. Cas9 from *Francisella tularensis* subsp. *novicida* (FnCas9) targets bacterial mRNA. In this investigation, FnCas9 was retargeted to HCV's ssRNA(+) genome in order to block both viral translation and replication. An RNA-targeting guide RNA complementary to a portion of the highly conserved HCV 5′ **un**translated **r**egion (UTR) was developed. Vectors encoding either this sgRNA or FnCas9 were transfected into human hepatocellular carcinoma cells and subsequently infected with HCV. Expression of the 5′ UTR-targeting sgRNA and FnCas9 together reduced the levels of viral proteins. Applications to other ssRNA(+) viruses such as flaviviruses, enteroviruses, and rhinoviruses as well as ssRNA(−) viruses, like filoviruses, paramyxoviruses, and orthomyxoviruses are anticipated. Because FnCas9 can target both negative- and positive-sense strands of RNA, it is likely that the FnCas9:sgRNA machinery will be used to target diverse viruses. The replication cycles of viruses with both RNA and DNA genomes require an RNA stage (generated during transcription, replication, or both). Taken together, the field of the CRISPR/Cas9 system holds much promise for future therapies against virus diseases. While there are many examples of the efficacious use of CRISPR/Cas9 in cell culture, application to the clinic or agricultural field will require methods of safe and efficient delivery.

FURTHER READING

Damborský, P., Švitel, J., Katrlík, J., 2016. Optical biosensors. Essays in Biochemistry 60, 91–100. Available from: https://doi.org/10.1042/EBC20150010.

Doerflinger, M., Forsyth, W., Ebert, G., Pellegrini, M., Herold, M.J., 2017. CRISPR/Cas9—The ultimate weapon to battle infectious diseases? Cellular Microbiology 19, e12693. Available from: https://doi.org/10.1111/cmi.12693.

Fakruddin, M., Mannan, K.S.B., Chowdhury, A., Mazumdar, R.M., Hossain, M.N., Islam, S., et al., 2013. Nucleic acid amplification: Alternative methods of polymerase chain reaction. Journal of Pharmacy & Bioallied Sciences 5, 245–252. Available from: https://doi.org/10.4103/0975-7406.120066.

Fredericks, D.N., Relman, D.A., 1996. Sequence-based identification of microbial pathogens: a reconsideration of Koch's postulates. Clinical Microbiology Reviews 9, 18–33.

Gootenberg, J.S., Abudayyeh, O.O., Lee, J.W., Essletzbichler, P., Dy, A.J., Joung, J., et al., 2017. Nucleic acid detection with CRISPR-Cas13a/C2c2. Science 356, 438–442. Available from: https://doi.org/10.1126/science.aam9321.

Hadidi, A., Flores, R., Candresse, T. & Barba, M. Next-generation sequencing and genome editing in plant virology. *Frontiers in Microbiology* 7:1325. https://doi.org/10.3389/fmicb.2016.01325.

Khalid, A., Zhang, Q., Yasir, M., Li, F., 2017. Small RNA-based genetic engineering for plant virus resistance: application in crop protection. Frontiers in Microbiology 8, 43. Available from: https://doi.org/10.3389/fmicb.2017.00043.

Li, Y., Li, L.-F., Yu, S., Wang, X., Zhang, L., Yu, J., et al., 2016. Applications of replicating-competent reporter-expressing viruses in diagnostic and molecular virology. Viruses 8, 127. Available from: https://doi.org/10.3390/v8050127.

Lindbo, J.A., Falk, B.W., 2017. The impact of "Coat Protein-mediated virus resistance" in applied plant pathology and basic research. Phytopathology 107, 624–634. Available from: https://doi.org/10.1094/PHYTO-12-16-0442-RVW.

Lipkin, W.I., Anthony, S.J., 2015. Virus hunting. Virology 479–480, 194–199. Available from: https://doi.org/10.1016/j.virol.2015.02.006.

Mahony, J.B., 2008. Detection of respiratory viruses by molecular methods. Clinical Microbiology Reviews 21, 716–747. Available from: https://doi.org/10.1128/CMR.00037-07.

Simmonds, P., Adams, M.J., Benkő, M., Breitbart, M., Brister, J.R., Carstens, E.B., et al., 2017. Consensus statement: virus taxonomy in the age of metagenomics. Nature Reviews Microbiology 15, 161–168. Available from: https://doi.org/10.1038/nrmicro.2016.177.

Souf, S., 2016. Recent advances in diagnostic testing for viral infections. BioscienceHorizons 9, Available from: https://doi.org/10.1093/biohorizons/hzw010hzw010.

Conclusion

It's a Viral World

Paula Tennant[1] and Gustavo Fermin[2]
[1]The University of the West Indies, Mona, Jamaica, [2]Universidad de Los Andes, Mérida, Venezuela

I think computer viruses should count as life.

Stephen Hawking

There are a few main points to draw out of these thirteen chapters and to reiterate in this conclusion. Viruses are ubiquitous. All kinds of cells can be infected with viruses. Our cells can be infected by viruses causing acute disease, such as respiratory diseases by influenza viruses, or chronic disease, such as hepatitis. However, most viral attacks are fended off, or the spread of viruses in an infected organism is restricted by immunity defenses and other defenses based on DNA and RNA immunity mechanisms. By virtue of triggering host defenses, viruses have been major drivers of host evolution, and have contributed an estimated 30% of the adaptive changes in the human proteome. One notable example of how a virus product that integrated into a host's genome has ended up serving the host's purpose is syncytin. Syncytin is essential for the placenta to properly connect with the uterus. Curiously, syncytin turned out be a retroviral gene. It appears that on at least six different occasions its ancestral form got incorporated into mammals' genomes and helped in the development of a placenta. Other viral stowaways are also very important. Viruses, for example, belong to the human body as a well-balanced ecosystem. They help all organisms cope with abiotic agents of stress; and in some instances, they keep other viruses and pathogens under check. What happens to viruses affects us; but the opposite is also true.

Added to this is the impact on microorganisms, including bacteria, archaea, and microeukaryotes, all of which are vital players in the global fixation and cycling of key elements such as carbon, nitrogen, and phosphorus. The enormous number of viruses associated microbial life, and their intimate relationship, suggest that viruses too play a critical role in the planet's homeostasis. Also, microeukaryotes are thought to be hosts of the recently described giant viruses. Giant viruses are characterized by disproportionately large genomes and virions. They encode several genes involved in protein biosynthesis, a unique feature which has led to diverging hypotheses on the origins of viruses and the diversity of life.

Having established that ours is after all a viral world, and that viruses are important in medicine, biology, biotechnology and ecology, we close with a very brief overview of recent reports which indicate that we have just touched the tip of the iceberg; there is much more to learn about viruses. The reports clearly illustrate this and point, at least, to three crucial features of viruses and the way we *deal* with them; what is a virus, how is host range defined and what measures should be taken to protect against the diseases that some cause. The latter intimately depends on how well we know the first two.

Virosphere viruses, unlike computer viruses, exist in the material world in the form of RNA or DNA molecules; both, however, consist of pieces of information capable of producing copies of themselves – with or without changes in their execution programs. Recently, it has been shown that the encryption that allows viruses to code for encapsidation *and* be able to be encapsidated are linked, and that this dual cipher can be cracked – and used to *improve* viruses in order to make them better (as viruses, that is). From our perspective, this means that we have to approach viruses in a more structure/information-oriented fashion if we want to better understand viruses, as well as manipulate and control them. Thus, might it be then useful to treat virosphere viruses in the way we treat computer viruses, for example? Would it not be worth the effort to deal with viruses in the same way that virus and antivirus creators work with these pieces of executable information? After all, if we have created computer viruses in our own image, as stated by Hawking, why not generate the way of managing those from the virosphere based on what we know from the *virtual* ones?

On the aspect of host range, we have long believed viruses to be host-specific. That is, they should infect a taxonomically narrow subset of species within a community at a given time in any given environment. But evidence in 2003 suggests that marine viruses may not be so picky after all, and are capable of infecting multiple microbial species and even distantly related organisms. Added to this, a 2017 study reported on a natural infection of the phytopathogenic fungus, *Rhizoctonia solani*, by a plant virus, *Cucumber mosaic virus* (CMV). The plant virus-infected fungus was isolated from a potato plant and researchers were able to demonstrate that *R. solani* can acquire and transmit CMV during plant infection. The study brings to question 1) how virus transmission between plant and fungal hosts contributes to the evolution and genetic diversities of plant and fungal viruses, and 2) where else can plant viruses spread beyond their known conventional hosts. But more importantly, is this an isolated case? Or should we anticipate discoveries of more cross infections with deeper probing into virus-host interactions? Will we have to eventually relook at the definitions of host range?

Finally, and unfortunately, vaccination against deadly virus infections is still a matter of heated debate and discussion. In November 2017, it was revealed that a promising vaccine against dengue was in jeopardy after a vaccination program in the Philippines was suspended amid fears about its safety. It appears that the vaccine can lead to more detrimental effects on the immunized patient if the person was not previously infected by *Dengue virus* when vaccinated. According to the **W**orld **H**ealth **O**rganization (WHO) "*While vaccinated trial participants overall had a reduced risk of virologically-confirmed severe dengue and hospitalizations due to dengue, the subset of trial participants who had not been exposed to dengue virus infection prior to vaccination had a higher risk of more severe dengue and hospitalizations due to dengue compared to unvaccinated participants, regardless of age*." As pointed out by WHO, the dengue vaccine effectively protects dengue-seropositive patients at the time of first vaccination, but not seronegative ones. Clearly, we need to better understand viruses and their interactions with their hosts in order to fulfill the needs, expectations and health and well-being of all.

In sum, all that has been presented in these chapters will not be the limit of what viruses are to be credited with. It is only the beginning after all.

FURTHER READING

Andika, I.B., Wei, S., Cao, C., Salaipeth, L., Kondo, H., Sun, L., 2017. Phytopathogenic fungus hosts a plant virus: A naturally occurring cross-kingdom viral infection. Proceedings of the National Academy of Sciences, USA 114, 12267–12272. Available from: https://doi.org/10.1073/pnas.1714916114.

Patel, N., Wroblewski, E., Leonov, G., Phillips, S.E.V., Tuma, R., Twarok, R., et al., 2017. Rewriting nature's assembly manual for a ssRNA virus. Proceedings of the National Academy of Sciences, USA 114, 12255–12260. Available from: https://doi.org/10.1073/pnas.1706951114.

World Health Organization. 2017. Updated Questions and Answers related to information presented in the Sanofi Pasteur press release on 30 November 2017 with regards to the dengue vaccine Dengvaxia®. http://www.who.int/immunization/diseases/dengue/q_and_a_dengue_vaccine_dengvaxia/en/. (Accessed on 18.12.17).

Index

Note: Page numbers followed by "*f*" and "*t*" refer to figures and tables, respectively.

B

C

D

E

I

J

K

L

M

N

O

P

Q

R

U

V

W

Printed in the United States
By Bookmasters